Integrating AI Techniques into the Design and Development of Smart Cyber-Physical Systems

Building on the knowledge of risks, vulnerabilities, and safety measures associated with cyber-physical systems, this book focuses on adapting artificial intelligence (AI) techniques to smart cyber-physical systems application development. The future is going to see cyber-physical systems in almost every aspect of life, so a book that focuses on shedding light on the design, development, and security aspects of cyber-physical systems in more crucial domains such as defense, healthcare, biomedical, smart city applications, is needed.

Integrating AI Techniques into the Design and Development of Smart Cyber-Physical Systems: Defense, Biomedical, Infrastructure, and Transportation offers an introductory exploration of the fundamental theories and concepts of artificial intelligence (AI) and machine learning (ML) that are utilized in the building of dependable cyber-physical systems. It brings the ideas of advanced design and development and empowered security measures to cyber-physical systems. By focusing on the application of AI in cyber-physical systems design as well as security aspects, an improvement in reliability and advancements can be explored. Also included are the latest findings and advancements as well as case studies and illustrative examples on the design and development of smart cyber-physical systems.

This resource is highly valuable for those employed in educational institutions, research laboratories, enterprises, and government agencies, as well as for students seeking novel ideas in the realm of smart cyber-physical systems design.

Prospects in Smart Technologies

Series Editor
Mohammad M. Banat

Mohammad M. Banat, Jordan University of Science and Technology, Irbid, Jordan
banat@just.edu.jo
Sara Paiva, Instituto Politécnico de Viana do Castelo, Viana do Castelo, Portugal
sara.paiva@estg.ipvc.pt

Published Titles

Emerging Technologies for Sustainable and Smart Energy
Edited by
Anirbid Sircar, Gautami Tripathi, Namrata Bist, Kashish Ara Shakil, and Mithileysh Sathiyanarayanan

CYBORG
Human and Machine Communication Paradigm
Kuldeep Singh Kaswan, Jagjit Singh Dhatterwal, Anupam Baliyan, and Shalli Rani

Emerging Technologies and the Application of WSN and IoT
Smart Surveillance, Public Security, and Safety Challenges
Edited by
Shalli Rani and Himanshi Babbar

WSN and IoT
An Integrated Approach for Smart Applications
Edited by
Shalli Rani and Ashu Taneja

Immersive Technologies
Navigating the Impacts, Challenges, and Opportunities
Sagaya Aurelia

Health 5.0
Concepts, Challenges, and Solutions
Hiren Kumar Thakkar, Chintan Bhatt, Victor C.M. Leung and Ilangko Balasingham

For more information on this series, please visit: www.routledge.com/Prospects-in-Smart-Technologies/book-series/CRCPST

Integrating AI Techniques into the Design and Development of Smart Cyber-Physical Systems

Defense, Biomedical, Infrastructure, and Transportation

D. Jeya Mala, Anto Cordelia Tanislaus Antony Dhanapal, Saurav Sthapit, and Anita Khadka

CRC Press
Taylor & Francis Group
Boca Raton London New York

CRC Press is an imprint of the
Taylor & Francis Group, an **informa** business

Designed cover image: Shutterstock

First edition published 2025
by CRC Press
2385 NW Executive Center Drive, Suite 320, Boca Raton FL 33431

and by CRC Press
4 Park Square, Milton Park, Abingdon, Oxon, OX14 4RN

CRC Press is an imprint of Taylor & Francis Group, LLC

ISBN: 9781032787701 (hbk)
ISBN: 9781032789613 (pbk)
ISBN: 9781003489986 (ebk)

DOI: 10.1201/9781003489986

Typeset in Times
by Newgen Publishing UK

Contents

SECTION IV Trustworthy AI in Smart Defence CPS Design

SECTION V AI in Infrastructure and Transportation Management

Foreword

It is a great honour to write this book's foreword. The authors are experts in their field. They have experience in teaching difficult and trending technical topics and explaining the concepts with examples. The contents of the book flow nicely and cover all topics in a beautiful sequence. The contents reveal the technical knowledge of the four authors. The book starts with an introduction to AI and leads to applications in defense, medicine, and others. Security, trustworthy designs, and autonomous vehicles are the latest important topics that are covered clearly in this book.

Graduate and undergraduate students will find the book very useful. This book is really appealing to practicing engineers, developers, and researchers.

Professor Subramaniam Ganesan
Director, Real Time & DSP Lab.
Electrical and Computer Engineering Department
Oakland University, Rochester, MI, USA

Preface

In today's connected world, the transformation has happened in almost every part of our life. Especially by means of the integration of cyber-physical systems (CPSs), this modern life has experienced several advancements. The way by which the computational aspects, communicational elements, and physical processes into CPSs has led to modernize our everyday life and they became the backbone for critical real-time domains such as defence, smart cities, healthcare, supply chain, and transportation.

Due to the convergence of AI with CPSs, the entire automation process becomes smart and paves ways for more advanced innovations and inventions. This has provided unprecedented potential for automation, decision-making, and efficiency. However, this integration has also magnified the challenges in ensuring reliability and protection against cyber threats and attacks.

This book addresses the design and development of smart and secure CPSs integrated with AI with their potential security challenges and addressing mechanisms across the key domains such as defence, healthcare, smart cities, supply chains, transportation, and supply chain. By weaving together theoretical foundations, practical strategies, and case studies, this book aims to provide a roadmap for researchers, practitioners, and policymakers to ensure the innovative and secure development of smart CPSs.

SMART CPS USING AI

The profound AI technologies, such as machine learning (ML), deep learning (DL), natural language processing (NLP), revolutionize the entire CPS development that turned CPSs into smart systems. This book deals with most important domains such as defence, healthcare, infrastructure, and transportation in which smart CPSs play a crucial role.

In the case of defence applications, the autonomous systems, such as drones, robots, under-water networks, satellite communications etc.; predictive analytics to meet the future warfare; rapid threat detection etc. are most important and critical areas of development using AI. In healthcare, AI-based CPS remote patient-monitoring systems, robotic surgery, and personalized healthcare play a crucial role to provide smart healthcare applications. For smart cities, the AI integration helps to optimize the traffic flows and manage energy consumption, logistics optimization, and public safety applications which are more crucial areas of innovative and secure development. For efficient supply chains, the CPSs provide predictive maintenance, optimization of logistic operations, and real-time tracking. CPS also plays a crucial role in transportation; especially in autonomous vehicles, smart mobility, automated infrastructure management etc.

Some of the potential challenges present in the design and development of smart CPSs include the following:

Defence: A typical AI-enhanced CPS for defence applications must include advanced threat intelligence and resilient architectures to effectively mitigate such kinds of threats. In the current and future warfare, the role of autonomous drones and robots will be so crucial. In those cases the fault-free and transformative mechanisms are required to ensure safety in operations and also will prevent adversarial attacks.

Biomedical: The increasing application of AI-integrated precision healthcare systems revolutionized the applications such as remote patient-monitoring systems, personalized nutrition systems, pacemakers, and robotic surgical tools. Further, these kinds of applications also are attacked by hackers and intruders through ransomware attacks which cause huge loss as well as breaching of personal data of the patients. Such kinds of breaches will also affect the connected devices such as ECG monitors, pacemakers, infusion pumps that endangers the lives. Hence, it is essential to provide a robust cyber security framework in order to provide safe and secure healthcare applications.

Infrastructure: The AI-integrated infrastructure management, especially for smart cities applications, provides efficient solutions for water, power, public transport etc. However, these kinds of systems are also vulnerable to be exploited by the hackers. Hence, it is essential to provide reliable, safe, and secure operations in order to serve critical infrastructure with essential failure-free operations.

Supply Chain: When AI-driven CPS is used in supply chain, the ransomware attacks disrupt various parts of the system. Secure AI frameworks with continuous monitoring, resource optimization techniques, predictive maintenance, and automated recovery mechanisms from intrusions and failures can minimize the impact of such attacks.

Transportation: The vulnerabilities in intelligent transportation systems have disrupted even the entire system operations. Especially autonomous vehicles, GPS-powered remote tracking systems, logistics operations with AI integration etc. are vulnerable to adversarial attacks by means of sensors or GPS signals. Hence, the design and development of secure AI algorithms will help provide resilient system architectures.

By considering all these challenges, this book adopts a holistic approach to design smart CPS with security as the primary concern. Some of the focusing areas of this book include

- **AI-Driven Innovative Approaches in CPS Design:** This book explores the transformative role of AI in making CPS systems to be smart in delivering applications for defence, infrastructure, biomedical, transportation, etc.
- **Resilience and Security:** By recognizing the vulnerabilities associated with AI and the interconnected systems of CPS, this book emphasizes the need for robust designs with various security mechanisms to develop secure CPS systems.
- **Predictive Analytics for Secure and Smart CPS:** This book has also dealt with the application of ML and DL models to provide security analytics solutions for safety critical smart CPS systems.

This book serves as a comprehensive guide to understand and address the various aspects related to the design and development of smart CPS. By means of various case studies, design principles, and insights from various resources, the authors aim to equip researchers, practitioners, students, and policymakers of CPS development industries with the knowledge to create systems that improve the trustworthiness as well as promise innovative development. By fostering interdisciplinary collaboration and security aspects, we ensure that the smart CPS will help transform the world with ethical guidelines and security measures.

About the Authors

Dr. D. Jeya Mala is currently working as a professor in the School of Computer Science and Engineering (SCOPE), Vellore Institute of Technology, Chennai, Tamil Nadu, India. She has more than 21 years of teaching and research and 4 years of industrial experience. She has completed a funded project from the Govt. of India and is doing several consultancy projects as well. Recently she is selected as a member of the National Level Workgroup on Quantum by TEC, Govt. of India, she is an Expert Evaluation Committee member for AICTE-NEAT to evaluate educational courses and software products for AICTE, the Govt. of India, and an evaluator for Smart India Hackathon- a mega event of AICTE, Govt. of India. Dr. Mala has been granted a design patent and has published a utility patent under IP, Govt. of India. She has published three edited books and one textbook and more than 60 papers in reputed, refereed journals and conferences along with 15+ book chapters. Her MOOC course for Udemy has more than 2900 learners' enrollment and her textbook has been placed in several national and international university curriculums. She has received "Best Faculty of the year award -2024 (under specific innovations category) from Computer Society of India, "Best Techno Faculty Award" from ICTACT, Govt. of Tamil Nadu, and several other laurels and recognitions from reputed Govt. bodies and organizations. She is a member of professional bodies like IEEE and ACM and she forms the editorial and reviewer boards of international journals and conferences. Dr. Mala is listed in the "Who's Who list of the SEBASE repository" of the University College of London, UK for her research work in search-based software engineering. Her area of research includes artificial intelligence (AI), machine learning, deep learning, explainable AI (XAI), software engineering, cyber physical systems, and cyber security.

Dr. Anto Cordelia Tanislaus Antony Dhanapal is an accomplished assistant professor in the Department of Chemical Science at Universiti Tunku Abdul Rahman (UTAR), Malaysia. With over two decades of academic and research experience across India and Malaysia, she specializes in food science, nutrigenomics, and AI-driven precision nutrition. Holding a Ph.D. in Food Science and Nutrition, Dr. Dhanapal is currently advancing the integration of artificial intelligence in precision nutrition, focusing on developing virtual digital twins and data-driven approaches to personalize dietary recommendations. She has spearheaded numerous research projects, securing 11 competitive grants in the last five years. Dr. Dhanapal has published over 60 research papers and book chapters, and her groundbreaking work has earned international recognition, including the UK-Malaysia Higher Institution Partnership Award. She is also a passionate advocate for public health, actively engaging in initiatives like the GENEY program to promote precision nutrition among Malaysian youth, further exemplifying her dedication to research, education, and societal well-being.

Dr. Saurav Sthapit is currently working as an assistant professor (research) at the Centre for Computational Science and Mathematical Modeling, Coventry University, UK. Before that, he was a research fellow in Cyber Systems Engineering at WMG at the University of Warwick. He received his B.E. in Electronics and Communication Engineering from Tribhuvan University, Nepal, his M.Sc. in Embedded Systems from the University of Kent, England, and his Ph.D. from the Institute for Digital Communications, within the School of Engineering, University of Edinburgh, Scotland. Dr. Sthapit has more than 15 years of experience in industry and teaching and has published more than 10 research papers in reputed international journals. His research interests include reinforcement learning, embedded systems, IoT, signal processing, simulation and modeling, machine learning, computer vision, game theory, operational research, optimization, automotive, and cyber security.

Dr. Anita Khadka is an Assistant Professor at the University of Warwick. Her research area involves Responsible AI investigating AI Resilience, and Assurance. She has contributed to several EPSRC-funded projects, including those on Robotic Artificial Intelligence in Nuclear (RAIN) and FAIRSPACE investigating the impact of adversarial attacks and developing defensive strategies to mitigate them. She focuses her research mainly on critical infrastructure including healthcare, and nuclear where the consequences of adversarial attacks can be high. She has published numerous papers in high-quality journals and conferences including *ACM Computing Surveys*, *AeroSpace*, and *RecSys* Conferences and has also secured grants funded by the University of Warwick and Sprite+. Dr. Khadka has taught and supervised students across various levels, including undergraduate, Master's, and M.Sc. apprenticeship programs, with a focus on Cybersecurity and Artificial Intelligence. She was an Assistant Professor at Northeastern University, London where she also held the Program Lead position for the M.Sc. in AI and Data Science degree. She obtained her M.Sc. in Intelligent Systems and Robotics from the University of Essex in 2012 and her Ph.D. in Computer Science from The Open University in 2020. She has extensive experience in both academia and industry.

Section I

Introduction

Role of AI in Smart CPS Design and Development

1 Role of AI Techniques in Smart Cyber-Physical Systems Development

1.1 CYBER-PHYSICAL SYSTEMS: AN INTRODUCTION

A cyber-physical system (CPS) can be regarded as a network of tangible objects, including sensors, actuators, control units, and others, that collaborate with computational elements to regulate and safeguard a physical infrastructure. These systems are highly interconnected and integrated, and they provide new functionalities that improve the quality of life and enable technological advancements in critical areas.

Some examples of these areas include defense and homeland security, personalized healthcare, emergency response, smart cities with traffic and infrastructure management, smart manufacturing, and energy supply.

They are made up of digital, analogue, physical, and human components that interact with one another and are designed to perform efficiently through the integration of software and hardware. They are basically the systems that have the collaborative working of a number of independent entities taking decisions to perform a particular task or functionality. They heavily depend on the integration of software programs with hardware elements.
Some typical examples are

- Robots for defense applications, industrial automation, healthcare, and other sectors;
- Drones equipped with software for specific purposes such as warfare, medicinal deliveries, wild-fire mitigation, smart agriculture, contagious disease preventions etc.;
- Healthcare devices equipped with decision-making algorithms; industrial internet of things (IIoT), medical internet of things (MIoT);
- Unmanned arial vehicles (UAVs), driverless cars, vehicle platoons;
- Intelligent and autonomous sensor systems with applications in weather forecasting, satellite positioning environmental monitoring etc.

CPSs are constructed from and depend on the seamless fusion of physical elements like robots and other equipment with computer algorithms. Such systems often include numerous decision-making entities. CPS is regarded as a smart factory in the industrial industryset-up.

DOI: 10.1201/9781003489986-2

CPS will have a revolutionary and pervasive impact; this is already obvious in technology such as autonomous vehicles, intelligent buildings, smart energy systems, robots, and smart medical equipment that are already in the process of being developed.

Generally, a CPS architecture, as shown in Figure 1.1, consists of the following elements:

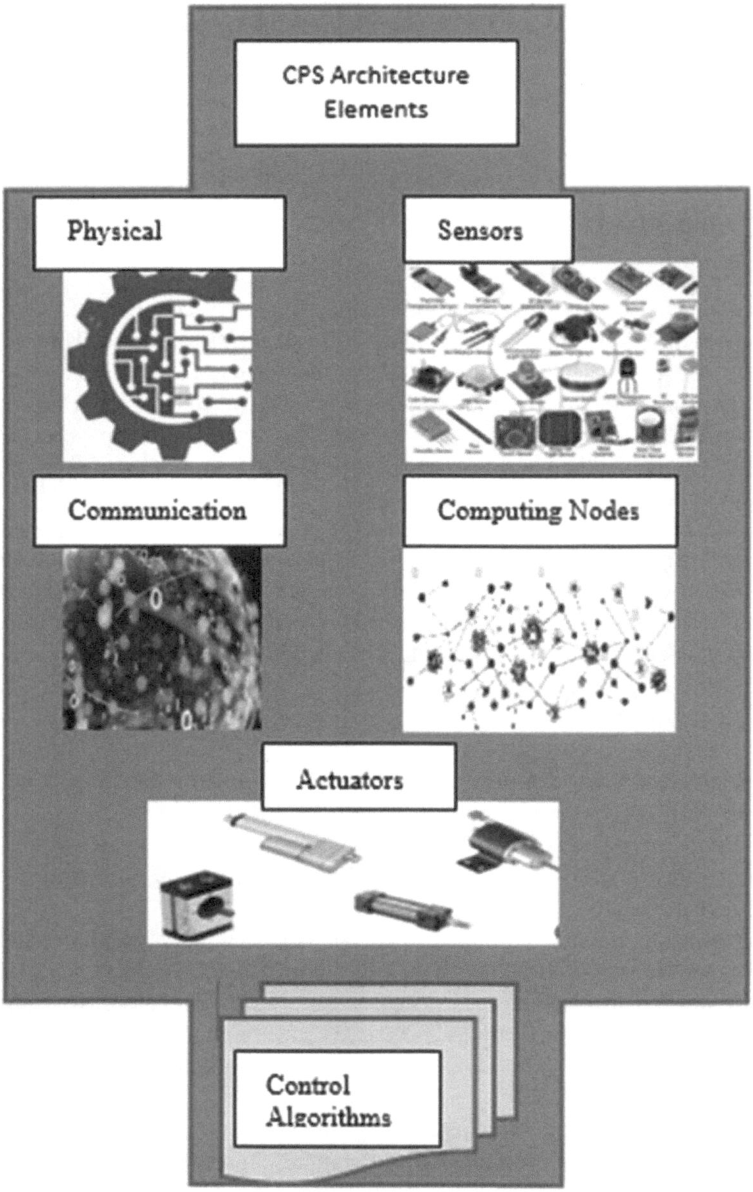

FIGURE 1.1 General CPS architecture with elements such as physical process, sensors, communication system, computing nodes, actuators, and control algorithm.

- **Physical Process:** It indicates the environment in which the CPS interacts to achieve its objective.
- **Sensors:** These are the elements that help to collect data from the environment. Some typical examples are pressure sensor, temperature sensor, humidity sensor etc.
- **Network**: The communication medium through which the elements interact with the other elements. It includes both wired as well as wireless communication medium.
- **Computing Nodes**: These are the microprocessors, microcontrollers, servers etc. to process the data collected by the sensors with their logic.
- **Actuators**: These are used to give outcome from the processed data. It indicates the physical process which may be switching off a control, raising up an alarm etc.
- **Control Algorithms (for computation):** These help in decision-making process of the CPSs. It has the internal logic specific to the application we are working on.

In order to fully realize the potential of CPS, it will be necessary to have interoperability between different types of components and systems. This will be enabled by new reference designs that make use of shared vocabularies and specifications.

In order to address the difficulties and opportunities that are associated with CPSs, it is necessary to have a widespread consensus in the fundamental principles, as well as a common understanding of the critical new capabilities and technologies that are specific to CPS. Thus, it is essential to refer the architecture (Griffor et al., 2017) provided by NIST USA, as shown in Figure 1.2. It has three elements:

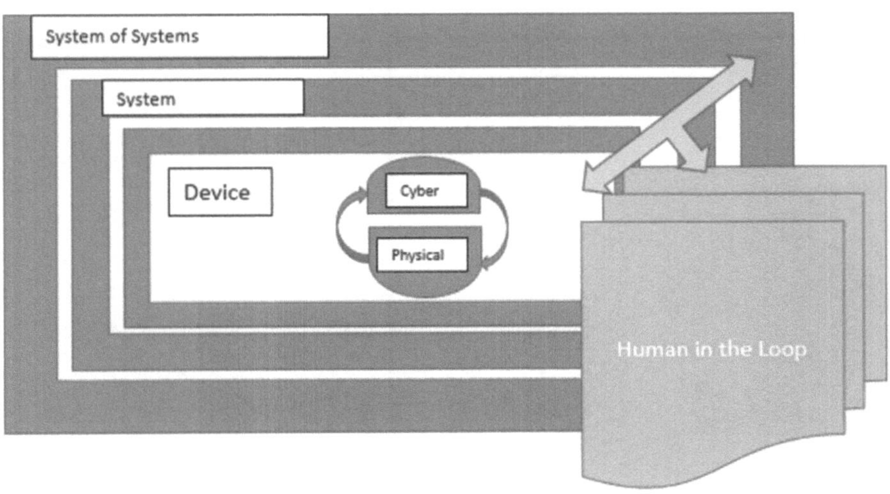

FIGURE 1.2 Cyber-physical system elements, as per NIST, USA (Griffor et al., 2017) showcased as system of systems at the outer layer, system at the next inner layer, device in the inner-most layer that has interactions with human and other systems. The device has a process cycle which has cyber part and the physical part.

- **Single CPS Device**
  - Refers to a singular device, such as a video camera, robot, or thermostat. The analysis will prioritize the robustness of the design to ensure its effectiveness as a valuable component of a CPS.
- **System or Subsystem**
  - A collection of distinct cyber, physical, and cyber-physical devices, such as heating, ventilation, and air conditioning (HVAC) system, which includes components including a thermostat, air handler, compressor, and furnace.
- **SoS (System of Systems)**
  - A network of interlinked systems, such as a power company's demand response program interacting with individual HVAC systems, in order to establish a harmonized energy system.

1.2 CPS FRAMEWORK AS PER NIST

Any CPS framework as per NIST (Griffor et al., 2017) is shown in Figure 1.3. It accommodates various sensor types, including basic to smart, static and adaptive, single-mode and multi-faceted, local, distributed, federated, or centralized control schemes, and input-fused loops. Interactions might be loosely or tightly tied, such as in repurposing distributed sensing in an existing CPS or in telemedicine or smart grid operations. Coupling presents both a chance to achieve CPS's ambition and a barrier to its certainty. Emergent behaviours may be desired or unwelcome in new services. To reduce complexity, CPS may be co-designed. Co-design involves integrating hardware and software design to explain choices between cyber and physical system components.

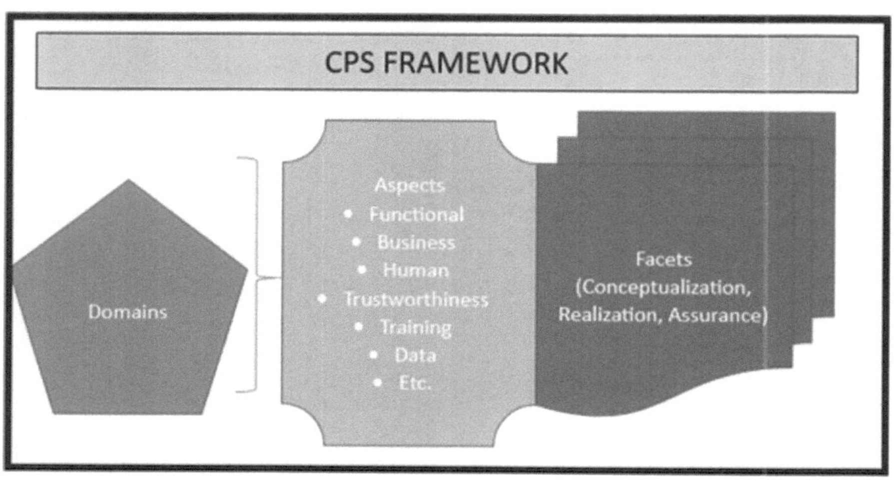

FIGURE 1.3 CPS framework, as per NIST, USA (Griffor et al., 2017) showcased as three major layers, namely domains, aspects, and facets.

Figure 1.3 illustrates the various application areas of CPS, which are represented by domains. Concerns, which are voiced by a wide variety of stakeholders in their individual and collective points of view, are a basic notion that drives the CPS framework.

Concerns that are conceptually identical or related are grouped together into *aspects*, and activities within the components of the facets are responsible for addressing these concerns.

They are addressed by properties, which are concrete assertions that address the worries. In addition to requirements, design elements, tests, and judgments, they also incorporate standards.

Aspects are collections of issues that are conceptually equivalent to one another or related to one another. It is possible that there are adjustments or alternative appropriate groups of concerns that could be of use to a particular application of the CPS framework in a particular setting. It is important to note that concerns and aspects are not considered to be orthogonal. Each component includes a collection of activities and artifacts (outputs) that are clearly specified for the purpose of addressing challenges. It has been determined that there are three distinct aspects: conceptualization, realization, and assurance.

1.3 COMPONENTS OF THE CPS FRAMEWORK

Based on Figure 1.3, the components are explained further:

- **Domains**
  - Manufacturing industries
  - Agriculture
  - Healthcare
  - Smart cities
  - Infrastructure
  - Defence
  - Education
  - Disaster management
  - Energy
  - Entertainment industry
  - Sports
  - Social media networks
  - Transportation
  - Weather
  - Scientific advancements, etc.
- **Facets**
  - Conceptualization
    - It includes the identification of devices, hardware components, sensors, and other things.
    - It also provides how composition and decomposition to be done on the CPS network of subsystems based on requirements and preparation of models for the final real-time development.

  - Realization
    - It shows the way the CPS components are performing their activities for the CPSs to deliver its expected functionality.
    - It includes the way of communication between the various components and the instance creation of new CPS components based on the need.
  - Assurance
    - It provides the confidence level of the CPS based on its expected performance.
    - It has the list of policies and guidelines to evaluate the performance of the CPS.
- **Aspects and Concerns**
  - It includes the internal activities to be done by the CPS such as security, communication, data collection through sensors, outcome execution through actuators etc. It has the tasks such as measuring CPS capability and other associated characteristics, performance of the CPSs, state changes, managing uncertainty, environmental aspects, regulatory concerns, time to market the product to improve the business possibilities, human usability of the CPSs, privacy, reliability, security and safety aspects to give assurance to all the stakeholders of the system.
  - The concerns include business, timing, human, assurance, life cycle, composition of components, etc.
  - Thus, it is necessary to describe the application of a set of multiple concerns that are considered pertinent to the CPS at hand. This is known as the composition of concerns, and it specifies how the subject should be interpreted and implemented.
  - Whether a concern is applied to a CPS has an effect is contingent on the aspect and activity under consideration. In general, the concern may associate that application with a particular set of properties or requirements that the CPSs must fulfil. Therefore, the application of two or more concerns necessitates the possession of every property specified by the set of concerns.
  - It is formally represented as,

$$\overrightarrow{\text{CPS}} = \text{Set of properties related to a concern in the CPS} \quad (1)$$

The integration of more than one concern can be represented as:

$$\overrightarrow{C1} * \overrightarrow{C2} = \overrightarrow{C1}\ \overrightarrow{C2} \quad (2)$$

- **Activities and Artifacts**
  - Some of the specific activities include business case analysis, managing lifecycle activities, prototypes design, implementation or production activities such as creation of functional components, test plans creation, integration and testing of components, deployment activities such as performance

assessment of the deployed product, quality assessment, version management of the CPS based on continuous evolution etc.
- Energy audit and assessments and energy recovery assessments to be done as part of the activities of CPS.
- Operations done on cyber-physical layer activities such as product-specific and domain-specific ontologies and realization of physical layer with respect to the life cycle activities are done.
- Some of the other artifacts include analysis report, business plan, strategy plan, assurance report, audit report based on configuration, validation reports, test reports and certifications based on regulatory compliance assessment.

1.4 PROCESS MODELS FOR CPS APPLICATION DEVELOPMENT

Generic life cycle phases of CPS application development are

- **Ideation**
  - Here, basic requirements with reference to the market need or based on the real-time system requirement to automate and to provide efficient and viable solution are collected. Based on that, the initial analysis will be done based on the real-time requirements and the market viability to proceed for the development of the CPS.
  - As the CPS involves high cost and resource requirements, this ideation phase initially conforms the expected product details.
  - Then, during this phase, the plan document outlines the expected product's expected functionalities, budget and timelines along with risk factors involved during the development and post deployment.
- **Proof of Concept**
  - Once the ideation phase is completed, a workable prototype or a simulated product should be developed to get the concrete evidence as the proof of concept.
- **Design**
  - Once the prototype is developed, the selection of appropriate hardware components, designing of software components, integrating the firmware will be done along with the design documents.
- **Development**
  - During this phase, the actual development of the entire CPS will be done by building the software components, and integrating the hardware and software components. It includes the development of firmware, integration of sensors, and developing the other associated infrastructure such as cloud services and other associated server-side functionalities.
  - During this development phase, the developers will do the testing of their developed software components in order to ensure their proper functionalities.
  - At this phase, the entire CPS will be integrated with the appropriate hardware components, software components, and firmware.

- **Testing**
  - During this phase, the entire CPS integrated with all the hardware, software, and firmware will be tested for its entire functionality, working of interfaces, and the quality attributes such as reliability, security, supportability, and safety. Both manual as well as automated testing will be conducted.
  - The objective here is to ensure that the developed CPS has met all the stated and quality requirements, as well as the software is ready for deployment.
- **Manufacturing**
  - During this phase, the CPS will be manufactured that involves the assembling of the entire system and testing the final manufactured product for reliability, quality, and performance.
- **Deployment**
  - The CPS needs to be deployed in the target environment or be placed in the market for sales. Here, the marketing and sales strategies will be formulated. The need for distribution channels will be identified in order for the product to reach the customers. Finally, a customer support system needs to be developed in order to provide continuous support throughout the lifetime of the product.
- **Support and Maintenance**
  - This is the final phase in the development life cycle of the CPS, in which the CPS will be monitored for its expected performance and functionalities and whenever there is a need, the appropriate updates and the maintenance activities will be performed.
  - It includes the support activities need to be provided whenever there is an issue or problem in the working of the system.

1.4.1 Waterfall Model

The steps involved are

- Conceptualization
- Realization
- Assurance

As waterfall life cycle model is a conventional life cycle model. It provides system engineering flow that facilitates a process driven by requirements that results in guaranteed and validated operation. It is important to understand that while this suggests a sequential progression through the different aspects, it is common to detect the ability to repeat and transfer modifications made in one aspect to the others.

1.4.2 Agile Methodology

It has the following stages

- Ideation
- Realization

- Assurance
- Iterative refinement to achieve increased levels of detail

Analysis in isolation might occasionally lead to a different outcome than what was first envisioned at a broad level. Therefore, an agile approach aims to achieve a swift realization and assurance by focusing on a minimal or fundamental conception.

After validating the initial assumptions about the CPS, the agile development process gradually adds more specific information to each aspect in order to ultimately complete the set of artifacts.

1.4.3 Service-Oriented Architecture (SOA)

It has the following stages

- Concept comprehension
- Selection of suitable manifestations as described
- Assurance

It is possible to imagine and implement dynamic services on current CPS.

1.4.4 Reverse Engineering

It involves the analysis of a product or system to understand its implementation, followed by the identification of its underlying concepts, and finally the verification of its integrity. In order to comprehend a deployed CPS and perhaps expand or improve it, reverse engineering examines the actualized CPS to identify its characteristics and examines its documentation to ascertain its guarantees. After analysis, tweaks and improvements can be initiated at any stage

In reverse engineering as well as in forward engineering, one has to conduct a gap analysis by examining a collection of CPSs, including their interconnected systems, in order to identify any deficiencies or redundancies in the crucial points of interoperability.

Gaining insight into the possibilities for integration or addressing gaps provides comprehensive guidance for making informed judgments regarding the integration of systems and capabilities.

1.5 TYPES OF CPS

- **Critical Tightly Coupled CPSs**
    - The framework would be utilized to construct a comprehensive and intricate procedure specifically tailored for critical infrastructures, such as the electricity grid. The focus on strict criteria and guarantees, as well as limitations from various perspectives, would be apparent.
- **Loosely Coupled CPSs**
    - This approach is particularly suitable for CPS applications that utilize current CPS capabilities and combine them in innovative ways. It places

less focus on strict requirements and places more importance on achieving functional objectives.

- **Semi-Coupled Using Superficial Analysis**
  - To discuss concepts or explore different approaches to CPS problems, it may be sufficient to use short portions of the framework. The utilization of the framework structure and language facilitates the comprehension of the concept's essence, as the framework establishes a framework for the discourse.

1.6 CPS VERSUS IOT

There are a number of terminologies and phrases in addition to CPS, define systems, and concepts that are comparable to or connected to one another. Some examples of these include the industrial internet, the internet of things (IoT), machine-to-machine (M2M), and smart cities (Radanliev et al., 2021).

There is a great amount of overlap between these concepts, particularly CPS and IoT, to the point where CPS and IoT are commonly used interchangeably. As a result, the strategy that is presented in this CPS framework ought to be considered to be equally applicable to IoT. It is shown in Figure 1.4.

Over time, the terms CPSs and IoT are coming closer together in their definitions to emphasize hybrid systems comprised of human, digital, analogue, physical, and interactive components that are function-engineered via integrated physics and logic.

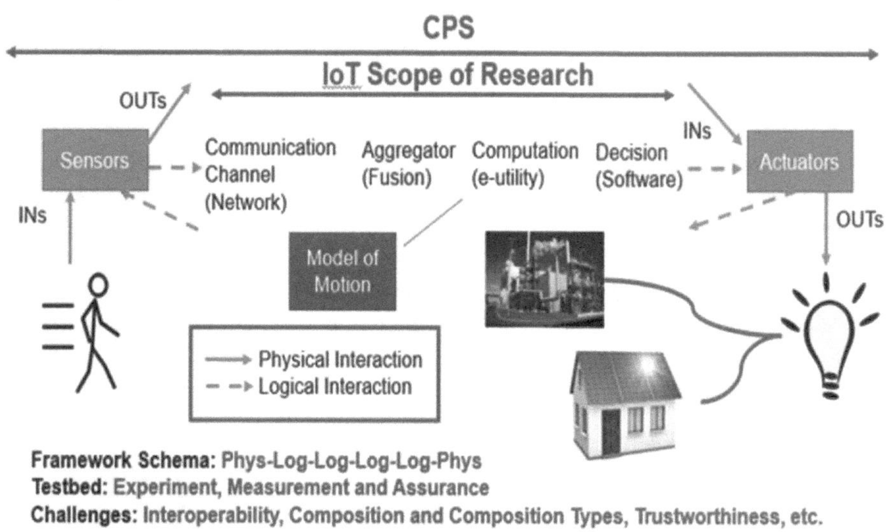

FIGURE 1.4 CPS with IoT scope of research by NIST, USA (Griffor et Al., 2017) given in this figure as IoT scope of research that has sensors to get the input from the human or from the environment and actuators through which the output is given out; for processing, there are communication channels, aggregator, computation and decision software present in the system that has both physical as well as logical interaction.

CPS and IoT, including, but not limited to, energy infrastructures, advanced manufacturing, building control, transportation, and healthcare, influence multiple economic sectors of the global economy. These sectors are enabled to develop innovative applications.

Existing domain-specific design and management approaches for these systems result in redundant efforts, limited interoperability and sharing across domains, and the absence of formal, robust methods for design, evaluation, verification, and validation.

The CPS framework presents a structured methodology for analysis that is founded upon the fundamental principles of facets (which encompass modes of system engineering: conceptualization, realization, and assurance) and aspects (which comprise clusters of concerns: business, human, trustworthiness, timing, data, composition, boundaries, and life cycle). These elements, in conjunction with integrated cyber-physical functional decomposition, serve as the underpinnings for both CPSs and IoT.

Physical components, their corresponding sensors and actuators, control systems, analytics, and ultimately user functionality and optimization, with assurance that requirements have been met, comprise the fundamental elements of a CPS or IoT.

Organized by groups of activities or facets (thinking modes: conceptualization, realization, and assurance) and aspects (collections of concerns: functional, business, human, trustworthiness, timing, data, composition, boundaries, and life cycle), the CPS framework delineates the labour and deliverables of analysis that facilitate the validation and assurance of CPS/IoT, as well as the system engineering analysis, design, development, operation, and vision.

While the majority of CPS/IoT currently operate within specific domains, it is indisputable that CPS/IoT of the future will involve extensive interoperability and communication across domains in order to provide coordinated functionalities. The methodology of this CPS framework establishes a strong premise for constructing consensus technical architectures that incorporate a shared lexicon and practical scenarios. These architectures serve as the foundation for collaboration and cooperation to advance all CPS/IoT applications.

1.7 CHALLENGES IN CPS DEVELOPMENT

CPS is likely to exhibit emergent behaviours due to its open nature. CPS requires a technique for interoperability, evolution management, and emergent consequences. Multiple manufacturers manage many subsystems in large-scale CPS like smart grid and smart city. Some of the general use cases of CPS include

- Cell phones in cars can serve as traffic sensors, and energy usage data can help diagnose equipment defects.
- Cross-domain apps for smart manufacturing, smart cities, smart healthcare, distribution of energy and sensor-based consumer analysis etc.

The potential impact of CPS on the real world and their interconnectivity raises trustworthiness concerns. We must prioritize security, privacy, safety, reliability, and resilience for interconnected devices and infrastructures. CPS networks may have third-party-owned and operated infrastructure, such as brokers and aggregators, leading to trust issues including publishing and subscribing communications, certificate authorities, and type and object registries.

CPSs can be made flexible so that the available components can be dynamically mixed and the system architecture can be adjusted during runtime to address new issues. However, there are still bottlenecks in terms of time required for composing the components, and if needed purchasing relevant resources to build a required system may have compatibility issues. Some services used from other service-providers may also be pay-per-use kind of resource sharing and utilization, which also poses some real-time implementation issues.

For the CPSs to work, there should be an integration of hardware and computational components. This will increase the complexity in terms of algorithmic as well as platform complexity.

The communication system in CPS should have both standalone as well as network-based one. The protocols will have both traditional as well as advanced objects transmission protocols. Some are very energy-constrained ones and hence it should also be kept as additional resource in the backup side. As CPS is a collection of different kinds of devices, and heterogeneous in nature, its complexity is dynamic based on its composition of components.

The next and most important challenge with respect to CPS is its sensitivity on time. Each time interval needs to be accommodated properly while designing the architecture of the CPS. Any lagging between data collection through sensors and the availability of data for processing will lead to hazardous impact on the environment, which relies on the outcome from CPS. In the case of time-sensitive real-time applications, the timestamps of each event and the availability of data on time are highly important as otherwise it will lead to huge life and material damages.

The environmental aspects assessment is a key factor for operating the CPS continuously. Generally, the operating environment includes not only the hardware components but also the human and hence the interactions with human and other hardware elements are essentially measured to take appropriate next action.

1.8 WHAT ARE SMART CPS?

The term "smart cyber-physical system" refers to intelligent systems that are comprised of engineered networks of physical and computational components that interact with one another.

These CPSs get smarter because of recent technical developments, which increase their efficiency and performance by incorporating new forms of information and communication. A key area of investigation, however, remains the question of how much intelligence CPSs should have to ensure the security of mission-critical systems.

When these CPS are integrated with advanced information and communication technologies, they become smarter and are present in a wide variety of everyday applications. These applications include education, where they are used to better support the learning and evaluation process; smart living, where they are used for efficient energy management, comfort, and security surveillance purposes; and the healthcare sector, where they are used to improve patient monitoring and treatment.

The supply chain of a smart factory facilitates equipment outage reduction, quicker equipment reuse, and shorter product development cycles with fewer defects. Such a smart industrial helping CPS denotes a cohesive and uninterrupted exchange of information and data among fully interconnected, adaptive systems operating at all levels of automation, which are capable of discerning and reacting to evolving circumstances.

In such a smart setup, the manufacturing system itself will be equipped with the smartness and the ability to expand the supply chain operations to make the factory's operations smart. In such a smart manufacturing system, it provides an autonomous execution system that has automation of the entire production process and helps to optimize the entire supply chain network. The monitoring units in such a system will be autonomous in the sense that they operate within the boundaries without any supervision and can interact with the external global network when a need arises.

The industrial robots are becoming common as they can perform dangerous and repetitive tasks, which may be hard for humans to carryout. However, several costly and still in its infancy, industrial robot technology is a current development. They frequently perform dangerous and repetitive tasks utilizing enormous caged devices in place of humans. Due to their limited dexterity in the areas of movement, manipulation, and grasping, these robots also suffer from vision impairments, which prevent them from detecting and navigating objects, including humans.

Nowadays, the hardware cost has become low and so the implementation of the more advanced CPS is possible with artificial intelligence (AI) and machine learning (ML) based smart actions in them. The objective of the smart CPSs is to provide an autonomous self-resilient solution that can make the environment become smarter whether its smart cities, smart infrastructure, smart transportation, smart education, smart healthcare, or smart agriculture etc.

1.9 AI IN CPSS: AN INTRODUCTION

AI is already causing changes in our economy and society, and the rising use of AI to make decisions has sparked a debate about the potential risks and the necessity of making AI decision-making more transparent.

First, let's take a look at the IIoT, which is equipped with AI in industrial automation. This is because AI plays a significant role in the development of smart CPSs.

The IIoT can be defined as the application of technologies such as the IoT and CPS to enhance manufacturing and industrial operations. At the same time, the term "Industry 4.0" (I4.0) represents a paradigm shift in industrial production, a generic designation for sets of strategic initiatives to boost national industries. This technical term helps to relate to new emerging business assets, processes, and services, and a brand to mark a very particular historical and social period.

After conducting a comprehensive evaluation of a substantial body of academic, government, and industrial literature, particular research questions have emerged from the research gaps that have been discovered following the review. The current research on how the integration of sophisticated and interconnected IoT along with

AI, coupled in CPSs, drives the inevitable and autonomous evolution of artificial cognition is lacking a considerable amount of information.

The literature review and taxonomic analysis take into consideration the significance of these research gaps in the context of the discussion on how the progression of technology leads to the unavoidable and independent growth of artificial cognition in socio-technical systems that are coupled and interconnected.

Autonomous vehicles, often known as self-driving automobiles, are yet another use of AI and ML in CPS. The Tesla automobile is a good illustration of how AI can be used in conjunction with IoT devices. AI is utilized in the vehicle, in order to ascertain the conditions of the road, the best speed, the weather, and to forecast the movement of both pedestrians and autos. The CPS can be utilized in smart transportation and smart home applications, which involve the utilization of sensors for turning on lights and opening doors. Additionally, when combined with AI, it has the potential to be utilized for anticipating the ideal time for heating or cooling the building.

In the future, AI in CPSs would encompass health and biological monitoring, robotics systems, intelligent edge devices, and a wide variety of other activities. It would also be used to remedy natural disasters, human errors, or malevolent actions, among other things. A change and evolution in the role of CPSs has occurred as a result of the growing number of linked devices and the IoT.

The incorporation of the IIoT, which has the potential to boost productivity, efficiency, and economic benefits, along with the evolving function of AI, has resulted in the development of CPSs that are more intelligent than they were in the past.

When we talk about CPSs, we are referring to computer–human networks that are in charge of directing physical processes. The capabilities for system analysis of the cyber and physical structures involved are provided by this integration, which includes advancements in real-time processing, sensing, and actuation between the physical domains and the IoT systems.

1.10 AI IN SMART CPS DEVELOPMENT

Manufacturing companies can now automate a greater number of their processes than ever before with the assistance of AI. A multitude of manufacturing processes, including planning, scheduling, and preventive maintenance, may benefit from the implementation of deep learning.

Implementing AI has the capacity to substantially enhance the effectiveness and efficiency of decision-making processes in environments replete with complex data. Additionally, it enables these intelligent systems to assimilate a more substantial volume of information.

As an illustration, John Deere is a manufacturer of driverless, extremely autonomous agricultural tractors at present. Utilizing ML and cloud services, robotic automation applications offer manufacturers novel opportunities to enhance productivity and dependability. AI has emerged as a force that surpasses the constraints of vegetation. While they have historically dominated production floors, contemporary robots no longer perform tedious mechanical tasks. They function as intelligent intermediaries connecting the capabilities of virtual reality and real-world smart manufacturing.

ML enables the optimization and rapid conversion of data into informed decisions and actions. ML methodologies utilize data for the following purposes: normative (including decision automation and assistance), predictive (forecasting future events), diagnostic (examining the cause of occurrences), and descriptive (analysing past events).

By utilizing training data sets, a subset of AI known as "machine learning" enables computers to acquire knowledge without requiring explicit programming. In the presence of new data, ML aids in the generation of predictions, which are subject to modification. The proliferation of data has become an enormous burden as a result of recent technological developments.

We must employ ML techniques to transform this data into an intelligent system. Consequently, ML and large data can be considered complementary. The proliferation of data has experienced a substantial surge in tandem with the advancement of IoT technologies. Efficient implementation of this data is necessary.

By employing ML, future forecasts can be generated.

AI is currently causing significant changes to both economies and society. The advent of AI decision-making has incited a discourse regarding the potential hazards involved and the imperative for increased transparency in AI decision-making.

This elucidates research inquiries regarding the societal and economic implications of AI decision-making's emergence, as well as potential strategies to enhance AI decision-making transparency.

In light of the fact that invasive self-construction processes resembling extremely intelligent human manifestations and the desires of numerous individuals initiate the collective creation, these inquiries serve to enhance comprehension of the subject matter. An example of such intrusive technology is the IIoT.

There is a significant dearth of current research examining how the integration of CPSs and the intricate and interconnected IoT initiates the self-evident and inevitable development of artificial cognition.

An example of the integration of AI and CPS with IoT devices is the Tesla automobile. The vehicle utilizes AI to predict the movements of pedestrians and other vehicles, as well as to evaluate the condition of the road and the optimal speed.

As sensors, the IoT and AI have the capability of not only controlling lighting and door access but also determining the optimal heating and cooling cycles for a structure, by leveraging these features, the CPS become smart enough to solve the real-world problems.

CPS of the future will be capable of performing an extensive array of functions, such as health and biological monitoring, robotic systems, intelligent peripheral devices, and much more. Additionally, they will be employed to rectify acts of nature-induced destruction, human error, and hostile actions.

These systems, which have emerged as captivating domains for AI applications, consist of "smart systems" and the IoT. In light of the elevated CPS standards, AI models, tools, and techniques must surmount a multitude of obstacles. Designers must take into account various non-functional requirements, including but not limited to resources (e.g., CPU and memory), power consumption, device and user safety, delay-tolerant connectivity, and autonomy in decision-making.

The AI-integrated CPS in sustainable smart systems development are listed:

- **Defence Applications**
  - Robots, drones, and other cyber warfare systems are integrated with AI to have highly integrated defence applications.
  - Reduced human casualties during war are the primary concerns in the current smart CPS-based defence applications.
- **Bio-Medical**
  - Smart healthcare integrated with personalization, emergency responder systems, intelligent health assistants etc. help revolutionize the healthcare and bio-medical applications.
  - Pharmaceutical industries are now focusing on CPS with AI in order to provide the corresponding solutions on need.
  - Satellite and underwater warfare are increasingly having adapted with smart CPS to meet out the constraints imposed by such environments.
- **Infrastructure and Transportation**
  - Smart Cities and Infrastructure
    - The energy distribution systems equipped with AI algorithms can optimize energy distribution, reducing waste and improving grid reliability.
  - Traffic Optimization
    - AI-powered traffic management systems can reduce congestion, improve traffic flow, and minimize emissions.
    - Smart CPSs for route optimization, supply chain management, and effective warehouse management will provide efficient transportation solutions.
  - Waste Management
    - AI-based CPSs can optimize waste collection routes and optimize recycling processes, reducing waste generation and landfill usage.
  - Renewable Energy Management
    - In renewable energy generation units, the AI can monitor renewable energy systems and predict maintenance needs, ensuring optimal performance and reducing downtime.
  - Demand Forecasting
    - AI models can analyse energy-consumption patterns and predict demand, enabling efficient energy production and distribution.
  - Grid Integration
    - AI facilitates the seamless integration of renewable energy sources into existing grids, ensuring a stable and reliable energy supply.
  - Sustainable Agriculture and Food Production
    - Under precision farming, AI-powered sensors and drones can monitor crop health, optimize irrigation, and minimize pesticide use.
  - Yield Optimization
    - AI algorithms can analyse data to predict crop yields and optimize planting and harvesting schedules.

- Sustainable Food System
  - AI can help reduce food waste by predicting demand and optimizing supply chains, ensuring food security while minimizing environmental impact.
- Effective Waste Management and Recycling
  - AI-powered systems can automatically sort waste, improving recycling rates and reducing landfill waste.
  - In smart bins, the AI can analyse waste data to identify patterns and develop strategies for reducing waste generation.
- Resource Recovery
  - In waste-segregation-based CPS, the AI can enhance resource recovery from waste, promoting circular economy principles and reducing reliance on virgin materials.
- Environmental Conservation
  - AI-integrated CPS can predict environmental changes, like deforestation rates or pollution levels, enabling proactive conservation efforts.
  - AI-powered drones and sensors can monitor wildlife populations and track endangered species, providing valuable data for conservation strategies.
  - AI algorithms in automated resource allocation systems can reduce waste, helping to protect ecosystems and ensure long-term sustainability.
- Climate-Change Monitoring and Mitigation
  - The smart weather-monitoring systems can analyse climate data and develop solutions for reducing greenhouse gas emissions and mitigating the impacts of climate change.

1.11 CASE STUDY: SAMPLE CPS DESIGN USING MATLAB'S SIMSCAPE

For case study, a Smart Manufacturing Robotics Cyber Physical System developed using MATLAB by Pieter Mosterman (2024) is discussed.

A pick-and-place machine simulates the service of repositioning blocks by one or two positions. Stacks of three coloured blocks must be moved from one spot to another in a specific order, with size and colour determining the final arrangement. Each block's local control determines the ordering, which is an emergent behaviour. So, this is a CPS in the form of a distributed Towers of Hanoi problem. In order to verify the model's resilience in the face of unreliable network conditions, a comprehensive industry-grade network model built on EtherCAT and TCP/IP enables control of the packet error rate (PER).

1.12 RESEARCH WORKS ON THE APPLICATION OF AI IN CPS DEVELOPMENT

CPSs and blockchain (BC) are seeing a meteoric rise in their respective application areas, as noted by Gupta et al. (2020). But due to their complexity, it is a very difficult task to frame robust and correct smart contracts (SCs) for these smart applications.

SC is bringing the conventional business, technical, and industrial procedures into the modern era. The elimination of trustworthy third-party systems is achieved by its self-executability, self-verification, and integration into the BC. This, in turn, saves money on administration and service costs. In addition to lowering security threats, it increases system efficiency. Despite SCs' enthusiastic support for Industry 4.0's new technical changes, several issues related to data privacy and security remain unresolved. This study presents an assessment of SC security flaws in the code that might be exploited by an attacker to compromise the entire BC network or just make it easier for them to infiltrate. According to the literature, authors from all around the globe do not go deeply enough into the difficulties associated with SC security and privacy. According to the current suggestions, creating complicated SCs will not solve their privacy and security problems. Consequently, this study explores numerous AI methods and resources for safeguarding SC privacy. Following that, we take a look at the current problems and obstacles with AI-based SCs. The paper concludes with a case study of retail marketing that employs SC and AI to protect customer privacy and security.

As indicated by Mavikumbure et al. (2024), advanced CPSs are more vulnerable to cyberattacks. Therefore, effective cybersecurity methods are needed to protect these important systems from cyberattacks. In recent years, generative AI (GenAI) has garnered prominence among AI technologies. Its unique capacity to autonomously generate original and diverse information across areas offers possibilities for fresh breakthroughs in various applications. Given GenAI's great success, its role in CPS cybersecurity must be investigated. Thus, in this work, we discuss GenAI's evolution and current condition, its benefits and dangers to CPS cybersecurity, protection techniques against threats, and future research potential. This thorough survey should help the community focus CPS cybersecurity research.

Nethani et al. (2023) have explored that the CPSs are becoming increasingly complex as IoT devices are integrated and decision-making procedures are needed. ML is a powerful tool for processing and analysing big data sets in CPS. Optimizing ML algorithms for fast, accurate decision-making remains a challenge. This work proposes using ML, PSO, and IoT to solve this problem. The proposed framework optimizes CPS ML algorithms using PSO, a natural optimization method. PSO mimics a swarm of particles to discover the best multidimensional answer. We use PSO and ML to increase CPS decision-making precision and effectiveness. IoT devices in CPS provide real-time data, allowing the collection of diverse and dynamic data from many sources. We employ ML algorithms to assess patterns, trends, and anomalies in this data for real-time decision-making. The technique we propose has two benefits. First, PSO optimization improves ML algorithm efficiency and reduces large-scale dataset computing complexity. Second, IoT devices enable real-time data use, improving decision-making speed and accuracy.

Liu et al. (2023) examined how explainable AI (XAI) might improve transparency and trustworthiness in medical image processing in medical cyber-physical systems (MCPS). Thus, this study presents an explainable paradigm combining ML with knowledge reasoning. When framework evolution target feature findings and reasoning results match and are dependable, the model is explainable. MIoT patient data security and privacy are new challenges with these devices. Attack detection is crucial to MCPS security. Based on sparse observability, the MCPS model with only sensor assaults has necessary and sufficient requirements for attack detection. Assuming some MIoT sensors are protected, the attack detector and state estimator are built. It is shown that protected MIoT sensors improve assault detection and state estimation. Experimental results suggest that XAI in medical image analysis within MCPS enhances lesion classification accuracy, removes low-quality medical images, and explains recognition results. This lets clinicians comprehend the mechanism's decision-making and decide whether to believe the results based on the framework's explanation.

Lu et al. (2022) have introduced a unique information-centric networking (ICN)-based system model in CPS that processes IIoT data closer to the edge than a content provider. Based on the ICN system paradigm, we present an edge-assisted authentication technique in CPS to prevent illegal access and reduce workload for resource-constrained devices. We use edge routers to delegate security operations and session handshake procedures to meet the growing issues of monitoring and securing IIoT devices in the ICN. Our scheme's security and practicality are proven by formal proof and performance study.

Almutairi et al. (2023) presented quantum dwarf mongoose optimization with ensemble deep learning based intrusion detection (QDMO-EDLID). The QDMO-EDLID technique uses feature selection (FS) and ensemble learning to detect intrusions. The QDMO algorithm selects feature subsets in QDMO-EDLID. In addition, intrusion categorization uses an ensemble of Convolution Residual Network (CRN), Deep Belief Network (DBN), and Deep Auto-Encoder (DAE) models. Performance of QDMO-EDLID was measured using benchmark intrusion datasets. QDMO-EDLID performed better in several performance indicators, according to simulations.

Lyu et al. (2023) have reviewed the complexity of CPS development and operation, which requires an effective engineering approach. In recent years, DevOps has promoted continuous integration of system development and operational deployment. Even though AI has great potential, it is presently limited in its usage to improve system design and integration. AIDOaRT is a three-year H2020-ECSEL European project involving 32 organizations in clusters from 7 countries that use AI-augmented automation to model, code, test, monitor, and build CPSs. Model-driven engineering (MDE) ideas and methodologies will be used to provide a framework with AI-enhanced approaches and tooling for constructing trustworthy CPSs. The framework fits DevOps principles, which combine software development with IT operations. The project aims to automate decision-making and system development with AI for IT operations (AIOps). This document describes the project's context, goals, and approach.

1.13 SUMMARY

In the rapidly changing paradigm of contemporary industry, there is a problem with integrating physical space with the enabling technology of Industry 4.0. As a result of the ever-changing nature of modern society, there is an increasing need for highly intelligent CPS equipped with cognitive capabilities, which allow these systems to function independently without human interaction.

The integration of CPS with AI and ML paves the way for smart factory-related highly automated processes in global supply and value chains. One of the main goals of developing fully or partially autonomous drones is to make it possible to conduct remote surveys and inspections, as well as rescue operations in hazardous environments.

In the past, people were too vulnerable to these dangers to enter these marketplaces. As a result of our rapid expansion, we require production-specific maintenance solutions and a plethora of control applications. Notably, AI has the potential to lessen the amount of human work needed to detect vulnerabilities or dangers in cyber/CPSs.

As these CPSs have operations or connections over the internet, they have cyber security risks and the cost of providing cyber security is very high. However, the combination of CPSs and AI with ML is vital in these contexts to have safe, secure, and reliable CPS for the domain under consideration.

PRACTICE QUESTIONS

I. Short Answers – Remember/Understand
   1. What are cyber-physical systems (CPSs)?
   2. What do you mean by aspects in the NIST framework? Give example.
   3. How do you think an industrial robot be considered as a CPS and not an IoT system?
   4. Name few CPSs that are used in healthcare sector.
   5. Classify the various components of CPS framework.
   6. List few aspects related to CPSs.
   7. Show the different artifacts derived from an "airplane auto-pilot" CPS.
   8. "Testing is the most critical phase in CPS development". Justify your answer.
   9. List the life cycle phases of agile methodology in CPS product development.
   10. Analyse the self-driving cars and examine each and every component of it with detailed explanation on why these are needed in developing such a system.

II. Descriptive Questions – Apply/Analyse
   1. Elaborate the NIST framework and the components in it.
   2. Why reverse engineering is needed in CPS design?
   3. Consider a drone-based fertilizer spray system. Identify its type and elaborate its characteristics.
   4. Classify the following as CPS or simple IoT:
      (i) Autopilot
      (ii) Temperature monitor
      (iii) Integrated sensor-based home automation system
      (iv) Weather forecasting application
      (v) Smart refrigerator with order booking
   5. Examine the challenges related to CPS design.

III. Higher Order Thinking Questions – Analyse/Create/Evaluate
   1. Explore the application of AI in smart CPS design and development.
   2. Evaluate the unique characteristics of CPS with reference to IoT.
   3. Explore the life cycle phases of software development in CPS design.
   4. Showcase the application of facets in the CPS design and development.
   5. Examine the components of factors and aspects with reference to "atomic power plant CPS".

BIBLIOGRAPHY

Almutairi, L., Daniel, R., Khasimbee, S., Lydia, E. L., Acharya, S., & Kim, H. I. (2023). Quantum dwarf mongoose optimization with ensemble deep learning based intrusion detection in cyber-physical systems. *IEEE Access,* 11, 66828–66837.

Dave, A., Bang, H., & Malikopoulos, A. A. (2024). A framework for effective AI recommendations in cyber-physical-human systems. *IEEE Control Systems Letters,* 8, 1379–1384.

Glaser, B. G., & Strauss, A. L. (2017). *Discovery of grounded theory: Strategies for qualitative research. Routledge:* IFR 2018 Positioning Paper, Robots and the Workplace of the Future, International Federation of Robotics. Retrieved from https://ifr.org/downloads/papers/IFR_Robots_and_the_Workplace_of_the_Future_Positi oning_Paper.pdf

Griffor, E., Greer, C., Wollman, D., & Burns, M. (2017). *Framework for Cyber-Physical Systems* (volume 1), Overview, Special Publication (NIST SP), National Institute of Standards and Technology: Gaithersburg, MD [online], https://doi.org/10.6028/NIST.SP.1500-201 (Accessed November 17, 2024).

Gupta, R., Tanwar, S., Al-Turjman, F., Italiya, P., Nauman, A., & Kim, S. W. (2020). Smart contract privacy protection using AI in cyber-physical systems: Tools, techniques and challenges. *IEEE Access,* 8, 24746–24772.

Jeschke, S., Brecher, C., Meisen, T., Özdemir, D., & Eschert, T. (2017). Industrial internet of things and cyber manufacturing systems. Springer.

Kammerer, F. (2020). Self-building technologies. *AI & Society*, 35(4), 901–915. doi:https://doi.org/10.1007/s00146-020-00962-8 1484

Koivisto, L. M. C., H., Iwakiri, I. G., Loureiro, J. M., Ribeiro, A. M., & Nogueira, I. B. (2020). Modelling of a pressure swing adsorption unit by deep learning and artificial intelligence tools. *Chemical Engineering Science,* 224, 115801. doi:https://doi.org/10.1016/j.ces.2020.115801

Koshijima, I., Niida, K., & Umeda, T. (1996). A micro module approach to the design and control of autonomous decentralized chemical plant. *Journal of Process Control,* 6(2-3), 169–176. doi:https://doi.org/10.1016/0959-1524(95)00047-X

Liberati, N. (2020). The Borg–eye and the We–I. The production of a collective living body through wearable computers. *AI & Society,* 35, 39–49. doi:https://doi.org/10.1007/s00146-018-0840-x

Lu, T., Zhao, J., Zhao, L., Li, Y., & Zhang, X. (2015). Towards a framework for assuring cyber physical system security. *International Journal of Security and its Applications*, 9(3), 25–40.

Liu, W., Zhao, F., Shankar, A., Maple, C., Peter, J. D., Kim, B. G., … & Lv, J. (2023). Explainable AI for medical image analysis in medical cyber-physical systems: Enhancing transparency and trustworthiness of IoMT. *IEEE Journal of Biomedical and Health Informatics,* 99, 1–12.

Lu, Y., Wang, D., Obaidat, M. S., & Vijayakumar, P. (2022). Edge-assisted intelligent device authentication in cyber–physical systems. *IEEE Internet of Things Journal*, 10(4), 3057–3070.

Lyu, D., Song, J., Zhang, Z., Wang, Z., Zhang, T., Ma, L., & Zhao, J. (2023). Autorepair: Automated repair for AI-enabled cyber-physical systems under safety-critical conditions. *arXiv preprint arXiv:2304.05617.*

Mackenzie Gavel, M. (2021). ABI research blog. Retrieved from www.prnewswire.com/.

Mavikumbure, H. S., Cobilean, V., Wickramasinghe, C. S., Drake, D., & Manic, M. (2024, July). Generative AI in Cyber security of cyber physical systems: Benefits and threats. *In 2024 16th International Conference on Human System Interaction (HSI)* (pp. 1–8). IEEE.

Nethani, S., Sivaranjani, L., Kumar, M. A., Lal, B., & Tiwari, M. (2023, August). Recognition and integration of AI with IoT for innovative decision making techniques in cyber physical systems. *In 2023 Second International Conference on Augmented Intelligence and Sustainable Systems (ICAISS)* (pp. 190–195). IEEE.

Nguyen, P. H., Sen, S., Cassoli, B. B., Jourdan, N., & Magnanini, M. C. (2023). *Software engineering and AI for data quality in cyber-physical systems/internet of things. SEA4DQ'22 Report.*

Nogueira, I. B., Ribeiro, A. M., Requião, R., Pontes, K. V., Koivisto, H., Rodrigues, A. E., & Loureiro, J. M. (2018). A quasi-virtual online analyser based on an artificial neural networks and offline measurements to predict purities of raffinate/extract in simulated moving bed processes. *Applied Soft Computing,* 67, 29–47. doi:https://doi.org/10.1016/j.asoc.2018.03.001

Pieter, M. (2024). *Smart Manufacturing Robotics Cyber Physical System.* MATLAB Central File Exchange. www.mathworks.com/matlabcentral/fileexchange/38515-smart-manufacturing-robotics-cyber-physical-system (Accessed September 20, 2024).

Pretorius, B. H. (2016). Cyber-security and governance for industrial control systems (ICS) in South Africa. Retrieved from http://hdl.handle.net/10413/15261.

Radanliev, P., De Roure, D., Van Kleek, M. et al. (2021). Artificial intelligence in cyber physical systems. *AI & Society,* 36, 783–796, https://doi.org/10.1007/s00146-020-01049-0

Shamila, M., Vinuthna, K., & Tyagi, A. K. (2019). *A review on several critical issues and challenges in IoT based e-healthcare system. Paper presented at the 2019 International Conference on Intelligent Computing and Control Systems (ICCS).*

Subraveti, S. G., Li, Z., Prasad, V., & Rajendran, A. (2019). Machine learning-based multiobjective optimization of pressure swing adsorption. *Industrial & Engineering Chemistry Research,* 58(44), 20412–20422. doi:https://doi.org/10.1021/acs.iecr.9b04173

Sun, R. (2020). Potential of full human–machine symbiosis through truly intelligent cognitive systems. *AI & Society,* 35, 17–28. doi:https://doi.org/10.1007/s00146-017-0775-7

Thomas, Gamer, Mario, Hoernicke, Benjamin, Kloepper, Reinhard, Bauer, & Isaksson, Alf J. (2019). The autonomous industrial plant – Future of process engineering. *Operations and Maintenance, IFAC-Papers Online,* 52(1), 454–460.

Yang, M., Khan, F., & Amyotte, P. (2015). Operational risk assessment: A case of the Bhopal disaster. *Process Safety and Environmental Protection,* 97, 70–79. doi:https://doi.org/10.1016/j.psep.2015.06.001

Ye, F., Ma, S., Tong, L., Xiao, J., Bénard, P., & Chahine, R. (2019). Artificial neural network based optimization for hydrogen purification performance of pressure swing adsorption. *International Journal of Hydrogen Energy,* 44(11), 5334–5344. doi:https://doi.org/10.1016/j.ijhydene.2018.08.104

Zhuang, W., Ye, Y., Chen, Y., & Li, T. (2012). Ensemble clustering for internet security applications. *IEEE Transactions on Systems, Man, and Cybernetics, Part C (Applications and Reviews),* 42(6), 1784–1796.

2 Security Challenges in the Development of Smart IoT

2.1 SMART IOT APPLICATIONS – AN INTRODUCTION

Internet of things, abbreviated as IoT, encompasses a decentralized network that integrates sensors, application software, and communication networks. The application software in IoT provides the functionalities related to various gadgets, tools, and appliances. The devices may range from simple devices such as temperature monitors to smart plugs, smart power grids that are connected over the internet.

IoT can be thought of as an interconnected network of various physical objects that have the communication ability with other devices through the internet.

IDC describes "an IoT solution as a network of distinct endpoints that communicate utilizing IP connectivity without human contact, whether locally or worldwide. The Internet of Things (IoT) provides organizations, governments, and consumers with inherent management, monitoring, and analytics, enhancing pervasive connection".

The IoT has proliferated globally due to affordable computing and cloud technology, resulting in billions of interconnected devices worldwide. The IoT connects the digital and physical realms through continuous, real-time communication in both consumer goods and industrial setups.

Smart IoT applications include technology stack with artificial intelligence (AI) and machine learning (ML) to deliver the smart functionalities. It can be visualized in Figure 2.1.

Some of the typical smart IoT applications are given as follows:

- Smart environmental monitoring and pollution control
- Smart traffic monitoring and regulation
- Smart fleet management in smart cities applications
- Smart agriculture
- Smart hospitality in hotels and cottages
- Smart healthcare
- Smart grid and energy saving
- Smart water supply and management
- Smart municipality maintenance management
- Smart homes automation
- Smart parking systems

DOI: 10.1201/9781003489986-3

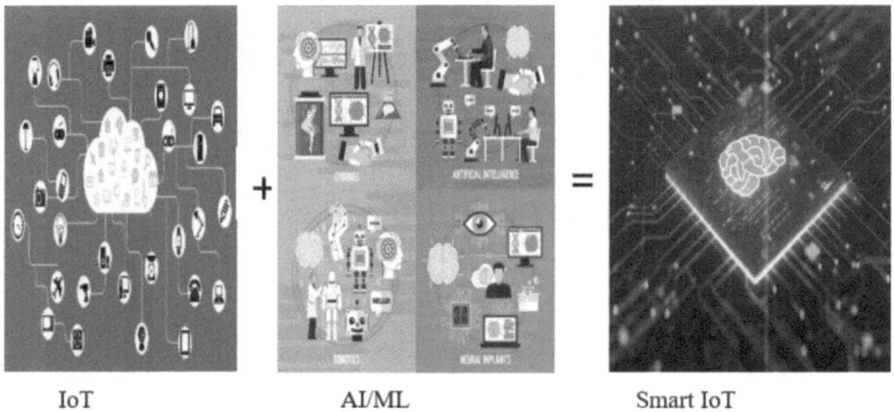

FIGURE 2.1 AI/ML integrated smart IoT framework has two major components: IoT and AI/ML which are integrated to provide smart IoT.

- Smart infrastructure management
- Smart disaster management
- Smart supply chain management

2.2 SMART IOT SYSTEMS WORKING MECHANISM

- **Smart Environmental Monitoring and Pollution Control**
  - Pollution affects most cities worldwide. IoT can assist reduce pollutants to breathable levels in such a case. Using sensors and IoT, city pollution data like car emissions, pollen levels, airflow direction, weather, traffic levels, etc. may be collected. ML systems can predict pollution in different metropolitan locations using this data, alerting city officials to potential issues. Then they may reduce pollution to make it safer. IBM China Research Lab's Green Horizons initiative illustrates this.
- **Smart Traffic Monitoring and Regulation**
  - Smart traffic management in major cities can benefit from the IoT. When we use our phones as sensors to collect and share vehicle data through apps like Waze or Google Maps, we are using the IoT to inform us and contribute to traffic monitoring, showing route conditions, and feeding and improving information on routes to the same destination, distance, and estimated arrival time.
- **Smart Fleet Management in Smart Cities Applications**
  - Sensors in fleet vehicles improve communication between managers and drivers. Drivers and managers/owners can access data collection, processing, and organization software to learn about vehicle status, operation, and needs and receive real-time maintenance alerts without driver detection. The IoT can help fleet managers with geolocation (and thus route monitoring and identification of the most efficient routes), performance

analysis, telemetry control, fuel savings, pollutant reduction, and driver improvement.

- **Smart Agriculture**
  - Smart farms help farmers who need information on good soil information to grow good crops, and the IoT gives them accurate soil information. IoT sensors can collect a lot of soil data. Soil moisture, acidity, nutrients, temperature, and other chemical characteristics help farmers control irrigation, use water more efficiently, determine the best sowing times, and detect plant and soil diseases.
- **Smart Hospitality in Hotels and Cottages**
  - IoT in hotels improves service quality in interesting ways. Electronic keys given to guests' phones can automate many encounters. Using integrated applications and IoT technology, guests' locations, offers, or information on activities of interest, room service orders, automatic room charges, and personal hygiene supply requests can be easily managed. Electronic keys automate the check-out procedure, disabling doors, providing room information promptly, and allocating housekeeping jobs to maintenance.
- **Smart Healthcare**
  - IoT has provided an efficient means to connect patients, doctors, nurses, attendants, pharmaceuticals, front-line workers, emergency care units etc.
  - Wearable devices equipped with sensors help the doctors to check the patients' status from a remote place. The cyber-physical system (CPS) equipped with appropriate software component will send alerts to the associated persons such as doctors, nurses, etc., based on the vital parameter measures.
  - This will instantly save human's life especially in the case of elders and those who have critical illnesses.
  - It can also provide a better way to store and retrieve the patients' medical records for easy access and helping to provide an easy access to the patients' details across the various departments of the hospitals.
  - Based on the sensors' data, the application of AI and ML techniques can assess the status of the patients and provide appropriate recommendations.
  - IoT-based pharmaceutical industries help to improve the entire manufacturing of medicines and the supply chain. The underlying methodology is the application of intelligence in the entire automation process.
  - Monitoring the glucose level of patients helps in keeping them to be assisted with proper drugs so that they will not get any further complications. Especially for pregnant women and elders, the historical patterns of their glucose level will help in providing precision healthcare.
- **Smart Grid and Energy Saving**
  - For industries, the energy conservation and energy audits will be useful in proper time intervals. However, the manual way of doing it will take a lot of time and resources and in several of the cases, this will never be considered as highly important.
  - To avoid this, the IoT devices are installed in energy meters, production plants, power distribution points, power transmission lines for continuous monitoring and reporting.

- Power grids incorporated with IoT are called smart grids. Here, the IoT devices equipped with sensors will monitor the energy consumption, nodes that are inactive at any point of time, and any abnormalities as and when needed.
- Using this, it is easier for the maintenance authorities to check for any inefficiency so that they can improve the performance and optimization of the power units accordingly.
- This will reduce the carbon footprints and which in turn will reduce the energy consumption costs.
- When there is an issue in the traditional power plants, then the sensor will detect, and using this information the actuators in the IoT device of the smart grid will switch over to wind or solar power plants to provide continuous power supply.

- **Smart Water Supply and Management**
    - For smart cities applications, the water supply is equipped with meters, and these are then connected to the centralized server via the internet. These smart meters will be equipped with the sensors, and the smart CPS will then collect, analyse, and process the data to get the behavioural analytics of consumers, the billing pattern, and any issues in the service etc.
    - These systems can continuously monitor the water consumption readings and then through the mobile app or web app, the consumers will receive the alerts based on the reading.
    - This will help the consumers to regularly check their readings and if needed they can control their water consumption.
- **Smart Home Automation**
    - The smart home automation systems are highly used in smart cities which provide home safety and security system, home surveillance, smart electric appliances monitor and control etc.
    - The smart CPSs in all these areas help to automate the home utilities in a better and intelligent manner.
- **Smart Parking Systems**
    - Smart parking solutions help the two wheelers and four wheelers to park in the slots based on their availability and the maximum time of parking.
    - This will assist the better way of managing and regulating traffic and to keep the customers be always connected.
- **Smart Infrastructure Management**
    - In smart cities, the infrastructure being maintained by the government often has a lot of issues in terms of street lamps, gas pipelines, roads, and parks management etc.
    - The monitoring and repairing of these infrastructure facilities need to be done seamlessly so that their functioning will not be affected. Hence, the smart infrastructure management systems will have IoT-based monitoring systems which will help identify the maintenance needed by means of analysis of the patterns and thereby providing the necessary support before the entire system gets affected.

- **Smart Disaster Management**
  - The natural calamities such as earth quakes, heavy rain falls, floods etc. can be monitored, and appropriate actions can be taken by means of smart CPSs employed in such disaster management.
  - A timely notification will help in providing immediate actions to save human lives in the affected areas.
- **Smart Supply Chain Management**
  - The smart CPSs employed in supply chain management has multiple layers of automation. The shipping organizations can use them to monitor the trucks in delivering the goods to the appropriate places.
  - The sensors can capture the route information and based on the need, the optimal route can be chosen so that the goods can reach the destination in less time and with less fuel.
  - Managing the fleets remotely will definitely reduce the delay and other related issues associated with transition. This smart fleet management using IoT devices help in monitoring the vehicles' health and pollution control.

2.3 IOT AND CPS SECURITY

To get better performance, the IoT devices must be optimized in their usage. These devices, due to their miniscule dimensions and reserved computational capabilities, incorporate fundamental authentication and risk mitigation systems, thereby enabling malevolent attackers to exploit them as entry points [1].

A white paper published by Microsoft discloses that by the early 2020s, 25% of cyber attacks will target IoT devices. Presently, there is a significant exponential growth in the adoption of IoT products across diverse industries, surpassing the ability of developers to keep up.

Exploitation of IoT devices, as cited in the most recent cyber report issued by the Canadian government, is one of the most significant threats to cyber security at present. Without considering the potential cyber hazards, organizations and sectors are integrating IoT devices into their networks [2]. Standards that presently exist and consist of advice such as "use strong passwords", "do not share your password", and "close unused ports" are insufficient to prevent another Mirai attack. However, these measures are insufficient to protect IoT devices.

In the case of IoT security, the protection of hardware components, software, data and information transfer, and network from any attacks or infiltrations is the need of the hour. As the entire IoT ecosystem needs to ensure the availability of services, reliability, integrity, and trustworthiness, it is essential to protect it from any outsiders' attacks. Hence, IoT security should include the security measures that comprise the set of methods, guidelines, processes, policies, software tools, and any supporting software and hardware systems [3].

IoT systems, which are highly vulnerable and may lead to data breaches especially in the case of healthcare devices, need to be protected from various array of attacks.

In the case of autonomous vehicles and self-driving cars, these kinds of security vulnerabilities will lead to major damages to the entire eco-system. Due to a security breach at a nuclear power plant in smart grid, the damage happened was huge. Hence, the IoT devices must be protected from these kinds of security breaches and vulnerabilities [3].

Also, IoT device access should be protected by means of authentication and authorization mechanisms so that these devices cannot be used as the pathway to access the backend data and processing elements by infiltrators.

Some of the malwares that are developed especially for IoT devices will have different attack patterns than for the other kinds of software. Hence, it is essential to develop appropriate security mechanisms to prevent the systems from such attacks.

Especially, the attackers and infiltrators can easily access these devices due to their accessibility over the network with less security. Hence, it is essential to find the attack types and provide appropriate security measures to reduce the impact of such attacks.

2.4 SECURITY CHALLENGES IN IOT APPLICATIONS

In 2022, there were 112 million cyber attacks on the IoT worldwide. In contrast, just 32 million comparable cyber incidents occurred in 2018. In addition, there was a 41% increase in the average number of weekly attacks from 2022 to 2023, which targeted IoT devices in the first half of 2023. Users and products are put at risk when developers fail to consider cybersecurity while creating IoT devices adequately [3, 4, 6].

As the IoT applications are developed primarily with functionality as the primary focus, the security aspects are given secondary level of importance. Hence, there are several vulnerabilities which cause the entire IoT ecosystem to be affected [5].

As there are only few standards and guidelines available for IoT security, there is a low level of focus on the risks associated with IoT devices. In several cases, the risks are not properly identified by the development team, the impact of such unforeseen risks will be very high and in certain cases it can't be revert back.
Some of the typical IoT security challenges are

- **Code Vulnerabilities**
  - Several of the IoT devices have open source firmware which leads to bugs and security threats.
- **Volume of Data**
  - The sensor-generated data will be very huge which can't be easily managed, sorted, and utilized for a specific purpose.
  - As the volume of data increases, the vulnerabilities associated with it will also increase.
  - Protection from unauthorized access really becomes the need of the hour in IoT security.
- **Security Integration**
  - As the IoT devices are very different, the integration of them to get a particular functionality really requires the analysis of security vulnerabilities during integration.

  - When the organizations have separate security systems, then integrating these systems with the actual application is also challenging as it requires generalized and secured operations.
- **Visibility Constraints**
  - As several of the IoT devices are deployed without the prior knowledge about each of its components, it will be highly difficult to protect them onsite as each of the components need individual monitoring.
- **Weak Security Testing**
  - Generally, the device security is considered only as the last priority task as the organizations are focusing on the functionality aspect. It is essential to perform effective testing on the IoT devices to ensure the reliability.
- **API Vulnerabilities**
  - The application program interfaces (APIs) are the vulnerable points for the attackers to indulge in injection attacks, denial of service (DoS) attacks, distributed denial of service (DDoS) attacks etc.
- **Password Hacking**
  - As many of the IoT devices are provided with very weak and same passwords across multiple applications, the intruders can easily enter into the device and the network to get the confidential data.
- **Vulnerabilities' Fixing Issues**
  - In several cases, after the vulnerabilities are identified, the fixing of bugs will be taken with less care, and several times the fixing task will be postponed due to other higher-priority tasks.

All these are posing challenges on the security issues of the IoT devices and smart CPSs [18]. IoT devices are susceptible to various vulnerabilities, which can compromise their security and expose users to risks.

2.5 TYPES OF ATTACKS IN IOT AND CPS

Generally, the IoT and CPS are facing the following kinds of attacks [18]:

Attacks on IoT Physical Devices: In these kinds of attacks, the attackers can manipulate devices, tamper sensors, take unauthorized access to the devices, which will lead to malfunctioning of the devices.

Zero-Day Attacks: Some of the vulnerabilities in IoT devices or software are completely unknown. These will help the attackers exploit the IoT functionalities without the knowledge of the developers, which in turn makes the developers face difficulty to defend against these attacks.

Middle-Man Attacks: These kinds of attacks are also called "Eves dropping attacks". In this, the attackers monitor the communication between IoT devices and communication channels. They may steal sensitive information or user credentials. Proper encryption and secure communication protocols are essential to thwart this threat. IoT devices with poor authentication, weak encryption, and/or insecure connections in the network are prime targets for this type of attack.

Injection Attacks: In the case of IoT device communication, there are ways to inject malicious data which will help the attackers to get access to the data storage or data processing code modules.

Replay Data Attacks: During data transfer from IoT device to other devices or to the centralized data storage or to the cloud, there may be chances to replicate a data packet with malicious content which will very much look like a normal data package. This is usually achieved by means of corrupting a legitimate data packet and as a response the source may try to resend it to get the communication success. These responses are used by the intruders to send their data packets instead of the actual data packets.

DDoS and DoS Attacks: The most critical services provided by the websites or cloud servers through the IoT devices' security protocols will be tampered by the intruders by means of excessive traffic. This in turn leads to disruption in critical and most important services such as healthcare alerts, manufacturing unit failures, coolant failures in chemical reactors, atomic power plants etc.

Malicious Botnet Attacks: The botnets created by the attackers prevent the IoT devices to provide the required functionality. These botnets act as agents that can perform malicious activities, such as launching DoS/DDoS attacks, sending spam emails, or spreading malware posing like a normal software functionality.

Malware Attacks: IoT devices are forced to fail in terms of their functionality by means of malicious software which is also called a malware. These kinds of attacks will also reflect on other devices in the network. Some of the most critical ones are the ransomware attacks on IoT devices. In these, the attackers can lock devices and demand payment for their release to perform their regular work.

These kinds of attacks need to be identified in the IoT devices, as otherwise it will lead to entire system failure.

2.6 BEST PRACTICES FOR IOT SECURITY

It is essential to follow some of the best practices in order to prevent the IoT devices from attacks. They are listed below [17]:

Update Credentials Regularly: The default credentials associated with the IoT devices are provided by their manufacturers at the time of manufacturing. Hence, it is essential to change these credentials. Common passwords across applications and easily predictable passwords should be avoided.

Firmware Update: IoT device's firmware must be regularly updated. As the firmware or any embedded software updates are to fix vulnerabilities in the IoT devices, it is essential to update them regularly.

Network Segmentation: IoT devices must be separated from the main network and if possible by creating a separate network for such devices. By applying this practice, even if one device is attacked, it won't affect other devices.

Firewalls Configuration: Configuring the router's firewall will help permit only necessary traffic to and from IoT devices. This will help to block unnecessary services and ports.

Unused Features Disabling: If a feature from an IoT device is not needed at that point of time, then it is better to disable them so that their use can be restricted. The amount of attack vectors will be reduced when the number of features are getting reduced.

Encryption Strengthening: The communication between devices and servers must be encrypted. It is advisable to always use protocols like TLS, HTTPS, or WPA3 for Wi-Fi.

Authentication: By means of two-factor authentication (2FA), the IoT devices can be protected from malicious attacks. It will improve the secured access to the IoT devices as it will provide an extra layer of security.

IoT Network Traffic Monitoring: It is important to use network-monitoring tools in order to detect the unusual traffic patterns while using IoT devices which will help identify potential attacks through the network.

Physical Security: It is better to lock the devices and keep them in less accessible areas in order to enhance physical security.

Privacy Settings: Always it is essential to review privacy settings in web interfaces and device apps. It is a best practice to limit data collection and sharing among the IoT devices and network.

2.7 SECURE SMART CPS DEVELOPMENT

2.7.1 Application of Intelligent Security Measures

Smart CPSs play a crucial role in various domains, from intelligent traffic control to healthcare and industrial automation etc. However, ensuring their security is of paramount importance.

Before exploring the application of intelligent security measures for developing secure smart CPSs, it is important to analyse the characteristics needed from such measures, as listed further [8]:

2.7.1.1 Understanding the Threat Landscape

Begin by comprehending the security threats specific to CPSs. These threats include manipulation of communication channels, hardware components, and associated software.

Recognize the complexity arising from the interaction among independent CPS domains.

2.7.1.2 Static and Adaptive Techniques

Employ static and adaptive detection and protection techniques across different layers of the CPS stack. Cross-layer and intra-layer connectivity should be considered when implementing security measures.

2.7.1.3 Challenges and Limitations in Applying Existing Techniques

Find the limitations of existing techniques. For instance, they may struggle to capture dormant or uncertainty-based runtime security attacks.

2.7.1.4 Open Research Problems in CPS Security

Explore open research problems related to intelligent security measures for CPSs. Understand the unique requirements of CPS domains such as autonomous vehicles, smart cities, and industrial automation.

2.7.1.5 Ongoing Projects and Analyses

It is essential to continuously analyse and adapt security measures to evolving threats as the current ones have limitations [2–6].

Based on this analysis, the next task is to address these challenges and issues to enhance the robustness of security measures using intelligent measures.

2.7.1.5 Intelligent Security Measures Using Machine Learning

Leverage ML-based techniques to enhance CPS security. Apply neural networks, deep learning, and other intelligent algorithms to detect and prevent attacks at various layers of the CPS stack. Explore the possibilities of using intelligent agents that run at the background to identify any threats or attacks.

2.7.2 Secure IoT Networks

Protecting connected devices via the IoT network is essential as attackers will use this as the medium to infiltrate into the IoT devices. Some of the effective measures are given [7]:

Zero-Trust Policy: It is always a best practice to separate IoT devices from the rest of the network. By implementing a zero-trust policy to grant only genuine operational access will provide next level of security to protect IoT networks from attacks.

Authentication of IoT Devices on the Network: Even though the devices are connected via a secured network, it is essential to enforce rigorous authentication processes before allowing connections with other IoT devices and external entities via the network.

Encryption Algorithms: The firmware and data must be encrypted while communication is established between the devices. Also, the firmware should be updated regularly. Encrypt firmware and data during manufacturing or firmware updates. Choose robust encryption algorithms like AES, RSA, or SHA.

Credential Update: Modify default usernames and passwords on IoT devices. Avoid common passwords like "admin" or "123456".

Firmware Updates with Security Patches: Keep device firmware up-to-date to address vulnerabilities. Regularly check for security patches [8].

Network Monitoring: Use antivirus/anti-malware software. Set up network firewalls to block unauthorized traffic. Detect and manage bot traffic.

Physical Security: Physically secure devices to prevent tampering. Disable remote management on routers. CPSs integrate physical processes, networking, and computation. Ensuring their security is crucial. Here are steps to protect CPS network's security.

- **Securing Access to Devices:** Proper authentication and authorization needs to be ensured in order to access CPS devices. If the authentication is poorly made, it will lead to unauthorized manipulation of the system by the attackers.
- **Securing Data Transmissions:** The data transmission in CPS communication networks while safeguarding it from imposters, unauthorized access, and disruption is of high importance. If these are blocked, the DoS attacks and routing interruptions can be detected and prevented.
- **Securing Applications:** In the application layer, user's privacy must be protected. By preventing private data leakage, we can safeguard information related to locations of the users and their social interactions.
- **Securing Data Storage:** By encrypting stored secret data on CPS devices, using security mechanisms at their edge itself by means of lightweight security, we can prevent major security attacks at the device level itself.
- **Securing Actuation:** Authorized issuance of actuation actions.

Secure Smart Home Networks: To safeguard your smart home network, follow these essential steps:

- **Secure the Wi-Fi Network:** Change the default credentials of your Wi-Fi router. Use the Wi-Fi Protected Access II (WPA2) protocol with a strong, complex password.
- **Ensure Strong and Unique Passwords:** Change default login credentials for smart devices (such as smart TVs, cameras, or voice assistants). Regularly update passwords and avoid using common or easily guessable ones [8].
- **Keep Firmware Up-to-Date:** Regularly update firmware for all connected devices. Manufacturers release security patches to address vulnerabilities [8].
- **Disable Unnecessary Features:** Disable any features on smart devices that may compromise network security. Enable only the essential functionalities.
- **Monitor and Secure Your Network:** Use network monitoring tools to detect unusual activity. Set up network firewalls to block unauthorized traffic.

2.7.3 Secure Data in IoT Networks

It is essential to secure the data that is transmitted in IoT networks as the attackers and intruders will have several mechanisms to get the data that leads to severe privacy and reliability concerns in critical IoT applications such as defense, healthcare, self-driving cars etc [9].

To enhance the security of the data transmission over the IoT networks, the following mitigation steps are provided:

1. Modify default usernames and passwords on IoT devices. When the IoT devices are purchased with in-built software, there will be default username and password. Many a times, the users will have a soft corner on these

credentials. They simply keep them when they bought for a specific purpose. This leads to serious backdoor attacks in the IoT networks with the intention of stealing the privacy data. Hence, it is advised to update the default username and password and also avoid using common credentials.

2. It is essential to keep the software and firmware be updated regularly as these updates will have the software patch files to enhance security in order to address the most recent security vulnerabilities.
3. It is always a best practice to encrypt the data in the IoT communication networks. This will help to improve the reliability, trustworthy, authenticated access, and integrity of data gathered from diverse IoT devices. It is also essential to manage the keys used for encryption to be both secure and confidential.
4. By implementing a secure network architecture for IoT devices, one can secure the highly sensitive data gathered from safety critical IoT systems. It is better to provide segmentation-based access to restrict the access and openness of the devices from the common access.
5. Employing the regular audit services on the IoT devices to monitor the behaviour of the IoT devices will help to identify any unusual and unauthorized access to such devices.

2.7.4 Secure Firmware

The most significant initiative to secure the firmware of the IoT devices is to provide a digital signature in the firmware image and then install it in the IoT devices [9].

The firmware signing procedure, as shown in Figure 2.1, consists of the following steps:

Step 1 – A cryptographic hash is created for the firmware image. To achieve this, the file having the firmware is hashed using a cryptographic hash algorithm. The output will be a fixed-size hash value which is also called a digest. Generally, SHA-3 and SHA-256 are applied for encryption.

Step 2 – The next step involves the generation of a private key. Here, an RSA or elliptic curve cryptography (ECC) is applied to generate the public and private keys. The manufacturer or developer will keep the generated private key securely. The public key corresponding to it will be written in the device's firmware or in the ROM which will be used while booting.

Step 3 – In this step, a digital signature is created by using the private key to encrypt the hash of the firmware image. This signature is unique with respect to that firmware and the private key, which was created during the creation time and is not common to all the firmware of all the devices. This process generates a signed firmware binary by embedding the digital signature to the image of the firmware.

Step 4 – The firmware verification during firmware update or booting is done by ensuring the software integrity. It is given in Figure 2.2. This is achieved by means of the following steps:

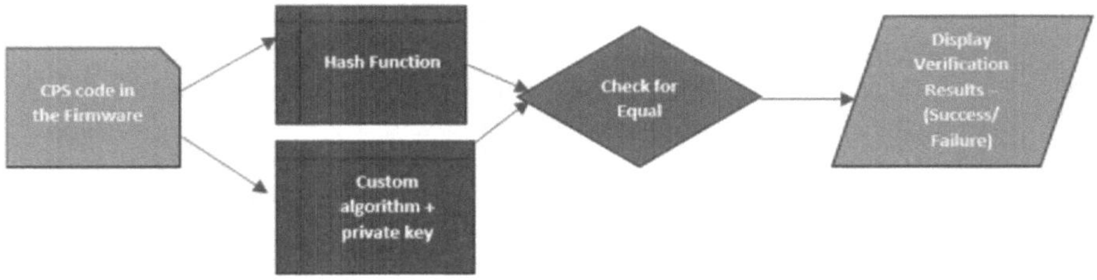

FIGURE 2.2 Firmware signing process that has, at the left side, the CPS code in the firmware which has two major functions: hash function and custom-defined algorithm with private key, to check whether the outputs from these two functions are equal, and based on that verification result it will be pass or fail (Courtesy: www.einfochips.com/blog/securing-firmware-in-iot-devices/).

Step 4.1 – The cryptographic hash of the firmware image is computed.
Step 4.2 – By applying the trusted embedded public key, the digital signature present in the firmware image will be deciphered.
Step 4.3 – To ensure that there are no inadvertent firmware update, the calculated hash value is compared with the decrypted hash value taken from the signature itself. If they are the same, then it is concluded that there are no alterations and can be signed with the private key.

The files that underwent security checks during invocation are as follows [10]:

1. Security Certificates: If there are no defined safety measures, the firmware itself will follow some of the security guidelines such as combining SSL or SSH security certificates with private keys. To get the certificates, a static analysis will be applied and then they are given in the final report.
2. Obsolete Components: The software components developed during earlier iterations are generally vulnerable to attacks. These components may be system libraries, bootloader files, kernel files, packages that consists of binary files etc. Thus, it is necessary to check these obsolete components to reveal locations that have high risks with openness.
3. Hard-coded Credentials: During debugging process, some of the developers may give the details like email addresses, IP addresses, host names, domain address, and some other application-specific information as the output when some error is raised by the application. These are used by the hackers to get the information to infiltrate into the application in order to get confidential information. This needs to be avoided in order to protect the system.
4. Password Saved Files: In the case of IoT-based systems, the passwords are saved in two files namely the password file and shadow. The technique to protect the system makes use of the reverse hashing in order to decrypt the passwords. These passwords are used to access the IoT devices through the Telnet interface. Once these are captured by the hackers, the IoT devices are

infiltrated by them without any difficulties. Hence, the device interface must be examined with the configuration files.

5. The vulnerability can be assessed by means of providing reverse certificate search. As many of the IoT devices are provided with the identical access credentials, they can be easily captured by the hackers. Hence, this reverse certificate search will help in protecting the devices from infiltrators..
6. The accessible ports and services have to be analysed for vulnerabilities. The open ports and services that are vulnerable to attacks using these ports must have been identified. And the end-users who are using these services must be guided on especially those services that are pretty old and must be identified and informed of the potential vulnerabilities on these devices.
7. By means of well-defined user interfaces in the web applications that are used to access the IoT devices functionalities, the threats can be reduced considerably. Especially the security verification steps, authentication, and session tracking with the entered credentials with proper reviews have to be applied.
8. For IoT devices, the Telnet and SSH are the most widely used network protocols, and their configuration and password data need to be scanned for any attacks [2,19]. The Zigbee protocol is still used in older IoT devices and the usage of bluetooth low energy (BLE) and Wi-Fi needs to be tested rigorously to assess the security measures in these technologies.

In spite of employing all these measures, still there are security concerns affecting the IoT devices as listed further [10, 11]:

1. Some of the botnets designed exclusively to infiltrate the IoT devices have made use of the default and hardcoded passwords used by the end-users.
2. As IoT devices gather sensitive data using the sensors, they need to be provided with adequate security mechanisms. Mostly, these devices do not have the encryption functionality applied on the collected data, these data are highly vulnerable and are open to access by the attackers when they are transmitted over the communication medium.
3. Due to the negligence in following the security software development best practices while developing the IoT software, the software is vulnerable to be attacked by the infiltrators by means of exploitation processes. Also, the network interfaces are also poorly secured which thus leads to higher level of vulnerability.
4. Due to the lack of application of security standards in the IoT devices, they lead to open doors for the infiltrators to attack. In several cases, these security standards are only at the theoretical capacity only and not at the level of implementation.

Hence, the vulnerabilities present in the IoT devices can be easily used by the infiltrators to easily take over confidentiality of the data by means of unauthorized login or even they can execute the attacks directly on these devices without any knowledge of the users.

2.8 IOT CHALLENGES IN SECURITY ENHANCEMENT

Some of the challenges faced by IoT (IoT) and the main security problems related to IoT are [11]

- Compromised SDLC in IoT software development: Due to negligence in adopting security as the first priority in software development life cycle of IoT system development, several vulnerabilities arise. Especially when there is high level of market competition, the manufacturers are pushed to release their product before the competitors, which in turn bypasses the security requirement.
- Customers who purchase the IoT systems for their usage may not be aware to change the default credentials provided by the manufacturers of the device. This will in turn help the attackers to easily get the access to the device. Also, weak login credentials also will lead to security vulnerabilities.
- The recent malware and ransomware attacks on IoT devices are increasing in multitude and also at critical level. Also some of the botnets have been used by the attackers to get the access to the devices.
- The collection, processing, transmission, and storage of data by the IoT devices have raised several privacy concerns. As the data from these IoT devices need to be shared with the other dependent devices, there exists potential vulnerabilities to exploit such data. It is highly important to understand how this data is shared among other devices. Hence, the users are required to check the terms and conditions of these devices and must have a complete knowledge about how their data is going to be used by these devices.
- In the worst-case scenario, the attackers are using the IoT devices which are compromised with security standards as the medium to perform DDoS kind of attacks. Using this, they can attack the connected devices such as PCs, laptops, mobile devices, and so on.
  - The usage of common interfaces to access the IoT devices lead to compromising of encryption and decryption standards. This will further lead to authentication failures in connected systems.
  - As the current scenarios lead to remote working in order to provide higher productivity and improved collaboration, especially due to the COVID-19 outbreak, the people need to work from home. The home networks generally don't have high level of security measures, it will bring the weaknesses in IoT security which needs to be addressed appropriately.

2.8.1 Instances of IoT Security Breaches

There have been several noteworthy cases of fraudsters successfully breaking into IoT devices in recent years. Included are the items [12, 13]:

- In 2016, the Mirai botnet assault took place. Using hundreds of thousands of compromised connected devices, a botnet known as Mirai was created in 2016. A botnet is a purposefully constructed network of malware-infected computers that is used to carry out automated online tasks without the owners' knowledge or approval. The Mirai attack caused a brief closure of well-known companies and websites, including PayPal, Netflix, and Spotify.

- Malware attack on VPNFilter: In 2018, more than 500,000 routers in more than 50 countries were successfully compromised by the VPNFilter malware. It is possible for the VPNFilter virus to install further harmful software. This malware is intended to intercept network traffic, steal passwords, and gather and collect information that is being communicated over the network.
- In 2020, an individual with cybersecurity experience penetrated the security of Tesla Model X. He exploited a bluetooth system flaw to gain unauthorized access to the vehicle in less than two minutes. Similar attacks on other cars that rely on wireless keys for ignition and opening have been noted.
- Hackers gained access to Verkada's camera feeds. Verkada is a business that focuses on manufacturing and selling security cameras. Swiss cybercriminals were able to breach and obtain illegal access to approximately 150,000 operational cameras. The functionality of these cameras was to record the actions that take place in public sector institutions such as schools, colleges, prisons, hospitals, and corporates.

There are security flaws in the IoT that affect a variety of products, such as wearables, smart home devices, cars, and smart grids. Researchers found that webcams might be hacked, allowing someone to access networks without authorization [13]. They also found security flaws in smartwatches that let hackers follow their movements and listen their chats.

2.9 IMPLEMENTATION OF IOT SECURITY STRATEGIES

Implementing and maintaining IoT security effectively requires an extensive strategy. It is essential to take into account a variety of tactics and tools in addition to taking linked systems like networks into account.

The ability to learn and gather data are three crucial components of a robust IoT security system.

- Make use of security solutions that provide network visibility to learn more about the ecosystem's elements and the risk profiles connected to various IoT device groupings.
- Guarantee the security or health of the IoT devices by keeping an eye on examining and enforcing IoT security regulations suitable for tasks carried out at different phases of the infrastructure.
- Use segmentation strategies based on policy groups and risk profiles to divide IoT systems, much like you would do with networks.

The following attributes are essential for guaranteeing the security of IoT devices:

- Application programming interface (API) security measures
- A more extensive and thorough inventory of IoT devices

- Constant software upgrades
- DNS filtering

It describes the procedure of using DNS manipulation to selectively ban or permit access to particular websites or online content.

- Information is safeguarded during transmission and storage by using data encryption. Programs that pose as honeypots are intended to trick would-be assailants. In order to mitigate these types of vulnerabilities, the following practical steps need to be implemented [13]:
- Authentication using many factors
- Security online
- Network traffic monitoring analysis
- Password management
- Management of software updates
- Devices for network security
- Informing developers, partners, and other interested parties of security flaws

2.9.1 Improve Internet of Things (IoT) Security to Achieve Enhanced Advantages

Individuals and businesses are increasingly using IoT devices. Not only are they here to remain, but they are also multiplying rapidly in an increasing number of variations. The outcome is a growing intricacy, which obstructs endeavours to effectively control the security of IoT devices.

The problems of IoT security encompass the task of countering threats posed by both internal individuals with evil intent and sophisticated attacks launched by nation-states. Due to the inherent susceptibility of IoT devices and the extensive deployment of these devices, assaults are increasing in both size and range.

Investing in the security of IoT devices is highly beneficial, despite the problems associated with IoT security. Enhancing the security of IoT devices is crucial in order to maximize their value and bring them up to the same level as other technologies. It will reduce risks and enhance returns.

2.9.2 Best Practices for Securing Internet of Things (IoT) Devices

A thorough comprehension of the devices that are connected is essential for ensuring the security of the IoT. To do this, a device identification and discovery solution that automates three crucial tasks for guaranteeing the security of IoT devices must be used [9].

The IoT devices connected to the network are automatically and reliably identified, described, and categorized by this system. Additionally, it maintains a current log of every device in real time. Keep an eye out for potential attack routes to deliver relevant risk information for every one of these asset classes.

The following are a few best practices for minimizing an organization's susceptibility to security risks related to the IoT:

- **Risk Assessment and Device Discovery:** IT personnel might not be aware of linked IoT devices. When identifying new IoT devices and assessing the potential security risks they pose to the organization, automated device discovery may be a useful tool.
- **Zero-Trust Network Access (ZTNA):** The goal of ZTNA is to stop hackers from utilizing IoT devices to access and traverse an organization's networks without authorization. By separating IoT devices from the rest of the network, ZTNA helps limit their access and lessens their capacity to access other systems and vital data.
- **Preventing IoT Threats:** It is possible for hackers to take advantage of security holes in IoT devices. However, because of things like different ecosystems, scarce resources, specialized functionality, and missing user interfaces, standard endpoint security solutions frequently don't work well on these devices. Furthermore, a number of IoT devices cannot be changed or updated by their owners. The goal of IoT threat prevention systems is to lessen the possibility that hackers may take advantage of these devices' weaknesses.

2.10 RISK MITIGATION IN SMART IOT

The following recommended practices should be considered in order to ensure the security of IoT networks and devices:

Maintain up-to-date hardware and software upgrades. Make sure the IoT gadget you are thinking about buying has updates, and make sure you always turn them on as soon as they are released. Software updates are essential to maintaining the security of IoT devices.

Device breach by hostile actors is made easier by the existence of erroneous IoT software. Your IoT device may receive updates automatically, or you may need to go to the manufacturer's website to find them. The predefined credentials on IoT devices have to be changed. A significant portion of users create an account and set a password for each device they use.

While this is easy to keep in mind, it also gives hackers a simpler way to get access. Always make sure that each registration is distinct, and on newly acquired devices, always update the default password. Using the same password on several devices is not advised.

Make sure that the Wi-Fi network and all devices have robust authentication setups.

A password that satisfies the robustness requirements must be a minimum of 12 characters long and contain a mix of characters, including numbers, symbols, lowercase and uppercase letters, and other characters. Never reveal anything that isn't absolutely necessary, such a series of digits (1234), or any personal information that may be guessed at by someone else, like your birthdate or the name of your friend.

An individual's login credentials can be more easily organized and monitored with the use of a password manager. Check the router's authenticity from your end. It may be possible for unauthorized persons to determine the brand or model of routers if the manufacturer's name is retained. Instead, think about changing your router; however, be sure the new name doesn't reveal any personally identifying information, such as your name or address. Apply a robust encryption technique for wireless networks.

One can improve the security of their network and communications by setting up a router to use a secure encryption mechanism (WPA2 or later). Brute force attacks can be used against WPA and WEP versions that are more outdated. Additional information on WPA versions is available on the website.

If the router has that capability, a person may want to set up a guest wireless network that uses WPA2 or later protocols and is password-protected. Use the guest network that is provided for guests as it is conceivable that relatives and friends used hacked or malware-infected devices before they connected to your network. A home network's security is improved by a guest network as a whole.

Ascertain the configurations for privacy on your IoT devices. It is probable that by default your IoT devices feature privacy and security configurations. It is prudent to thoroughly examine these and make any required adjustments to the parameters in order to ascertain that they are at a satisfactory level for oneself. Similarly, it is advisable to review privacy policies to obtain knowledge regarding the process by which the service-provider stores and utilizes your personal data.

Ensure a record of the available features on the device and deactivate any that are not being utilized. Conduct a thorough assessment of the functionalities available on your devices and deactivate any that you do not actively employ to reduce the likelihood of potential security breaches. For the purpose of illustration, consider a chronometer whose primary function is to display the time. Nevertheless, the implementation of bluetooth, near-field communication (NFC), and voice activation is also probable. Neglecting to employ these functionalities allows an unauthorized individual to compromise the device via supplementary methods, without providing any further benefit to the user. The threat posed by cyberattacks can be mitigated by deactivating these features.

Implement multi-factor authentication (MFA) whenever possible. User access to an online account necessitates the provision of two or more verification methods through the implementation of MFA. MFA augments the standard username and password system by necessitating supplementary data, such as an email address or mobile phone number that is transmitted by the authentication servers of the website in the form of a one-time password. If supported by your mobile device, utilize MFA for identification of IoT devices connected to a residential network. Obtain a thorough comprehension of the operation of every device that is transmitting data across your network. There is a potential for certain devices to be of an older model at this time; if so, upgrading to a more recent device could furnish the IoT with more robust security features.

Restrict your access to public Wi-Fi. While traveling, you might find it convenient to manage your IoT devices through your mobile device, even in a coffee shop, airport, or shopping mall. Comprehending the security vulnerabilities that are linked to the utilization of public Wi-Fi is of the utmost importance. One way to partially mitigate these risks is through the implementation of a virtual private network (VPN).

By maintaining a cognizance of IoT cybersecurity and adhering to IoT security best practices, it is possible to mitigate risks.

2.11 FUTURISTIC SOLUTIONS FOR IOT SECURITY

2.11.1 Security Analytics for IoT

IoT security vulnerabilities can be effectively handled, if they are analysed using analytics algorithms. In security analytics, the data sources from multiple resources are analysed so that it is effective to identify and protect the devices from threats and attacks.

In the case of security analytics, the malicious attacks on the network by means of traffic can be analysed by taking into consideration of data from multiple domains. The results received from it can be used to find the correlated data so that the security team in the organization can address these anomalies so that it will not create cascading effects on the connected devices in the network.

2.11.1.1 Network Access Control Mechanism

The solution to the IoT security challenge is the use of dedicated visibility tools by IT personnel that can have network access control (NAC). This provides insight on the availability of all the devices connected to the IoT network.

When a new device gets connected, NAC will automatically update list of available devices and verifies it monthly. This kind of monitoring of IoT devices' availability helps organizations respond to security threats and take appropriate security-related actions.

2.11.1.2 Endpoint Detection and Response (EDR)

Due to the constant streaming of data, some data may be lost. This is because the IoT devices are not under control all the time.

To address this issue, the EDR technology helps by identifying attacks in real time and avoiding any data loss. By employing EDR, the security personnel can identify malicious attempts as quickly as possible and get direct access to the IoT devices with real-time visibility and appropriate alerts when they are tampered.

Further, the EDR technology has the ability to detect and prevent any suspicious activities in real time. The threat intelligence of EDR helps to detect suspicious attempts on an IoT endpoint and give effective responses, which are not identified by the human security personnel in the software development team.

2.11.2 Software Development Considerations

2.11.2.1 Secure APIs

Generally, the APIs are used as the connecting link in many security strategies. The IoT devices are dependent on APIs to get data from other systems, and through them, they can share the collected data.

Hence, the organizations must ensure that hackers cannot gain access to their IoT devices via such APIs due to poor configuration or authentication issues. This can be ensured by means of continuous security testing.

2.11.2.2 Encrypted Communication

The attackers compromise IoT communication to gain access to IoT devices. However, such data breaches can be prevented by encrypting communication between IoT devices and web or mobile interfaces. The most commonly known web-based encryption protocol for data transfer is SSL/TLS.

2.11.2.3 Authentication

Comprehensive device authentication helps to reduce the vulnerability of IoT devices from hackers as they always try to gain access and obtain personal information,

Some of the authentication mechanisms available for IoT devices include digital certification, multifactor authentication, and biometrics.

2.11.3 Enforcement of Laws and Standards

Nowadays, cybersecurity laws, regulations, and standards help smart devices to securely handle a large amount of sensitive personal data. The application of legal actions and financial penalties are imposed when private data is disclosed. Some of the indirect impacts of such breaches include loss of consumers' trust and reputation.

It is important to consider various rules and regulations pertaining to privacy, consumer rights, and security while developing an IoT ecosystem, by prioritizing security and privacy in the IoT by the government.

In 2017, a US regulation was approved that forbids the use of default passwords on any IoT devices sold in the United States. The lawmakers included these and other mandates in the IoT Cybersecurity Improvement Act. The information privacy was protected by the California legislature in 2018 through the IoT security bill SB-327.

The General Data Protection Regulation (GDPR) and the privacy by design idea taken into consideration when the target markets are from the European Union. The privacy by design of IoT safeguards user information and lays forth guidelines for how developers and manufacturers can guarantee the security of IoT devices. Further, the Cyber Resilience Act, which is a new act specifically addresses IoT devices.

The Cybersecurity Framework given by National Institute of Standards and Technology (NIST) and the International Organization for Standardization (ISO) 27001: Information Security Management have offered suggestions on how to safeguard networks and devices connected to the IoT.

The CIA triad is considered as the gold standard in the IoT industry, but it doesn't always account for new threats that may arise when the team setting is done.

2.11.4 Ways to Improve Consumer IoT Security

Some of the ways by which the customers can keep their personal information safe in order to improve the security of the IoT:

1. **Physical Protection Measures:** Protecting IoT devices requires attention to physical security, which is frequently overlooked in spite of its critical significance. Customers can take physical security as an indentation further by not

letting their IoT devices stand in plain prospect or in an area where unauthorized people can easily access them. To lessen the likelihood of depiction, it is recommended to place IoT devices in less noticeable locations and stay away from openings. By installing physical locks or security measures, customers can take additional precautions against theft and tampering.

2. **Restrict Data Communication**: Restricting the data that IoT devices communicate is crucial for privacy protection. To acquire information more about the data gathered and sent by their IoT devices and related services, consumers should check the settings given in their application dashboard. By removing permissions or functionalities that aren't needed, users can limit the quantity of personal data disclosed. Essential data can be shared with trusted devices and services which can help reduce the likelihood of privacy breaches.
3. **Creation of Separate IoT Network:** An effective technique to increase the security of the IoT is to create a separate network that is exclusively for IoT devices. By creating a dedicated Wi-Fi network or virtual local area network (VLAN), customers can protect their IoT devices from interference on their home network. Due to this network segmentation, IoT devices cannot access other devices or private information. The security of other devices and information on a user's home network can be protected against a compromised IoT device by taking this precaution.
4. **Applying MFA:** To protect the IoT devices and accounts, it is advised to apply MFA. By means of this MFA, users can add a code sent to their mobile device as an extra layer of security, replacing the need for a password entirely. The MFA settings need to be updated regularly in order to get better security. This will ensure the users that their IoT devices are much safer from intruders since only authorized users will be able to access or manage sensitive features. In addition, to provide a strong foundation for IoT security, it is suggested to use hardware-based safety measures such as hardware root of trust and private enclaves. These features can help prevent unauthorized access and manipulation.

Secure boot procedures such as regular update of boot and cloud can help to make sure that only reliable firmware and applications may run on IoT devices. Some of the approaches such as securing "over-the-air (OTA)" update approaches allow for remote, cloud-implementable provisioning of fixes and updates without compromising security. Enforce encryption by encrypting data at rest and during transmission using strong encryption methods.

Further, by employing data anonymization and tokenization, it will be efficient to safeguard sensitive information. It is always better to stay informed about the ever-evolving regulations pertaining to the security of the IoT and the compliance requirements in different industries and places.

2.11.5 Continuous Monitoring of IoT Systems

Continuous monitoring helps to find and address security incidents instantly. Based on that, prepare a plan for incident response exclusively designed for IoT environments.

From the beginning of the conceptualization and manufacturing processes, it is better to bring together cybersecurity from the design process itself. Making threat assessment and security verification as critical steps in the product life cycle will address IoT security problems.

Some of the examples of security measures are surveillance, biometrics, MFA, and stringent password policies which can help keep unauthorized users out of the IoT devices and systems. It is essential to fix security vulnerabilities and build a strong foundation using authentication, encryption, and machine-driven monitoring.

The connectivity to commonplace appliances and devices such as Wi-Fi-enabled lightbulbs and thermostats through voice commands have been made possible by IoT. However, they become an easy target for cybercriminals, as in several cases no or less importance is given to security.

Some of the critical hacking activities of IoT devices includes hacking of baby monitors, smart cars, and even Wi-Fi-connected chastity belts. The devices that we allow into most private environments pose a threat to privacy, data, and security if proper protections are not taken to prevent their exploitation.

In order to mitigate such kinds of security threats, it is essential to have proper analysis about the vulnerabilities of IoT technologies, so that consumers can make decisions appropriately especially in the case of smart homes and workplaces.

2.12 WAYS TO IMPROVE ENTERPRISE IOT SECURITY

The practical knowledge of organizations on the implementation of IoT devices and functionalities is intricately associated with security. The IoT solutions provided by the industries are different from each other. Some of the sectors in industries that can significantly get benefits from implementing basic or even erudite IoT security solutions are discussed in the subsequent sections [14].

2.12.1 The Manufacturing Sector

The product manufacturing companies can prove the application of security measures in thwarting a wide range of attacks associated with the IoT, whether due to cyberattacks or other threats that can take information pertinent to their operations and work. A machinery-related security violation may result in a temporary cessation of production until the issue is resolved in the manufacturing sector. In such cases, the plant can utilize IoT devices to monitor processes and regulate activities of any nature.

2.12.2 The Healthcare Sector

With reference to the usage of IoT devices in healthcare, patient privacy is one of the most critical considerations that must be taken into account. For instance, medical devices, when linked through online platforms and configurations, become vulnerable to attacks that have the potential to interrupt patient care.

In this regard, IoT security is critical for the healthcare industry, as medical facilities rely on IoT applications and technology for a variety of functions, including patient monitoring, inventory management, and more.

2.12.3 Critical Establishments

In cities and urban areas, the critical infrastructures established as large structures and complexes comprise the bare necessities for daily operations. Some of such services are communication through internet, security services, communication and internet connections, water and electricity systems, and waste disposal.

Due to the fact that intelligent devices and sensors are typically employed to regulate IoT security, it is crucial to protect our critical infrastructure because an attack has the potential to generate extensive interferences and apprehensions for individuals [14].

2.12.4 The Energy Sector

The energy sector comprises infrastructure and production firms operational across diverse domains. Industries such as oil and gas, coal, power plants, or grids nowadays have the integration of IoT devices and functionalities into their operations in order to get a steady and consistent process. These intelligent devices are employed to oversee their manufacturing measures and aid in the guideline or monitoring and analysis of diverse components of the entire enterprise.

As with any critical infrastructure, many in the energy sector will be vulnerable to attacks that disrupt the flow of electricity to and from the plants or cause blackouts when we don't have proper IoT security measures.

2.12.5 The Transport Industry

The transportation industry frequently employs IoT devices due to its widespread digitalization. With reference to various application areas of transport industry such as infrastructure maintenance, traffic management, video surveillance cameras, tolls, or traffic control etc., there are numerous prospects for others to intercept and disrupt the system. Hence, IoT security has become crucial in the transportation industry, as numerous nations rely on intelligent technologies to control traffic and ensure the security and safety of pedestrians and vehicles on roads. Accordingly, it is critical to apprehend the threats and problems associated with these technologies. Hence, it is important to take every necessary measure to mitigate them prior to their impacting one's IoT devices or network.

2.12.6 Managing an Extensive Ecosystem

Adopting IoT into one's operations could be as straightforward for numerous businesses spanning diverse industries as integrating intelligent monitors or sensors into the machinery and equipment utilized in those operations. For alternative enterprises, the scope may span from minor modifications to extensive IoT solutions encompassing a multitude of devices and equipment.

As IoT is an ever-expanding landscape for businesses, it is quite natural for them to have a variety of interconnected devices for their functioning purposes. The fact that IoT for a business may consist of numerous devices or solutions with varying capabilities and issues that can make a whole ecosystem.

IoT security management is a challenge in these circumstances as many individuals are unable to conquer on their own. Further, the industries frequently require additional support to address the innumerable security structures and vulnerabilities that they need.

The software organizations that wish to ensure that their devices and machinery equipped with IoT solutions regularly update the firmware may encounter difficulties in accomplishing this. This means that the devices have extremely fundamental processing and memory capabilities, which has two primary effects on firmware and software updates:

(i) The hardware constraints hamper the implementation of robust IoT security features, which are frequently an absolute necessity for the organization, and the process of upgrading may be significantly sluggish due to the diminished processing capabilities, which can lead to prolonged downtimes and susceptibilities throughout the procedure;
(ii) Privacy concerns are indistinguishably linked to any data collection, but particularly with regard to user data. IoT devices depend on the collection of said data for various operational purposes irrespective of the industry in which an organization or company implements them. This has been predictable as a challenge for IoT security, as maintaining the confidentiality and integrity of this data may necessitate erecting numerous safeguards, which can ultimately impede work efforts or render the system less reliable than projected.

It should be understood that not all IoT solutions and devices are capable of performing data encryption and decryption as they need to perform variety of other major functionalities such as patient monitoring, atomic reactor monitoring, oil and gas pipeline monitoring etc. Hence, this further complicates the task of confirming the security of data through encryption, rendering it more susceptible to breaches.

2.12.7 Risks to the Supply Chain and Network

Businesses that depend on IoT applications to accomplish their supply chain and logistics may also meet this exertion. The supply chain may experience vulnerabilities as a result of insufficient security measures. Additionally, malicious entities or cybercriminals may target these supply chains with the intention of causing disruptions or financial losses to the organization. Hence, it is absolutely necessary to include security measures during supply chain operations and logistics.

2.12.8 Key Considerations for Enhancing Enterprise IoT Security

(i) **Application of Blockchain for Enhanced IoT Security**
Blockchain technology can provide enterprise IoT deployments with enhanced security by means of secure and translucent data management, trustworthy transaction records, and enhanced device identity management due to its decentralized and immutable characteristics. By incorporating blockchain technology into the IoT ecosystem, the potential for data breaches and unauthorized manipulation can be reduced [15].

(ii) **Enhanced IoT Security by Design**
By adopting a "security by design" methodology, security measures are assimilated at each stage of the life cycle of an IoT device development. This includes encryption protocols, secure device provisioning, and vigorous authentication mechanisms. By means of the integration of security measures during the design and development phases, industries can construct IoT solutions that possess an inherent resistance to threats related to cyber security [16].

(iii) **Application of Encryption in Enhanced IoT Security**
It is extremely important to safeguard data communicated between IoT devices and backend systems by implementing healthy encryption protocols. Encryption guarantees the conservation of confidentiality and security of sensitive data, even in the event of transmission interruption. Industries can enhance data privacy and minimize the likelihood of unauthorized access by implementing industry-standard encryption algorithms [16].

(iv) **Application of IoT Access Guidelines**
To effectively manage and secure IoT environments in enterprises, stringent access controls must be implemented. It is recommended that organizations implement granulated access policies that limit interaction with IoT devices and access to sensitive data to approved personnel only. This demands the implementation of robust authentication techniques, such as MFA, routine evaluation and revocation of access privileges when needed [16].

(v) **Application of Enterprise Security Measures**
It is imperative for organizations to embrace an inclusive security policy that not only encompasses IoT devices but also their wider IT infrastructure. Intrusion detection and prevention systems, security information and event management (SIEM) solutions, and robust firewalls are all components to achieve this. Further essential elements of an enterprise-wide security strategy include the implementation of incident-response plans, the provision of employee training, and the regular monitoring and updating of security procedures. In the realm of IoT security, it is imperative for industries to establish collaborations with weathered and loyal vendors [17].

2.13 SUMMARY

Due to economic considerations or hardware and software availability, it is occasionally possible that IoT applications need to be integrated into older, legacy systems. As these systems or extant infrastructure are outdated, they are susceptible to a variety of speed and connectivity issues, as well as vulnerabilities and exploits. An additional difficulty for IoT security arises from the possibility that certain manufacturers may prioritize system functionality for the sake of restrictions or similar considerations.

Furthermore, when utilized concurrently, the more recent IoT applications may malfunction improperly, resulting in compatibility and security concerns. Regardless of the domain, numerous industries have begun to encounter or at the very least become conscious of the problems that IoT security encounters during the implementation of intelligent devices and solutions.

These challenges effectively require organizations to implement continuous monitoring, application of robust policies, and security development practices. With the growth of IoT device usage to comprehend the needs of enterprises in addition to individual users, the implementation of robust security protocols becomes extremely necessary. Thus, enterprise IoT security requires the implementation of a proactive strategy in order to protect critical information, impede cyber threats, and preserve operational integrity.

PRACTICE QUESTIONS

I. Short Answers – Remember/Understand
   1. Define IoT.
   2. Name any five smart IoT systems.
   3. List any four security challenges associated with smart IoT systems.
   4. What do you mean by zero-day attacks?
   5. Write the working of replay data attacks.
   6. What is zero-trust policy?
   7. What is security firmware?
   8. What do you mean by hard coded credentials?
   9. List the attributes essential for security in IoT devices.
   10. Explain the application of blockchain in security of IoT-based transactions.

II. Descriptive Questions – Apply/Analyse
   1. Analyse the architecture of CPSs and examine each of its components.
   2. Explore the life cycle phases of CPS application development.
   3. A city power supply department wants to automate their facilities in order to give an efficient power distribution to the households. Apply the appropriate type of CPS to provide the required functionality.
   4. Differentiate CPS and IoT.
   5. Which characteristics distinguish normal CPS from a smart CPS?
   6. For an Industry 4.0 application, an industry is in need of a smart CPS that can take decisions appropriately to improve productivity. Examine the application of AI in such kind of application development.

III. Higher Order Thinking Questions – Analyse/Create/Evaluate
   1. A smart home safety system needs to be developed with all the appropriate safety mechanisms of embedded authentication and authorization. In order to transfer the information a sophisticated cryptography mechanism needs to be employed. The customer of the organization wants everything to be part of the CPS as a firmware. Examine the scenario and elaborate the steps involved to create a firmware with digital signature [13].
   2. Consider the recent Mirai botnet attack that caused the closure of PayPal and other most important financial organizations. Analyse the situation and suggest some AI-based smart solutions to mitigate such kinds of risks.

REFERENCES

1. Dibaji, S. M., Pirani, M., Flamholz, D. B., Annaswamy, A.M., Johansson, K.H., & Chakrabortty, A. (2019). A systems and control perspective of CPS security. *Annual Reviews in Control,* 47, 394–411.
2. Humayed, A., Lin, J., Li, F., & Luo, B. (2017). Cyber-physical systems security—A survey. *IEEE Internet of Things Journal*, 4(6), 1802–1831. doi: 10.1109/JIOT.2017.2703172
3. Yaacoub, J-P.A., Salman, O., Noura, H.N., Kaaniche, N., Chehab, A., & Malli, M. (2020). Cyber-physical systems security: Limitations, issues and future trends. *Microprocessors and Microsystems*, 77(C), 1–33, 103201.
4. Kumar, C., Marston, S., & Sen, R. (2020). Cyber-physical systems (CPS) security: State of the art and research opportunities for information systems academics. *Communications of the Association for Information Systems,* 47, pp. 1–14, https://doi.org/10.17705/1CAIS.04731
5. Z. Wang, W. Xie, B. Wang, J. Tao, & E. Wang (2021). A survey on recent advanced research of CPS security. *Application Science,* 11, 3751. https://doi.org/10.3390/app11093751
6. Mbiriki, A., Katar, C., & Badreddine, A. (2018). Improvement of security system level in the cyber-physical systems (CPS) architecture. *2018 30th International Conference on Microelectronics (ICM)*, Sousse, Tunisia, 40–43. doi: 10.1109/ICM.2018.8704100
7. Cai, X., Han, K., Li, Y., Wang, H., Zhang, J., & Zhang, Y. (2020). Research on security estimation and control of cyber-physical system. *2020 IEEE 39th International Performance Computing and Communications Conference (IPCCC)*, Austin, TX, 1–5. doi: 10.1109/IPCCC50635.2020.9391573
8. Thakur, A. (2021). Securing cyber physical systems I. https://medium.datadriveninvestor.com/securing-cyber-physical-systems-i-d508543a5187
9. Gohil, H. (2023). Securing Firmware in IoT devices, Balbix white paper, Internet of Things (IoT) Biggest Security Challegences Niharika Chaurasia, IoT security challenges and best-practices – overview.
10. Gloukhovtsev, M. IoT Security: Challenges, solutions & future prospects, https://learning.dell.com/content/dam/dell-emc/documents/en-us/2018KS_Gloukhovtsev-IoT_Security_Challenges_Solutions_and_Future_Prospects.pdf
11. Dreiblatt, R. Challenges for cyber security, www.linkedin.com/pulse/idea-4-part-two-challenges-cybersecurity-reuben-dreiblatt/
12. Dang, T. Internet of Things Security Issues: You Won't Believe How Hackable Your Smart Devices Are! Huichen Lin and Neil W. Bergmann. *IoT Privacy and Security Challenges for Smart Home Environments*, 7, 44. https://doi.org/10.3390/info7030044
13. Lin, H., & Bergmann, N.W. (2016). IoT privacy and security challenges for smart home environments. *Information*, 7(3), 44, https://doi.org/10.3390/info7030044
14. Göçoğlu, V. (2019). Akıllı Şehirlerdeki Kritik Altyapıların Siber Güvenliği [Cyber Security of Critical Infrastructures in Smart Cities], *Uluslararası Yönetim Akademisi Dergisi Yıl,* 2 (1), 51–63. https://doi.org/10.33712/mana.548909
15. Almarri, S., & Aljughaiman, A. (2024). Blockchain technology for IoT security and trust: A comprehensive SLR. *Sustainability*, 16, 10177. https://doi.org/10.3390/su162310177
16. Kokila, M., & Srinivasa Reddy, K. (2025). Authentication, access control and scalability models in Internet of Things security–A review, *cyber security and applications*, 3, 100057. https://doi.org/10.1016/j.csa.2024.100057
17. Whitfield, B. *11 Key Enterprise IoT Security Recommendations, Industrial report.* www.altexsoft.com/blog/11-key-enterprise-iot-security-recommendations/

3 Protecting Cyber-Physical Systems (CPSs) from Cyber Attacks Using Artificial Intelligence (AI)

3.1 INTRODUCTION TO CYBER-PHYSICAL SYSTEMS

Cyber-physical system (CPS) is a growing field of engineering that integrates physical and computational components. These systems coupled the physical systems with the cyber world using components like sensors, actuators, and computational elements to monitor and control physical processes. Let us take an example of an autonomous vehicle, first sensors (e.g., camera) collect data on the vehicle's surroundings such as the position of obstacles (e.g., kerb, other vehicles, walls, and pedestrians). This data is then processed by a system (e.g., learning models) which makes decisions about when to stop, how to steer the steering wheel etc. The system then sends signals to the vehicle actuators, which control the steering wheel, brakes, and accelerator. Therefore, the importance of CPSs lies in its ability to augment and manage complex systems, offering significant improvements in efficiency, safety, and functionality [1, 2].

The utilization of CPSs is embraced by various sectors from energy sectors to digital health. For example, integrating renewable energy sources, optimizing power generation and distribution, and ensuring grid stability are adopted in smart grids. In terms of autonomous vehicles, enabling self-driving cars to sense their environment, make driving decisions, and navigate safely are some of the applications in which CPSs are utilized. Digital health uses CPSs in the form of wearable and implantable devices that monitor vital signs and deliver personalized healthcare. There are diverse sectors that use CPSs such as industrial automation and smart manufacturing: systems that optimize production processes, improve efficiency, and reduce downtime.

The development of CPSs has the potential to revolutionize many aspects of human lives, for example, it can make transportation systems efficient and safer, improve the quality of healthcare etc. However, it can also raise a number of challenges, one challenge is security. They are vulnerable to cyber attacks. If a hacker were to gain control of a CPS, they could cause significant damage. For example, a hacker could take control of a self-driving car and cause it to crash. CPSs are complex systems and this complexity can make them difficult to design, develop, and test.

DOI: 10.1201/9781003489986-4

CPSs include several key components:

- Sensors for data collection from the physical environment
- Actuators to exert changes on the physical system based on computational decisions
- Networking capabilities to connect devices and systems for data exchange and coordination
- Computational units (e.g., hardware and software) for processing data, making decisions, and executing control actions

CPSs are distinguished by several defining characteristics, for example, CPSs seamlessly integrate hardware and software with physical processes, enabling new types of feedback loops where computational processes affect physical ones and vice versa. It often requires real-time or near-real-time responses to physical world inputs, necessitating precise timing and synchronisation [3]. CPS components communicate and exchange data, enabling them to function as a unified system. This often involves internet connectivity and real-time data exchange. Many CPSs can operate autonomously, making decisions based on sensor data without human intervention and most importantly they are designed to adapt to changing environments and to be resilient to disruptions, whether from physical failures or cyber threats.

On the other hand, when CPSs are embedded with AI there can be a change in the characteristics of CPSs. The systems will be able to do advanced decision-making, AI techniques provide CPS with the ability to analyse complex data, identify patterns and make dynamic decisions in response to changing situations. This can enable intelligent and adaptable systems behaviour [4]. Additionally, CPS can also learn from data, predict future states of the physical system, and adapt their behaviour to optimise outcomes. This includes adjusting to unforeseen changes in the environment. AI allows for the real-time analysis of data, enabling immediate responses to changes in the physical system. This can help improve efficiency and effectiveness. By automating decision-making and learning from operational data, artificial intelligence (AI)-enabled CPSs can reduce the need for constant human monitoring and intervention. Incorporating AI into CPSs transforms them from automated systems into intelligent entities capable of learning, adapting, and operating autonomously.

Let us sum up the characteristics of CPSs with and without embedding of AI in Table 3.1.

TABLE 3.1
Characteristics of CPS with and without Integration of AI

Characteristics of CPS	Characteristics of CPS with AI
Deterministic behaviour	Autonomous decision-making
Limited adaptability	Adaptive and predictive capabilities
Manual data analysis	Real-time data analysis
Predefined functions	Increased efficiency and optimization
Human intervention	Reduced need for human intervention

3.2 ATTACKS ON CYBER-PHYSICAL SYSTEMS

Attacks on CPSs represent a significant security challenge due to the integration of computational and physical processes within these systems. CPSs play crucial roles in critical infrastructures including industrial control systems, smart grid networks, transportation systems, and healthcare devices. These attacks can have severe consequences, ranging from economic losses and privacy breaches to threats to human safety and life. Understanding the nature of these attacks is crucial for developing effective defence mechanisms.

This section discusses different types of attacks including cyber threats that are likely to affect CPSs. In Section 3.4, we discuss the use of AI for protecting CPS.

- **Malware:** Malicious software designed to cause damage or gain unauthorized access to the CPSs. Some of the examples are virus (attach themselves to clean files and infect other clean files and spread uncontrollably damaging systems' core functionality), Trojan horse (disguises as legitimate applications to destroy systems), rootkit (enables remote control of a system without the users' knowledge; they are challenging to detect and can hide other malware), ransomware (encrypts data, making it inaccessible until a ransom is paid; attackers can use ransomware to lock out legitimate users from critical control systems, demanding payment to restore access) etc.
- **Man-in-the-Middle (MitM):** Attackers secretly intercept or alter communications between two parties who are directly communicating with each other. In terms of CPSs, this can lead to incorrect operation commands being sent to different components of CPSs, for example, between physical systems, potentially causing physical damage, such as spoofing where an attacker deceives the system by impersonating another device, or user, leading to unauthorized access, manipulation of the system, or disruption on its normal operations.
- **Data Attacks:** Attackers alter the data sent from sensors to control systems or spoof sensor data to trick the system into making unsafe decisions. For example, to hide an overheating condition that could lead to an explosion, attackers can tamper with the data from a temperature sensor in a chemical plant, attacker can alter data either stored or transmitted by CPS to cause incorrect system behaviour, they can even steal confidential information from the system. Data attacks are serious threats in the era of machine learning (ML), as advanced AI technologies have been widespread.
- **Network Attacks:** Attackers disrupt by attacking communication networks of CPS to intercept, modify, or block the flow of information between system components. Momentum iterative method (MIM), spoofing and packet sniffing attacks are some of the examples of network attacks.
- **Software Exploitation:** Exploiting software vulnerabilities within the CPS to gain unauthorized access or control. For example, exploiting zero-day vulnerabilities where attackers use unknown software flaws before they are

patched, and SQL injection where attackers insert malicious SQL queries into an input field to manipulate or corrupt database content.

- **Insider Threats:** Threats from individuals within the organization who have access to the CPS. For example, deliberate actions to damage or disrupt the system, and unauthorized access for exploiting legitimate access privileges for malicious purposes.
- **Supply Chain Attacks:** Compromising the security of CPS through vulnerabilities introduced in the supply chain. For example, introducing vulnerabilities through malicious hardware or software components, exploiting vulnerabilities in third-party services or software used by the CPS.

With the understanding of various attacks that can affect CPS, Table 3.2 presents several real-world threats that occurred in different industries.

3.3 FUNDAMENTALS OF ARTIFICIAL INTELLIGENCE IN CYBERSECURITY

The integration of AI into cybersecurity represents a transformative shift in how organizations defend against cyber threats, detect vulnerabilities, and respond to incidents [13]. AI can enable more efficient, accurate, and automated security processes. This section explores the fundamentals of AI in cybersecurity, covering key concepts, applications, and the impact of AI technologies.

3.3.1 Key Concepts of AI in Cyber Security

- **Machine Learning (ML):** As a subset of AI, ML enables systems to learn from data, identify patterns, and make decisions with minimal human intervention. For implementing ML in CPS, ML algorithms can learn from historical security data to predict and identify potential threats [14]. As the progression of the AI technologies has increased rapidly, the development of learning algorithms has also been improved exponentially – for example, the introduction of deep learning technology. Deep learning is an advanced and complex ML technology. It comprises neural networks with many layers to analyse data. It is beneficial for detecting anomalies in large datasets, making it valuable for identifying sophisticated cyber threats [15].
- **Natural Language Processing (NLP):** This AI technology helps computers understand, interpret, and produce human language. In cybersecurity, NLP can analyse and understand the content of emails, social media, and other text-based data to identify phishing attempts in text-based data or malicious communications [16].
- **Artificial Neural Networks (ANNs):** Inspired by the human brain's network of neurons, ANNs are crucial for deep learning models. They excel in processing complex patterns in data, useful for detecting malware or anomalous behaviour within network traffic [15].

TABLE 3.2
Examples of Real-World Threats/Attack in Several Industries

Threat	Threat Type	When	Scale	Sectors	Adversaries	Victims	Reference
Stuxnet	Worm/Zero-day exploit	2010	Physical damage to Iranian nuclear centrifuges	Nuclear sector	Suspected nation-state	Iran Nuclear Programme	[5]
Ukraine Power Grid	Cyber attack on physical system	2015	Left 230,000 residents without power	Energy/power grid	Suspected nation-state	Ukraine power companies	
WannaCry Ransomware	Ransomware	2017	Infected over 200,000 computers worldwide	Healthcare, transportation, government	Cyber-criminal	Health sector – NHS	[6]
NotPetya	Ransomware	2017	Caused over $10 billion in total damages	Diverse sectors such as shipping, manufacturing	Suspected nation-state	Global including Maersk, Merck	[7]
Triton/Trisis	Cyber attack on physical system	2017	Targeted safety systems, could have caused physical harm	Petrochemical	Suspected nation-state	Middle Eastern Petrochemical Plant	[8]
Atlanta Ransomware Attack	Ransomware	2018	Estimated	$17 million in recovery costs government/ municipal services	Cyber-Criminals City of Atlanta		[9]
Solarwind Orion Software Supply Chain Attack	Cyber espionage	2020	Affected thousands of organizations globally, including US government agencies	Government, IT, cybersecurity	Allegedly Russian government-backed group	US government agencies, private companies, and other entities worldwide	[10]
Maroochy Water Services Attack	Insider threat/ physical sabotage	2000	Millions of litres of raw sewage spilt into local parks, rivers, and even a hotel	Water utility	Disgruntled former employee	Residents and environment of Maroochy Shire, Australia	[11]
Iran's Railroad System	Cyber attack	2021	Trains delayed or cancelled across the country	Transportation	Unknown, possibly politically motivated hackers	Iranian National railway company and passengers	[12]

3.3.2 Application of AI in Cybersecurity

- **Threat Detection:** AI systems can analyse vast quantities of data in real time to detect known and unknown threats, significantly reducing the time to detection compared with traditional methods [17].
- **Anomaly Detection:** AI can identify deviations that may indicate a cyber attack by learning what normal behaviour looks like, such as unusual network traffic or unexpected access patterns [18].
- **Phishing Detection:** AI and NLP techniques can analyse emails to detect phishing attempts, flagging suspicious links or language indicative of fraud.
- **Predictive Analytics:** AI can forecast future threats based on current trends in data, allowing organizations to proactively strengthen their defenses.
- **Automated Incident Response:** AI can automate responses to certain types of security incidents, such as isolating affected systems, thereby speeding up containment and reducing damage.

3.3.3 Impact of AI on Cybersecurity

AI is changing the traditional technological way of handling tasks, especially in detecting and responding to (emerging) threats. Due to the ability to learn and predict from the historical data, the use of AI can have a distinct impact on threat detection, response time efficiency, and emerging threats adaptation ensuring scalability of security measures [17, 19].

For example, advanced AI algorithms can better differentiate between benign activities and genuine threats, decreasing the number of false positives that security teams need to investigate.

It can accelerate the speed and accuracy of threat detection and response time, enabling organizations to mitigate risks more effectively. It can handle increasing volumes of data and more complex networks, ensuring security measures scale with organizational growth. However, at the same time, challenges such as data quality, model transparency, ethical considerations, and adversarial AI use necessitate careful management [19].

3.4 AI TECHNIQUES FOR PROTECTING CYBER-PHYSICAL SYSTEMS

This section discusses how AI has started to integrate into CPSs, and how researchers and practitioners can study and improve the security of CPSs.

Integrating AI into the security of CPSs offers innovative solutions to protect critical infrastructure from cyber threats. The integration can strengthen the systems with the capability to predict, detect, and respond to threats in real time. This can enhance the resilience and reliability of the systems. Table 3.3 presents some key strategies and how to integrate AI into CPS security.

3.4.1 Threat Detection and Prevention Using AI

With the use of AI for diverse tasks from marketing to surgery, it can be used for providing safety to CPSs as well. It can involve several strategies that are aimed

TABLE 3.3
Key Strategies for AI Integration into CPS Security

Strategies	Detail	Reference
Data-driven security	AI algorithms require vast data to learn and make informed decisions. Implementing robust data collection and preprocessing strategies ensures high-quality data for AI models. This will facilitate effective threat detection and response mechanisms.	[20]
Real-time monitoring and anomaly detection	Leveraging AI for continuous monitoring allows for the real-time detection of anomalies and potential security breaches. ML models can analyse patterns and predict potential threats before they escalate.	[21]
Automated response systems	AI can automate the response to security incidents, reducing the time from detection to response. This includes isolating affected systems, applying patches, and even counteracting attacks in progress.	
Collaborative AI systems	Integrating AI systems across different sectors and components of CPS can enhance overall security. Sharing intelligence about threats and vulnerabilities allows for a coordinated defence strategy.	[22]
Adaptive and evolving Security measures	AI systems are capable of learning and evolving, which means they can adapt to new threats over time. Regularly updating AI models with new data ensures that the security measures remain effective against evolving cyber threats.	[23]
Ethical and privacy considerations	Integrating AI into CPS security raises ethical and privacy concerns. Ensuring transparency in how AI makes decisions and maintaining the privacy of data within AI-driven security systems are paramount. Implementing ethical guidelines and privacy-preserving algorithms is essential for maintaining public trust.	[24]

at enhancing the security, resilience, and reliability of the interconnected systems. The interconnection includes critical infrastructure including power grids, water supply networks, manufacturing, healthcare and transportation systems. This section discusses key approaches that can be utilized to protect CPSs using AI.

1. **Anomaly Detection:** AI can analyse vast amounts of data generated by CPSs to identify patterns and detect anomalies in real time. ML models can be trained to deduce system behaviour and flag any deviations, which could indicate potential security threats or system malfunctions [25]. Several works have integrated AI in defense security.
2. **Predictive Maintenance:** AI can help in scheduling maintenance activities by predicting equipment failures before they occur. This can avoid potential vulnerabilities that attackers could exploit. ML algorithms can be used to analyse historical data to predict when and where maintenance is needed [26].

3. **Resilience and Self-Healing Mechanisms:** Enhancing the resilience of CPSs through self-healing mechanisms is another effective strategy. These mechanisms allow systems to automatically detect, isolate, and recover from faults. The use of intelligent attack and anomaly detection methods can be useful to improve the robustness of CPSs against cyber attacks [27].
4. **Physical Security Enhancements:** AI can also be used to improve the physical security of infrastructure through advanced surveillance systems, facial recognition, and motion detection technologies. These systems can identify unauthorized access attempts and trigger alerts for timely intervention [28].
5. **Incident Response:** In the event of a security breach or system failure, AI can assist in automating the incident response process. AI systems can help in quickly isolating affected areas, assessing the impact of the breach, and implementing remedial actions to restore normal operations.
6. **Data Protection:** AI algorithms can help protect the integrity and confidentiality of data through encryption methods, secure data storage solutions, and intelligent access control systems that learn and adapt to potential threats.
7. **Simulation and Training:** AI can be used to simulate cyber attacks and other threat scenarios to train CPS operators in recognizing and responding to security incidents. These simulations can help in identifying vulnerabilities and improving the overall security posture of the system.
8. **Compliance and Risk Management:** AI tools can assist in ensuring that CPSs adhere to regulatory requirements and standards by continuously monitoring compliance and assessing risks. These tools can help in identifying potential compliance issues and suggesting corrective actions.

Implementing AI in the protection of CPSs requires a comprehensive approach that includes the development of robust AI models, continuous training with up-to-date data, and integration with existing security protocols. Collaboration between cybersecurity experts, AI researchers, and CPS operators is essential to address the complex challenges posed by the integration of physical and digital systems.

3.4.2 Future Trends and Challenges

The continuous progress in the trends and challenges in the AI technologies has led to the protection of CPSs also evolving. This section discusses such future trends and challenges of the AI technologies in CPS protection.

1. **Advanced Predictive Analytics:** The availability of data is increasing at an unprecedented level with the advancement in data storage, data processing, and computational power. AI can predict systems' failures and security breaches by analysing historical data patterns. This trend will likely be useful and advance with more innovative algorithms that can anticipate a wider range of issues based on historical and real-time data.
2. **Autonomous Response Capabilities:** AI systems are moving towards autonomous operations where they can not only detect anomalies but also

take immediate corrective actions without human intervention. This includes automatic patching of vulnerabilities, shutting down compromised systems, or rerouting network traffic to avoid attacks.

3. **Integration of AI with IoT Devices:** As the internet of things (IoT) becomes more prevalent, AI is increasingly used to manage and secure the vast networks of connected devices, from industrial sensors to smart home products.
4. **Blockchain for Security:** Integrating blockchain technology to enhance the security of CPSs ensures data integrity and traceability. This is particularly relevant for distributed systems where coordinated attacks can be more challenging to detect and mitigate.
5. **AI-Powered Security Frameworks:** Development of comprehensive AI-driven security frameworks can be adopted for various sectors, ensuring that security measures are robust and capable of evolving with new threats.
6. **Ethical AI and Privacy:** As AI systems handle more sensitive data, ensuring these systems are developed and used ethically while maintaining privacy will be a key focus. This includes addressing bias in AI algorithms and ensuring they do not inadvertently compromise user privacy.

On the other hand, as the CPSs and their defence mechanisms become more sophisticated, attackers also evolve their strategies. This includes AI-powered attacks that can learn and adapt to bypass AI-driven security measures. Integrating AI security solutions across diverse hardware and software platforms poses significant challenges, especially in sectors with legacy systems. As discussed earlier, AI technologies require large volume of data to train, and the extensive data required to train AI models raises concerns about data privacy and protection, particularly when sensitive information is involved. There is a pressing need for industry-wide standards to ensure that AI-driven security solutions are robust and interoperable across different sectors and geographical regions. At the same time, balancing the benefits of autonomous AI systems with the need for human oversight is complex, especially considering the potential consequences of failure in critical systems. Implementing AI solutions requires significant computational resources, which might not be feasible for all organizations, particularly small- to medium-sized enterprises. As we look into the future, it is evident that while AI offers tremendous potential to enhance the security and efficiency of CPSs, these advancements will also need to address significant challenges to ensure they are safe, effective, and equitable.

3.4.3 Practice Questions

3.4.3.1 MCQs

1. Which AI technique is particularly effective for anomaly detection in CPS?
   (A) Supervised learning
   (B) Unsupervised learning
   (C) Reinforcement learning
   (D) Decision trees
   Answer: (B) Unsupervised learning

Feedback: Unsupervised learning is ideal for anomaly detection because it does not require labelled data to identify patterns or anomalies. It can learn the normal operation pattern and detect deviations, which are potential threats.

2. What role does AI play in intrusion detection systems (IDS) for CPS?
   (A) Decreases the accuracy of threat detection
   (B) Increases the false positive rate
   (C) Enhances the detection of complex and novel attacks
   (D) Simplifies the attacks on the system
   Answer: (C) Enhances the detection of complex and novel attacks
   Feedback: AI significantly improves the ability of IDS to detect complex and novel attacks by analysing and learning from vast amounts of data, thus reducing the chances of missing sophisticated cyber threats.
3. Which AI technique is commonly used for anomaly detection in CPS?
   (A) Linear regression
   (B) Decision trees
   (C) Convolutional neural networks
   (D) Reinforcement learning
   Answer: (D) Reinforcement learning
   Feedback: While various AI techniques can be used for anomaly detection, reinforcement learning is particularly effective in adapting to new threats and anomalies by learning from interactions with the environment.
4. Which of the following is not a benefit of using AI in protecting CPSs from cyber attacks?
   (A) Real-time threat detection
   (B) Automated response to threats
   (C) Complete elimination of human error
   (D) Pattern recognition and anomaly detection
   Answer: (C) Complete elimination of human error
   Feedback: AI significantly reduces human error but cannot completely eliminate it. AI systems themselves can be subject to flaws and require human oversight.

3.4.3.2 True/False

1. Predictive analytics can only analyse past security incidents, not predict future ones.
   (A) True
   (B) False
   Answer: False
   Feedback: Predictive analytics uses past and current data to identify patterns and predict potential future security incidents, allowing proactive defence measures.
2. Automated response systems can reduce the time to mitigate a cyber attack on CPS.
   (A) True
   (B) False
   Answer: True

Feedback: Automated response systems, powered by AI, can immediately react to detected threats, significantly reducing the mitigation time and potential damage.

3. AI can predict and mitigate all types of cyber attacks on CPS.
   (A) True
   (B) False

Answer: False

Feedback: While AI significantly improves the ability to predict and mitigate many types of cyber attacks, no system can guarantee prevention against all possible attacks, especially new and unknown (zero-day) attacks.

3.4.3.3 Case Studies

Case Study 1: Smart Grid Protection

Background: A smart grid manages electricity demand in a sustainable, reliable, and economical manner. However, it is also a target for cyber attacks that can disrupt power supply and distribution. The power grid is a complex CPS that delivers electricity from producers to consumers. It has become a prime target for cyber attacks due to its critical importance and potential for high-impact disruptions. A power grid operator implemented an AI system to safeguard against potential cyber attacks aimed at destabilizing the grid.

Your task is to analyse how AI can be used to protect a smart grid from a coordinated cyber attack aimed at disrupting power distribution.

Model answer: AI was deployed to monitor network traffic and operational data in real time. ML algorithms were trained on historical data to recognize normal operational patterns and detect anomalies. During the attack, the system identified unusual patterns in power requests that deviated significantly from typical behaviour, suggesting a cyber attack. The AI system then automatically rerouted power distribution to stabilize the grid and alerted human operators for further investigation and response.

This case study demonstrates the effectiveness of AI in recognizing and responding to cyber threats in real time, ensuring the continuity of critical infrastructure services. The ability of AI to learn from historical data and identify anomalies can significantly enhance the resilience of CPSs like smart grids against cyber attacks.

Case Study 2: AI-Enhanced Cybersecurity in Industrial Control Systems

Background: An industrial control system (ICS) manages the operational processes within a manufacturing plant. These systems are increasingly connected to the internet for efficiency and data analysis purposes, which exposes them to cyber threats. A threat actor targeted the ICS with malware designed to manipulate the control processes, potentially causing physical damage to the plant.

Your task is to analyse the deployment of AI in detecting and responding to the cyber threat, focusing on how the AI system distinguishes between normal operations and malicious activities.

Model answer: The deployed AI system uses a combination of supervised and unsupervised ML models that have been trained on historical data from the plant's operations. These models developed a baseline of "normal" patterns and could detect deviations indicative of a cyber intrusion or malfunction. When the malware begins to alter the control commands, the AI system could detect anomalies in the operational data that do not match the available set patterns. The AI system then automatically initiates a safe shutdown procedure to prevent damage and issues alerts to the cybersecurity team. The swift response allows the team to isolate the affected systems and remove the malware before any physical damage occurs.

This case study illustrates the effectiveness of AI in monitoring complex systems for signs of cyber threats. The use of both supervised and unsupervised learning is a robust strategy, as it leverages the strengths of both approaches. However, relying solely on AI could lead to potential risks, such as the AI system being manipulated by sophisticated threats including adversarial attacks. Therefore, integrating human oversight and a multi-layered defence strategy is crucial for optimal security.

REFERENCES

1. R. Rajkumar, I. Lee, L. Sha, and J. Stankovic, Cyber-physical systems: The next computing revolution, in *Proceedings of the 47th Design Automation Conference, DAC'10*, pp. 731–736, date: 13-06-2010 Through 18-06-2010.
2. F.O. Olowononi, D.B. Rawat, and C. Liu, Resilient machine learning for networked cyber physical systems: A survey for machine learning security to securing machine learning for CPS, *IEEE Communications Surveys & Tutorials* 23 (2021), pp. 524–552.
3. E.A. Lee, Cyber Physical Systems: Design Challenges, in *2008 11th IEEE International Symposium on Object and Component-Oriented Real-Time Distributed Computing (ISORC)*. 2008, pp. 363–369.
4. P. Radanliev, D.D. Roure, M.V. Kleek, O. Santos, and U. Ani, *Artificial Intelligence in cyber physical systems – AI & society*. Springer London, 2020. Available at https://link.springer.com/article/10.1007/s00146-020-01049-0#citeas.
5. K. Zetter, *An unprecedented look at Stuxnet, the world's first digital weapon* (2014). Available at www.wired.com/2014/11/countdown-to-zero-day-stuxnet/.
6. Cloudflare, *What was the WannaCry ransomware attack?* (2024). Available at www.cloudflare.com/en-gb/learning/security/ransomware/wannacry-ransomware/, accessed: 2024-07-08.
7. A. Hurst, *Notpetya: The most devastating cyberattack in history* (2022). Available at www.information-age.com/notpetya-five-years-on-cyber-security-lessons-learned-by-organisations-19999/, accessed: 2024-07-08.
8. Martin Giles, *Triton is the world's most murderous malware, and it's spreading* (2019). Available at www.technologyreview.com/2019/03/05/103328/cybersecurity-critical-infrastructure-triton-malware/, accessed: 2024-07-08.
9. S. Subramanian, P. Renneker, D. Powers, J. Mariani, A. Keyal, and A. Routh, *Breaking free from government ransomware attacks* (2018). Available at www.deloitte.com/global/en/our-thinking/insights/industry/government-public-services/government-ransomware-attacks.html, accessed: 2024-07-08.
10. L. Abrams, *The solarwinds cyber-attack: What you need to know* (2020). Available at www.bleepingcomputer.com/news/security/the-solarwinds-cyberattack-the-hack-the-victims-and-what-we-know/, accessed: 2024-07-08.

11. G. Cohen, *Throwback attack: An insider releases 265,000 gallons of sewage on the Maroochy shire* (2021). Available at www.industrialcybersecuritypulse.com/facilities/throwback-attack-an-insider-releases-265000-gallons-of-sewage-on-the-maroochy-shi, accessed: 2024-07-08.
12. K. Townsend, *Details emerge on Iranian railroad cyberattack* (2021). Available at www.securityweek.com/details-emerge-iranian-railroad-cyberattack, accessed: 2024-07-08.
13. Artificial intelligence (AI) cybersecurity (2024). Available at www.ibm.com/ai-cybersecurity?utm_content=SRCWW&p1=Search&p4=43700068027043008&p5=p&p9=58700007551028531&gad_source=1&gclid=CjwKCAiA6aW6BhBqEiwA6KzDc0a_cMCbSwqd7y3kjtYxda_t450Kk9GOSRKZvXFl4pjNz0kncB5lrxoCuGcQAvD_BwE&gclsrc=aw.ds.
14. A. Handa, A. Sharma, and S.K. Shukla, Machine learning in cybersecurity: A review. *WIREs Data Mining and Knowledge Discovery,* 9 (2019), e1306, Available at https://wires.onlinelibrary.wiley.com/doi/abs/10.1002/widm.1306.
15. S. Mahdavifar and A.A. Ghorbani, Application of deep learning to cybersecurity: A survey. *Neurocomputing,* 347 (2019), 149–176, Available at www.sciencedirect.com/science/article/pii/S0925231219302954.
16. D.O. Ukwen and M. Karabatak, *Review of NLP-based systems in digital forensics and cybersecurity* (2021), pp. 1–9.
17. K. Dhanushkodi and S. Thejas, (2024). AI enabled threat detection: Leveraging artificial intelligence for advanced security and cyber threat mitigation. *IEEE Access* 12 (2024), 173127–173136.
18. D. Li, D. Chen, B. Jin, L. Shi, J. Goh, and S.K. Ng, MAD-GAN: Multivariate anomaly detection for time series data with generative adversarial networks, in *Artificial Neural Networks and Machine Learning – ICANN 2019: Text and Time Series*, I.V. Tetko, V. Kůrková, P. Karpov, and F. Theis, eds., Springer International Publishing, Cham, (2019), pp. 703–716.
19. K. Kaushik, A. Khan, A. Kumari, I. Sharma, and R. Dubey, Ethical Considerations in AI-Based Cybersecurity, in *Next-Generation Cybersecurity: AI, ML, and Blockchain*, K. Kaushik and I. Sharma, eds., Springer Nature Singapore, Singapore (2024), 437–470, Available at https://doi.org/10.1007/978-981-97-1249-6_19.
20. A. Paullada, I.D. Raji, E.M. Bender, E. Denton, and A. Hanna, Data and its (dis) contents: A survey of dataset development and use in machine learning research. *Patterns,* 2 (2021), 100336, Available at www.sciencedirect.com/science/article/pii/S2666389921001847.
21. C.M. Ahmed, G.R. M R, and A.P. Mathur, Challenges in machine learning based approaches for real-time anomaly detection in industrial control systems, in *Proceedings of the 6th ACM on Cyber-Physical System Security Workshop*, New York, NY. Association for Computing Machinery, CPSS'20, (2020), pp. 23–29, Available at https://doi.org/10.1145/3384941.3409588.
22. M. Robertson, MARS—machine automated response system, in *Proceedings of the 21st Annual ACM SIGUCCS Conference on User Services*, New York, NY. Association for Computing Machinery, SIGUCCS'93, (1993), pp. 16–20, Available at https://doi.org/10.1145/263814.263821.
23. J.A. Wong, A.M. Berenbeim, D.A. Bierbrauer, and N.D. Bastian, Uncertainty-quantified, robust deep learning for network intrusion detection, in *Proceedings of the Winter Simulation Conference*. IEEE Press, WSC'23, (2024), pp. 2470–2481.
24. A. Thekkilakattil and G. Dodig-Crnkovic, Ethics aspects of embedded and cyber-physical systems, in *2015 IEEE 39th Annual Computer Software and Applications Conference*, 2 (2015), 39–44.

25. J. Inoue, Y. Yamagata, Y. Chen, C.M. Poskitt, and J. Sun, Anomaly detection for a water treatment system using unsupervised machine learning, in *2017 IEEE International Conference on Data Mining Workshops (ICDMW)*. (2017), pp. 1058–1065.
26. Q. Cao, C. Zanni-Merk, A. Samet, C. Reich, F. de Bertrand de Beuvron, A. Beckmann, and C. Giannetti, KSPMI: A knowledge-based system for predictive maintenance in Industry 4.0. *Robotics and Computer-Integrated Manufacturing,* 74 (2022), 102281, Available at www.sciencedirect.com/science/article/pii/S0736584521001617.
27. J. Sakhnini and H. Karimipour, *AI and Security of Cyber Physical Systems: Opportunities and Challenges,* in *Security of Cyber-Physical Systems: Vulnerability and Impact*, H. Karimipour, P. Srikantha, H. Farag, and J. Wei-Kocsis, eds., Springer International Publishing, Cham (2020), pp. 1–4, Available at https://doi.org/10.1007/978-3-030-45541-5_1.
28. P. Radanliev, D. De Roure, M. Van Kleek, O. Santos, and U. Ani, *Artificial intelligence in cyber physical systems,* 36 (2021), 783–796, Available at https://doi.org/10.1007/s00146-020-01049-0.

Section II

Application of AI and ML in Smart CPS Analytics

4 Performance Analytics of Cyber-Physical Systems (CPSs)

4.1 INTRODUCTION

Cyber-physical systems (CPSs) are characterized by the integration of computational systems with physical processes, creating a connection between the digital and physical systems. These systems are central to modern applications such as autonomous vehicles, smart healthcare devices, and industrial automation, where performance, safety, and security are important. As CPSs become more pervasive, understanding and optimizing their performance through rigorous analytics is essential ensuring their reliability and effectiveness. This chapter explores the methodologies, metrics, and tools for analysing CPS performance, highlighting their importance across various domains.

4.1.1 Key Metrics for CPS Performance

The performance of CPSs is multifaceted, involving metrics that collectively determine the system's effectiveness in integrating computational and physical processes. Understanding these metrics is crucial for designing, analysing, and optimizing CPSs across different application domains. We discuss some of the key performance metrics of CPSs in detail:

Latency and Real-Time Performance:
Latency refers to the time delay between the input and the corresponding output in a system. In CPSs, low latency is important because these systems often interact with dynamic physical environments requiring immediate reactions. Real-time performance maintains the system's processing time for the inputs and generates outputs within an acceptable time frame for the specific application. For example, let us consider an autonomous vehicle navigating through traffic. The vehicle's sensors continuously collect data about the surrounding environment, such as the positions of other vehicles, pedestrians, and obstacles. The onboard processing unit must analyse this data in real time to make decisions about steering, acceleration, and braking. Here, even a slight delay in processing can lead to accidents, highlighting the critical nature of low latency in such CPS applications [1]. Likewise, in industrial robotics, for example, robots on an assembly line must coordinate their actions with precision

DOI: 10.1201/9781003489986-6

timing. High latency can result in misalignment of components or even damage to the products being assembled [1].

To quantify latency in CPS, the formulation is

$$\text{Latency} = T_{\text{Processing}} + T_{\text{Communication}} + T_{\text{actuation}} \tag{4.1}$$

where $T_{\text{Processing}}$ is the time taken to process the input data, $T_{\text{Communication}}$ is the time taken to transmit data between components. $T_{\text{Actuation}}$ is the time taken for the system to perform the physical action.

Throughput and Efficiency:
The rate at which a system can process tasks or data over a given period is throughput. In CPSs, high throughput is essential for handling large volumes of data and ensuring efficient system operation, especially in environments where multiple processes occur simultaneously. For example, in a smart manufacturing setting, sensors and actuators are deployed across the factory floor to monitor and control various processes. The CPSs must process data from numerous sources, such as temperature sensors, pressure gauges, and machine status indicators, to optimize production efficiency. High throughput ensures that the system can handle this data influx without delays, maintaining optimal operation [2].

Throughput (θ) can be calculated as:

$$\Theta = \frac{N_{\text{Tasks}}}{T_{\text{Total}}} \tag{4.2}$$

where N_{Tasks} is the number of tasks processed, T_{Total} is the total time taken.

For example, in smart grids, high throughput is necessary to process real-time data from distributed energy resources and consumers to balance supply and demand effectively [3].

Reliability and Fault Tolerance:
While reliability refers to the ability of a system to perform its required functions under stated conditions for a specified period, fault tolerance is the system's capacity to continue operating properly in the event of the failure of some of its components. In CPSs, reliability and fault tolerance are crucial due to the potential consequences of system failures, which can range from operational downtime to catastrophic events. For instance, in healthcare CPS, such as remote patient monitoring systems, reliability ensures that critical health data is accurately captured and transmitted without interruption. On the other hand, fault tolerance mechanisms, like redundant sensors of failover communication channels, are implemented to prevent single points of failure that could jeopardize patient safety [4].

The reliability ($R(t)$) of a system can be expressed using the exponential reliability function:

$$R(t) = e^{-\lambda t} \tag{4.3}$$

where, λ is the failure rate and t is time.

To enhance reliability and fault tolerance, CPSs often employ techniques such as:

- **Redundancy:** Incorporating multiple instances of critical components.
- **Diversity:** Using different methods or technologies to achieve the same function.
- **Error Detection and Correction:** Implementing algorithms to detect and correct errors in data transmission and processing.

Table 4.1 lists common fault tolerance techniques used in CPSs.

In energy systems, reliability and fault tolerance are critical to prevent blackouts. Smart grids incorporate self-healing mechanisms that automatically detect and isolate faults, rerouting power to maintain service [3].

- **Scalability** is the system's ability to handle increasing amounts of work to accommodate that growth. In CPS, scalability ensures that as the number of connected devices or the volume of data grows, the system continues to perform efficiently.

  A scalable CPS architecture allows for adding new sensors and nodes without significant redesign or performance degradation [5]. For example, environmental monitoring networks were initially deployed in a limited area, these networks may need to expand to cover larger regions or include more parameters like air quality, humidity, and noise levels as expansion, therefore CPS should be designed considering scalability from the start.

  Scalability considerations involve

  - **Hardware scalability:** Ability to add more physical devices.
  - **Software scalability:** Software's ability to handle increased load.
  - **Network scalability:** Network infrastructure accommodating more nodes and data traffic.

  Scalability ensures that as more machines and sensors are connected, the CPS can manage the increased data flow and maintain performance levels [6].

- **Safety and Security:**

While safety in CPS refers to the system's ability to operate without causing an unacceptable risk of physical injury or damage to people or the environment, security involves protecting the system against malicious attacks that could compromise its integrity, availability, or confidentiality.

TABLE 4.1
Fault Tolerance Techniques in CPS

Technique	Description
Redundant hardware	Multiple hardware units performing the same function
Checkpointing	Saving system states periodically for recovery
Watchdog timers	Monitoring system operation and triggering resets
Failover systems	Switching to backup components upon failure
Software redundancy	Diverse software algorithms for the same task

TABLE 4.2
Safety and Security Metrics in CPS

Metric	Description
MTBF	Average operational time between failures
Safety integrity level (SIL)	Level of risk reduction provided by safety functions
Detection rate	Percentage of attacks correctly identified
False positive rate	Incorrect identification of benign activity as malicious
Response time	Time taken to respond to a security incident

In CPSs, safety and security are connected. For example, a security breach in an autonomous vehicle's control system could lead to unsafe operating conditions. Therefore, performance metrics must consider both aspects to ensure the system not only functions correctly under normal conditions but also remains robust against threats. For example, the Stuxnet worm attack on industrial control systems, which manipulated the physical processes of nuclear centrifuges, causing physical damage while appearing normal to operators [7]. This incident highlights the importance of incorporating security measures into CPS performance metrics. Further, in smart healthcare systems, ensuring patient safety involves protecting medical devices from unauthorized access that could alter their operation, such as insulin pumps or pacemakers [8]. Table 4.2 shows the performance metrics and their description to be used by CPS designers.

It is important to note that these performance metrics are often interdependent. For instance, enhancing security measures might introduce additional processing overhead, potentially increasing latency. Therefore, CPS designers must balance these metrics to meet the specific requirements of their applications.

4.2 PERFORMANCE MONITORING AND DATA COLLECTION

Performance analytics in CPSs rely on robust performance monitoring and efficient data collection mechanisms. These processes are critical for understanding system behaviour, detecting anomalies, and optimizing performance in real time. This section explores the methodologies and technologies used in performance monitoring and data collection for CPS.

- **Sensor Networks and Data Acquisition:** Sensor networks play essential part in CPS performance monitoring. It consists of distributed sensors that capture data about the physical environment and system states. These sensors collect various types of data, including temperature, pressure, motion, and more, enabling the CPS to interact intelligently with the physical world. For example, in a smart agriculture system, soil moisture sensors, temperature sensors, and humidity sensors work together to monitor crop conditions. This data allows for precise irrigation and fertilization, improving crop yields while conserving resources [9].

TABLE 4.3
Common Sensors in CPS and Their Applications

Sensor Type	Measured Parameter	Application
Temperature sensor	Temperature	Climate control in buildings
Pressure sensor	Pressure	Industrial process control
Accelerometer	Acceleration	Vehicle dynamics monitoring
Gyroscope	Orientation	Drone flight stabilization
Proximity sensor	Distance to objects	Robotics and automation
Light sensors	Light intensity	Smart lighting systems

Table 4.3 presents common types of sensors used in CPS and their typical applications.

Typically sensors feed data into the CPS, where it is processed and analysed to make informed decisions. Next section elaborates on the importance of data and how to collect such data for the CPS-related systems. It will explore the trade-offs between processing data locally (at the edge of the network) and using centralized cloud computing resources.

4.2.1 Edge Computing versus Cloud Analytics

This section will explore the trade-offs between processing data locally and using centralized cloud computing resources. A significant consideration in CPS data collection is the choice between edge computing and cloud analytics for data processing.

Edge computing involves processing data locally, near the source of data generation. This approach reduces latency, conserves bandwidth, and enhances data privacy. For instance, in autonomous vehicles, edge computing allows for immediate processing of sensor data to make real-time driving decisions [10]. The vehicle's onboard computers analyse inputs from cameras, LiDAR, and radar sensors to detect obstacles and navigate safely.

Cloud analytics, on the other hand, leverages centralized data centres to perform complex computations and store large datasets. This approach is suitable for tasks that are not time-sensitive but require substantial processing power, such as training machine learning (ML) models. In smart city applications, data from traffic sensors can be aggregated in the cloud to analyse congestion patterns and optimize traffic light sequences [11]. The choice between edge and cloud depends on factors such as latency requirements, data volume, and computational needs.

4.2.2 Data Aggregation Techniques

As CPSs generate vast amounts of data, managing it can be a challenging task. Hence, efficient data aggregation techniques are essential. Data aggregation reduces the volume of data transmitted and stored, conserving network bandwidth and storage resources. One common method is *hierarchical aggregation*, where data is combined at various levels before reaching the central system. For example, in a wireless sensor network monitoring environmental conditions, sensors may send data to a local aggregator node, which computes summary statistics like averages or maximum values before forwarding the data upstream [12]. Another approach is *in-network processing*, where data processing tasks are distributed across the network nodes. This technique minimizes the need for raw data transmission by performing computations at the source [13]. We present several data aggregation techniques in Table 4.4.

These techniques help in optimizing network performance and extending the lifespan of battery-powered devices.

4.2.3 Real-Time Monitoring Systems

Real-time monitoring is critical for CPSs that require immediate responses to changing conditions. It enables prompt detection and response to issues, enhancing safety and efficiency. Systems like supervisory control and data acquisition (SCADA) and distributed control systems (DCS) are employed to monitor and control industrial processes in real time [14]. For example, sensors in a chemical processing plant monitor variables such as temperature, pressure, and flow rates. The SCADA system collects this data, displays it to operators, and can automatically adjust control valves or shut down equipment if unsafe conditions are detected [15].

4.2.4 Data Fusion and Sensor Fusion

Sensor fusion combines data from multiple sensors to improve the accuracy and reliability of the information obtained. This is especially important when individual sensors may be noisy or have limited accuracy. The fusion of these data sources compensates for the weaknesses of individual sensors. In robotics, for instance, combining data from global positioning system (GPS), accelerometers, gyroscopes, and vision systems allows for precise localization and navigation [16]. Equation (4.4) demonstrates a simple sensor fusion model using a Kalman filter:

$$\hat{x}_k = \hat{x}_{k|k-1} + K_k (z_k - H\hat{x}_{k|k-1}) \tag{4.4}$$

TABLE 4.4
Comparison of Data Aggregation Techniques

Technique	Advantages	Disadvantages
Hierarchical aggregation	Reduces data transmission, scalable	Potential loss of data granularity
In-network processing	Lowers communication overhead	Increases processing burden on nodes
Compressed sensing	Efficient for sparse data	Requires complex algorithms
Time-series compression	Preserves important patterns	May miss anomalies in compressed data

where

- $\hat{x}$ is the estimated state at time k
- $\hat{x}_{k|k-1}$ is the prior estimate
- K_k is the Kalman gain
- z_k is the measurement at time k
- H is the measurement matrix

This mathematical model helps in predicting the system's state by minimizing the estimation error.

4.3 CHALLENGES IN DATA COLLECTION

Despite technological advancements, data collection in CPS faces several challenges. We discuss some challenges in this section:

- **Data Volume:** The high volume of data can overwhelm storage systems and make real-time processing difficult.
- **Data Velocity:** The speed at which data is generated requires systems capable of handling rapid data streams.
- **Data Variety:** Heterogeneous data types from different sensors complicate data integration and analysis.
- **Energy Constraints:** Battery-powered sensors need energy-efficient data transmission methods.
- **Security and Privacy:** Protecting sensitive data from unauthorized access is crucial.

4.4 STRATEGIES FOR EFFECTIVE PERFORMANCE MONITORING

Effective performance monitoring is crucial for ensuring that the system operates efficiently, reliably, and securely in CPSs. However, with the integration of physical processes and computational resources, monitoring becomes a dynamic challenge. This section presents some of the widely adopted strategies to overcome these challenges.

4.4.1 Adaptive Sampling

A strategy that adjusts the frequency of data collection based on the observed variability in the system is called adaptive sampling. When the system is stable, the sampling rate can be reduced to conserve resources, such as bandwidth and energy, particularly in resource-constrained environments like wireless sensor networks. Conversely, when rapid changes are detected, the system can increase the sampling rate to capture more granular data. For example, soil moisture sensors may collect data less frequently during stable weather conditions in a smart agriculture system. However, suppose the system detects changes in weather patterns, such as rainfall or sudden temperature shifts, in that case, the sampling rate can increase to provide more detailed monitoring of the soil's condition [17]. This technique helps to optimize resource use while maintaining system responsiveness. Some of the benefits include

- **Resource Efficiency:** Reduces unnecessary data collection, conserving energy in battery-powered sensors.
- **Focused Monitoring:** Increases data collection precision when needed, providing higher fidelity during critical events.

4.4.2 Data Compression

It is another technique that is particularly important in systems where vast amounts of data are generated continuously. Compression algorithms reduce the size of data without significant loss of information, thus optimizing both bandwidth usage and storage. One common approach is *lossless compression*, where data can be perfectly reconstructed from the compressed version. This is important in applications like healthcare monitoring, where critical data, such as patient vitals, must not be compromised. *Lossy compression*, on the other hand, can be used in less critical applications where minor data loss does not affect system performance, such as environmental sensing [18]. For example, smart meters generate voluminous amounts of data about energy consumption in smart grids, compression techniques can help reduce the amount of data transmitted back to the control centre, making data storage and transmission more efficient [3]. Compressing data provides several benefits including (i) less data needs to be transmitted over networks, which is critical in systems with bandwidth limitations; (ii) reduction in the demand for storage resources, allowing for more data to be stored over time without requiring large storage capacities.

4.4.3 Anomaly Detection

Anomaly detection is crucial in identifying abnormal behaviour or unexpected events that may indicate system faults, security breaches, or other issues. CPSs often employ ML algorithms to analyse data streams and detect deviations from normal patterns in real time [19]. For instance, in smart grids, anomaly detection can help identify unusual electricity consumption patterns, which may indicate equipment malfunction or energy theft. By flagging anomalies early, the system can trigger alerts or corrective actions before the issue escalates [20]. Common anomaly detection techniques include clustering, where data points are grouped based on similarity, and statistical methods like Z-score, which measures how far a data point is from the mean. More advanced techniques involve unsupervised ML algorithms, which can autonomously detect anomalies without needing labelled data for training. Some of the benefits of using anomaly detection techniques are that they are capable of detecting and responding to issues instantly, minimizing potential downtime.

4.4.4 Distributed Processing

Distributed processing involves delegating computational tasks across multiple nodes or devices in the system, rather than relying on a single central unit. This strategy improves scalability and reduces bottlenecks, especially in large-scale CPS deployments such as smart cities or industrial automation systems [21]. For example, in an industrial CPS, distributed processing allows each machine to handle some portion of the data analysis locally. By processing data close to the source, latency is reduced, and the system can make faster decisions without waiting for centralized processing. With distributed processing, decisions can be made closer to the source of the data, improving system responsiveness. It also helps to scale the system horizontally by adding more devices without overwhelming central processors.

4.4.5 Predictive Analytics and Maintenance

Predictive analytics uses historical data to predict future system behaviour, particularly focusing on identifying potential failures before they occur. By using ML models to analyse historical trends, CPSs can anticipate issues such as equipment breakdowns, network congestion, or energy spikes [22]. In predictive maintenance, CPSs can monitor key performance indicators (KPIs), such as vibration levels or temperatures in industrial machines, to predict when a part is likely to fail. This allows for maintenance to be performed at an optimal time, reducing the risk of unexpected failures and costly downtime. For example, wind turbines in a smart energy system can use predictive analytics to forecast when their blades or mechanical components will require maintenance, ensuring that repairs are made during planned outages rather than during operation [23]. It can facilitate a proactive maintenance approach which reduces downtime by addressing issues before they cause failure and is able to minimize repair costs by preventing catastrophic failures and optimizing maintenance schedules.

4.4.6 Data Redundancy and Backup Systems

In mission-critical CPSs, where system failure could result in significant consequences, data redundancy and backup systems are essential. Redundancy ensures that even if one component or sensor fails, the system can continue to operate using backup data from other sources. For example, in autonomous vehicles, having redundant sensor inputs (such as multiple cameras, radar, and LiDAR) ensures the system can still function if one sensor fails [24]. Similarly, backup systems ensure that critical data is always available, even in the event of network outages or hardware failures. These systems periodically back up data to secure locations, either locally or in the cloud, ensuring that data integrity is maintained even during catastrophic events [24].

4.4.7 Encryption and Security Measures

In many CPS applications, especially those dealing with sensitive information (e.g., healthcare, finance, or critical infrastructure), security is paramount. Encrypting data during transmission and storage ensures that unauthorized entities cannot access or tamper with it [25]. For example, in smart healthcare systems, encryption protocols (such as AES) are used to secure patient data transmitted from monitoring devices to healthcare providers. In addition to encryption, multi-factor authentication and network firewalls are used to protect against unauthorized access to the system. They can provide several benefits including

- **Data Privacy:** Ensures sensitive information is protected from unauthorized access.
- **System Integrity:** Prevents tampering and ensures that only trusted parties can modify or access critical data.

The summarization of benefits and challenges of each strategy is listed in Table 4.5.

4.4.8 Conclusion

The strategies for effective performance monitoring in CPSs are critical for ensuring that the system operates efficiently, reliably, and securely. From adaptive sampling and data compression to anomaly detection and predictive maintenance, these techniques allow CPSs to handle the complexities of real-world environments. By employing distributed processing, redundancy, and robust security measures, CPSs can not only enhance performance but also minimize the risks of failures or attacks [24]. Implementing these strategies effectively will lead to more resilient and capable systems across industries like healthcare, energy, transportation, and manufacturing. These strategies contribute to more efficient and secure data collection processes.

4.5 TECHNIQUES FOR PERFORMANCE ANALYSIS

Various analytical techniques are employed to assess CPS performance. Statistical methods are often used to detect anomalies in the data, which may indicate potential system failures or inefficiencies. For example, time-series analysis can be applied to monitor temperature data in industrial processes, where deviations from normal patterns may signal overheating or equipment malfunction [27, 28]. Increasingly, ML is being used to optimize CPS performance by enabling predictive analytics and adaptive control systems. ML models can predict system behaviours based on historical data, allowing for preemptive adjustments to prevent performance degradation. In autonomous vehicles, for instance, ML algorithms are trained to predict traffic patterns and optimize routing decisions in real time [29].

Another powerful tool for performance analysis in CPSs is modelling and simulation. Digital twins – virtual replicas of physical systems – allow engineers to simulate various scenarios and analyse how changes in one part of the system affect the overall performance. For example, in smart manufacturing, digital twins can simulate different production workflows to identify the most efficient configurations, thereby enhancing both throughput and reliability [30]. Analysing the performance of CPSs requires a multidisciplinary approach that incorporates methods from statistics, ML, control theory, and simulation modelling.

TABLE 4.5
Benefits, and Challenges, of Widely Adopted Strategies for Effective Performance Monitoring in CPS

Strategy	Benefit	Challenges	Reference
Adaptive sampling	– Conserves energy and bandwidth – Dynamically adjusts to system conditions – Focuses on critical events	– May miss short-lived events when sampling rate is low – Complexity in determining optimal sampling rate	[26]
Data compression	– Reduces bandwidth usage – Optimizes storage requirements – Can enable longer-term data retention	– Lossy compression can lead to data loss – Decompression may add to processing overhead	[18]
Anomaly detection	– Enables real-time fault detection – Prevents system downtime through early detection – Improves system reliability	– False positives can lead to unnecessary interventions – Training models for accurate detection may require significant data	[19]
Distributed processing	– Reduces latency by processing data closer to the source – Improves scalability as additional devices are added – Prevents bottlenecks at central processing nodes	– More complex to manage and coordinate distributed nodes – Risk of inconsistent data across nodes	[21]
Predictive analytics	– Reduces unplanned downtime – Enables proactive maintenance, extending equipment life – Optimizes maintenance schedules	– Requires large amounts of historical data to train models – Can be computationally expensive	[22]
Data redundancy	– Ensures high availability and reliability – Protects against data loss in the event of sensor failure	– Increases resource usage (e.g., extra sensors, storage) – Managing redundant data may introduce complexity	[24]
Encryption and security measures	– Ensures data privacy and integrity – Protects sensitive information from unauthorized access	– Can increase system overhead (e.g., encryption/decryption time) – May impact system latency due to additional security processing	[25]

4.5.1 Statistical Methods

Statistical analysis is fundamental for interpreting the vast amounts of data generated by CPSs. Techniques such as time-series analysis, regression, and hypothesis testing help in identifying patterns, trends, and anomalies. One common statistical method is time-series analysis, which examines sequences of data points collected over time. For instance, in a smart grid CPS, analysing the time-series data of electricity consumption can help predict future demand and optimize energy distribution [31].

Figure 4.1 illustrates a toy example of showing time-series plot of electricity consumption over a week, highlighting peak usage periods.

4.5.2 Regression Analysis

It is another statistical technique used to model the relationship between dependent and independent variables. In CPS, regression can predict system behaviour under varying conditions. For example, in analysing the performance of solar panels, regression models can determine how temperature and sunlight intensity affect energy output [32].

Table 4.6 presents an example of regression coefficients indicating the influence of environmental factors on solar panel efficiency.

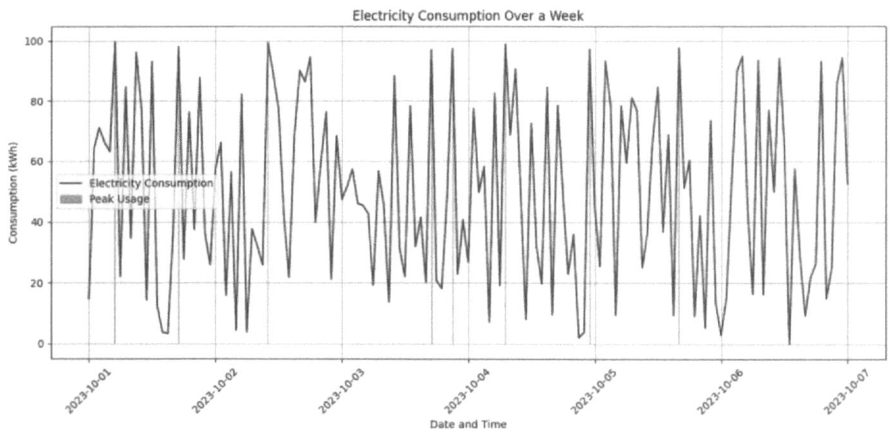

FIGURE 4.1 Toy example of time series of electricity consumption over one week.

TABLE 4.6
Regression Coefficients for Solar Panel Efficiency

Variable	Coefficient
Temperature (°C)	–0.0045
Sunlight intensity (W/m^2)	0.0009

These coefficients suggest that higher temperatures slightly decrease efficiency, while increased sunlight intensity improves it.

4.5.3 Machine Learning (ML) for CPSs

ML techniques enable CPSs to learn from data, make predictions, and improve over time without being explicitly programmed. ML is particularly useful for pattern recognition, anomaly detection, and predictive maintenance. In predictive maintenance, ML models like neural networks and support vector machines analyse sensor data to predict equipment failures before they occur [23]. For example, vibration and temperature data from industrial machines can be used to predict bearing failures, allowing for timely maintenance. Especially effective in anomaly detection, unsupervised learning algorithms identify patterns that deviate from normal behaviour. For instance, clustering algorithms can detect unusual network traffic in CPSs, indicating potential security threats [19].

Table 4.7 compares different ML algorithms used in CPS performance analysis.

4.5.4 Modelling and Simulation

Modelling and simulation allow for the virtual representation and analysis of CPSs, enabling testing and optimization without affecting real-world operations. Digital twins are virtual models that simulate their physical counterparts in real time. In manufacturing, a digital twin of a production line can simulate operations, identify bottlenecks, and test improvements without disrupting actual production [33].

Figure 4.2 shows typical concept of a digital twin in CPSs [34] where different layers are combined together to form a connected system.

Simulation tools like MATLAB/Simulink and NS-3 help model the behaviour of CPS components and their interactions. For example, NS-3 can simulate network protocols in CPS to evaluate performance metrics like latency and throughput [35]. While performance profiling tools help identify bottlenecks

TABLE 4.7
ML Algorithms for CPS Performance Analysis

Algorithm	Application	Advantages
Neural networks	Predictive maintenance, pattern recognition	Handles complex nonlinear relationships
Support vector machines	Fault detection, classification	Effective in high-dimensional spaces
Clustering algorithms	Anomaly detection, data segmentation	Simple implementation, scalability
Principal component analysis	Dimensionality reduction	Reduces data complexity, highlights key variables

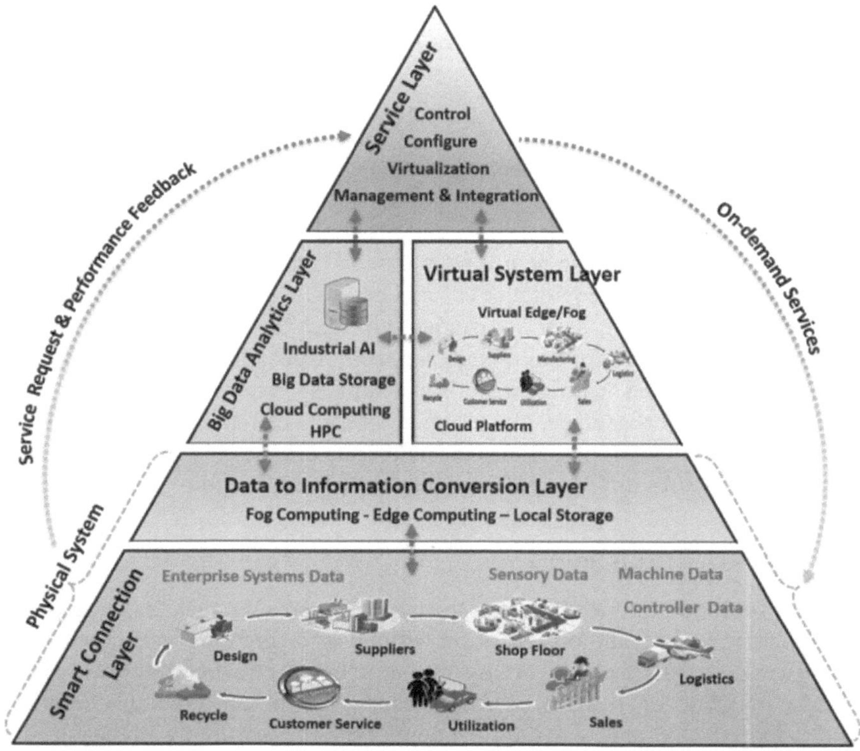

FIGURE 4.2 Digital twin concept in cyber-physical systems [34].

TABLE 4.8
Performance Profiling Tools for CPS

Tool	Application	Features
Wireshark	Network analysis	Packet capture, protocol decoding
Perf	CPU profiling	Monitors CPU usage, performance counters
Valgrind	Memory debugging	Detects memory leaks, profiling
MATLAB Profiler	Code optimization	Measures execution time, identifies bottlenecks

and optimize system performance, tools like Wireshark, Perf, and Valgrind are instrumental in CPS analysis. (i) Wireshark captures and analyses network traffic, useful for diagnosing latency and packet loss in networked CPS [36]. (ii) Perf profiles CPU performance, identifying processes that consume excessive resources. (iii) Valgrind detects memory leaks and errors, which is crucial for ensuring system reliability. Table 4.8 summarizes these tools and their applications.

4.6 CHALLENGES IN CPS PERFORMANCE ANALYSIS

Despite the advances in performance analytics, several challenges remain. One major issue is the heterogeneity of systems, as CPSs often involve a mix of hardware and software components from different vendors, each with its own standards and protocols. This lack of uniformity complicates performance monitoring and optimization, as integrating and analysing data from disparate sources can be difficult [37].

Another challenge is ensuring data privacy and security while performing analytics. Since CPSs often deal with sensitive data – such as health information or industrial control systems – there is a risk that performance monitoring systems could inadvertently expose vulnerabilities to malicious actors. Thus, ensuring that data analytics frameworks incorporate robust security measures is crucial for maintaining system integrity [38].

Finally, computational resources can be a limiting factor, particularly in resource-constrained environments. While essential for real-time processing, edge computing may not have the computational power required for advanced analytics. Balancing the trade-offs between computational efficiency, accuracy, and timeliness is a key challenge in the CPS performance analytics [10].

4.7 CASE STUDY: INDUSTRIAL INTERNET OF THINGS (IIOT)

CPSs are adopted widely to enhance processes including manufacturing in industrial internet of things (IIoT) through improved performance monitoring and data analytics. A manufacturing plant might deploy sensors on machinery to monitor parameters like vibration, temperature, and operational speed. The plant can implement predictive maintenance, identifying potential equipment failures before they occur from this data [22]. Figure 4.3 illustrates a typical example of an IIoT architecture with data flow from sensors to analytics platforms.

The image highlights the vertical flow of data and control from the perception layer, where raw data is collected, to the application layer, where actionable insights are used to optimize industrial processes. (i) *The perception layer* is responsible for sensing and collecting real-world data through various devices such as cameras, robots, sensors, and meters. These devices constantly monitor physical conditions (e.g., temperature, pressure, motion) and relay this data upward through the IIoT architecture. In a smart manufacturing plant, for example, machines are equipped with sensors that monitor operational data like vibration, temperature, and output levels. *Cameras and sensors* provide real-time monitoring of equipment health, identifying potential faults before they lead to breakdowns. Robots in manufacturing lines can also be equipped with sensors that allow them to self-monitor, ensuring that any deviations from normal operation are flagged immediately. Meters capture usage data such as energy consumption, which is crucial for predictive maintenance and optimization.

The system will gain an accurate and detailed picture of real-time operations by collecting this data. This will enable the next layer – the network layer – to transmit it for further analysis.

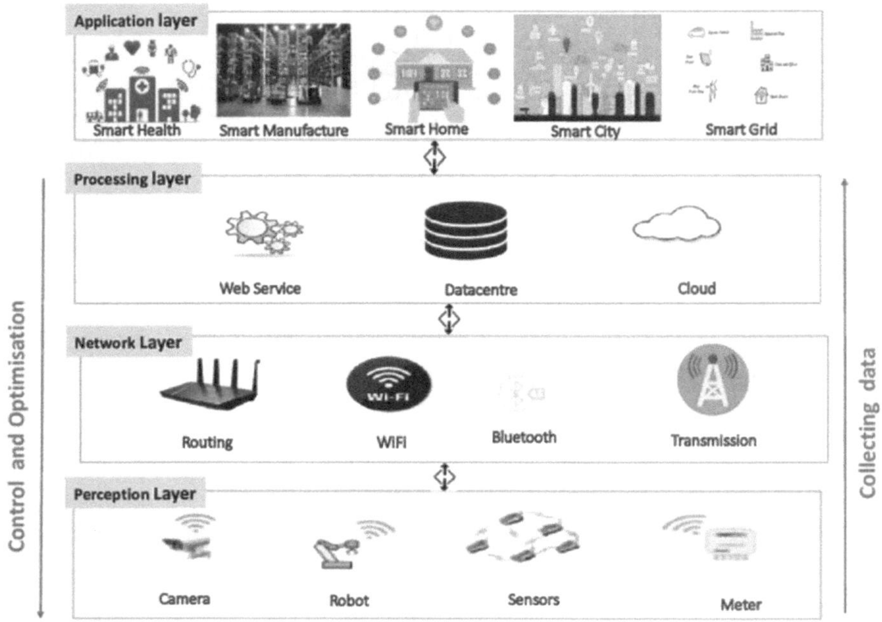

FIGURE 4.3 Data flow in an IIoT architecture [39].

(i) In the **Network Layer**, the collected data is transmitted to the processing layer through networking technologies like Wi-Fi, bluetooth, and other transmission protocols. Effective communication within this layer maintains the timely and efficient delivery of sensor data to the cloud or data centres for processing. For example, Wi-Fi and routing are used in factories where a wireless infrastructure can enable data flow without wired connections. These networks facilitate the transmission of data between machines, robots, and the centralized system. Bluetooth is often used for short-range data exchange, particularly useful in industrial setups where proximity-based device communication is needed. Next, transmission towers or other long-range communication protocols can connect the IIoT infrastructure to remote data centres or cloud systems, enabling monitoring over vast geographical distances, such as in smart grids or smart city projects. *The network layer ensures* that the massive amount of data generated by the perception layer is relayed efficiently to the next stage, the processing layer, for deeper analysis and control.

(ii) The **Processing Layer** is where the real power of IIoT comes into play. In this layer, collected data is aggregated and analysed to generate insights that can drive decision-making. This layer can be hosted on web services, data centres, or cloud infrastructures depending on the specific application requirements. For example, cloud computing offers scalable storage and processing power, which is critical for handling the vast amounts of data generated by IIoT systems. In a smart factory, for instance, data from hundreds of machines is uploaded to the cloud where advanced analytics and ML algorithms can predict when equipment is likely to fail, triggering maintenance alerts before an actual breakdown occurs. Web services enable real-time interaction between devices and users. For example, a plant operator can monitor machine performance via a web-based interface, receiving live updates from sensors installed on the factory floor. Data centres provide centralized processing capabilities, often used in scenarios where sensitive data, such as in healthcare applications, needs to be processed locally for security or compliance reasons. By leveraging the processing layer, IIoT systems can harness the power of big data analytics and predictive modelling to optimize industrial processes, minimize downtime, and reduce operational costs.

(iii) At the top of the IIoT architecture is the **Application Layer**, which translates the insights generated from the lower layers into actionable outcomes across industries such as smart health, smart manufacturing, smart homes, smart cities, and smart grids. In healthcare settings, IIoT devices such as wearable sensors collect patient data like heart rate or glucose levels in real time, which is then transmitted via secure networks to healthcare providers. This enables early diagnosis, remote monitoring, and timely interventions. Likewise,

Smart Manufacturing: in manufacturing plants, IIoT enables predictive maintenance by analysing sensor data to predict when a machine will need repairs. This helps to avoid unscheduled downtime and increases operational efficiency.

Smart Homes: In smart homes, IIoT devices like smart thermostats and security cameras collect environmental data that is analysed and used to adjust settings automatically, improving energy efficiency and security.

Smart Cities: IIoT enables cities to monitor everything from traffic flow to environmental conditions in real time. For instance, traffic sensors can alert traffic control systems to adjust signal timings to reduce congestion.

Smart Grids: Smart meters and sensors in the power grid enable more efficient energy distribution by adjusting loads and predicting demand, helping utilities prevent outages and optimize energy delivery.

In all these applications, the IIoT system operates in a loop: data is collected at the perception layer, transmitted through the network, processed for insights, and then fed back to the application layer for control and optimization. The implementation of such systems can lead to reduced downtime, lower maintenance costs, and increased productivity.

4.8 CONCLUSION

Performance analytics play a pivotal role in ensuring the effectiveness, safety, and reliability of CPSs. By monitoring key metrics such as latency, throughput, reliability, and scalability, and by leveraging tools such as ML and digital twins, CPSs can be optimized for a wide range of applications. However, challenges such as system heterogeneity, data security, and computational limitations must be addressed to fully realize the potential of CPSs in the future.

PRACTICE QUESTIONS

I. Understanding Key Performance Metrics

Consider a CPS deployed in an autonomous drone used for aerial surveillance. The drone must process real-time video feeds to detect objects and avoid obstacles.

1. **Latency Requirement:** Explain why low latency is critical in this CPS application. Calculate the maximum acceptable latency if the drone travels at a speed of 15 m/s and must detect obstacles at least 10 meters ahead to avoid collisions.
2. **Throughput Consideration:** Discuss the importance of throughput in processing high-definition video feeds for real-time object detection. Suggest methods to optimize throughput without compromising data quality.
3. **Reliability and Fault Tolerance:** Identify potential failures that could occur in this CPS and propose fault tolerance mechanisms to address them.

Answer Guide:

(a) Latency Requirement:

Explanation: Low latency is crucial because the drone must quickly process sensor data to react to dynamic environments. Delays can result in the drone failing to detect and avoid obstacles in time, leading to collisions.

Calculation: The time available to detect and react is

$$t = \frac{d}{v} = \frac{10\,\text{m}}{15\text{m/s}} \approx 0.667\,\text{seconds} \tag{4.5}$$

Hence, the total system latency (including sensing, processing, and actuation) must be less than 0.667 seconds.

II. Performance Monitoring Strategies

A smart factory uses a network of sensors to monitor machine performance. The data collected includes temperature, vibration, and output quality metrics.

1. **Edge versus Cloud Computing:** Evaluate the pros and cons of using edge computing versus cloud analytics for this application.
2. **Data Aggregation:** Propose a data aggregation technique suitable for this scenario and justify your choice.

3. **Real-Time Monitoring:** Design a simple real-time monitoring system architecture for the factory, highlighting the key components and their interactions.

Answers: (1) Edge versus cloud computing:

Edge computing pros: Low latency, immediate response, reduced bandwidth usage, enhanced privacy.

Edge computing cons: Limited computational resources, potential scalability issues.

Cloud analytics pros: High computational power, advanced analytics capabilities, easy scalability.

Cloud analytics cons: Higher latency, increased bandwidth usage, potential security concerns.

REFERENCES

1. M. Sun, S. Quan, X. Wang, and Z. Huang, Latency-aware scheduling for data-oriented service requests in collaborative IoT-edge-cloud networks. *Future Generation Computer Systems* 163 (2025), p. 107538, Available at www.sciencedirect.com/science/article/pii/S0167739X24005028.
2. V.K. Annanth, M. Abinash, and L.B. Rao, Intelligent manufacturing in the context of industry 4.0: A case study of Siemens industry. *Journal of Physics: Conference Series* 1969 (2021), p. 012019, Available at https://dx.doi.org/10.1088/1742-6596/1969/1/012019.
3. X. Fang, S. Misra, G. Xue, and D. Yang, Smart grid—The new and improved power grid: A survey. *IEEE Communications Surveys & Tutorials* 14 (2012), pp. 944–980.
4. N. Dey, A.S. Ashour, F. Shi, S.J. Fong, and J.M.R.S. Tavares, Medical cyber-physical systems: A survey. *Journal of Medical Systems* 42 (2018), p. 74.
5. X. Mao, X. Miao, Y. He, X.Y. Li, and Y. Liu, CitySee: Urban CO_2 monitoring with sensors, in *2012 Proceedings IEEE INFOCOM*. (2012), pp. 1611–1619.
6. E.L.X. Li Da Xu and L. Li, Industry 4.0: state of the art and future trends. *International Journal of Production Research* 56 (2018), pp. 2941–2962, Available at https://doi.org/10.1080/00207543.2018.1444806.
7 R. Langner, Stuxnet: Dissecting a cyberwarfare weapon. *IEEE Security & Privacy,* 9 (2011), pp. 49–51.
8 D. Halperin, T.S. Heydt-Benjamin, B. Ransford, S.S. Clark, B. Defend, W. Morgan, K. Fu, T. Kohno, and W.H. Maisel, Pacemakers and implantable cardiac defibrillators: Software radio attacks and zero-power defenses, in *Proceedings of the 2008 IEEE Symposium on Security and Privacy*, USA. IEEE Computer Society, SP'08, (2008), pp. 129–142, Available at https://doi.org/10.1109/SP.2008.31.
9. N. Zhang, M. Wang, and N. Wang, Precision agriculture—A worldwide overview. *Computers and Electronics in Agriculture* 36 (2002), pp. 113–132, Available at www.sciencedirect.com/science/article/pii/S0168169902000960.
10. W. Shi, J. Cao, Q. Zhang, Y. Li, and L. Xu, Edge computing: Vision and challenges. *IEEE Internet of Things Journal* 3 (2016), pp. 637–646.
11. H. Schaffers, N. Komninos, M. Pallot, B. Trousse, M. Nilsson, and A. Oliveira, *Smart cities and the future internet: towards cooperation frameworks for open innovation*, in *The Future Internet*, J. Domingue, A. Galis, A. Gavras, T. Zahariadis, D. Lambert, F. Cleary, P. Daras, S. Krco, H. Müller, M.S. Li, H. Schaffers, V. Lotz, F. Alvarez, B.

Stiller, S. Karnouskos, S. Avessta, and M. Nilsson, eds., Springer, Berlin, Heidelberg, (2011), pp. 431–446.
12. I. Akyildiz, W. Su, Y. Sankarasubramaniam, and E. Cayirci, Wireless sensor networks: A survey, *Computer Networks* 38 (2002), pp. 393–422, Available at www.sciencedirect.com/science/article/pii/S1389128601003024.
13. L. Krishnamachari, D. Estrin, and S. Wicker, *The impact of data aggregation in wireless sensor networks, in Proceedings 22nd International Conference on Distributed Computing Systems Workshops*, Vienna, Austria (2002), pp. 575–578, doi: 10.1109/ICDCSW.2002.1030829 (2002), pp. 575–578.
14. S.A. Boyer, *SCADA: Supervisory Control and Data Acquisition*, ISA, 2009.
15. G. Clarke and D. Reynders, *Practical Modern SCADA Protocols: DNP3, 60870.5 and Related Systems*, Newnes, 2004.
16. B. Khaleghi, A. Khamis, F.O. Karray, and S.N. Razavi, Multisensor data fusion: A review of the state-of-the-art. *Information Fusion* 14 (2013), pp. 28–44, Available at www.sciencedirect.com/science/article/ pii/S1566253511000558.
17. H. Luo, J. Wang, Y. Sun, H. Ma, and X.Y. Li, *Adaptive sampling and diversity reception in multi-hop wireless audio sensor networks, in 2010 IEEE 30th International Conference on Distributed Computing Systems*, Genoa, Italy (2010), pp. 378–387, doi: 10.1109/ICDCS.2010.32
18. C.M. Sadler and M. Martonosi, Data compression algorithms for energy-constrained devices in delay tolerant networks, in *Proceedings of the 4th International Conference on Embedded Networked Sensor Systems*, New York, NY. Association for Computing Machinery, SenSys'06, (2006), pp. 265–278, Available at: https://doi.org/10.1145/1182807.1182834.
19. V. Chandola, A. Banerjee, and V. Kumar, Anomaly detection: A survey, *ACM Computing Surveys* 41 (2009), Available at https://doi.org/10.1145/1541880.1541882.
20. P.K. Reddy Shabad, A. Alrashide, and O. Mohammed, *Anomaly detection in smart grids using machine learning, in IECON 2021 – 47th Annual Conference of the IEEE Industrial Electronics Society* (2021), pp. 1–8.
21. Z. Jia, L. Wang, J. Zhan, L. Zhang and C. Luo, Characterizing data analysis workloads in data centers, in *2013 IEEE International Symposium on Workload Characterization (IISWC)*, Portland, OR, USA (2013), pp. 66–76, doi: 10.1109/IISWC.2013.6704671.
22. J. Lee, H.A. Kao, and S. Yang, *Service innovation and smart analytics for industry 4.0 and big data environment, Procedia CIRP* 16 (2014), p. 3–8.
23. L. Xiang, P. Wang, X. Yang, A. Hu, and H. Su, Fault detection of wind turbine based on scada data analysis using CNN and LSTM with attention mechanism. *Measurement* 175 (2021), p. 109094, Available at www.sciencedirect.com/science/article/pii/S026322412100124X.
24. D.I. Curiac, C. Volosencu, D. Pescaru, L. Jurca, and A. Doboli, Redundancy and its applications in wireless sensor networks: A survey, *World Transactions on Computing* 8 (2009), pp. 705–714.
25. J. Zhou, Z. Cao, X. Dong, and A.V. Vasilakos, *Security and privacy for cloud-based IOT: Challenges, IEEE Communications Magazine* 55 (2017), pp. 26–33.
26. X. Yu, Y. Fu, J. Li, J. Mao, T. Hoang, and H. Wang, *Recent advances in wireless sensor networks for structural health monitoring of civil infrastructure, Journal of Infrastructure Intelligence and Resilience* 3 (2024), p. 100066, Available at www.sciencedirect.com/science/article/pii/S2772991523000415.
27. B. Yang, J. Ye, S. Coshatt, W. Song, and F. Zahiri, Data-driven approach for detection of physical faults and cyber attacks in manufacturing motor drives, in *2022 IEEE Energy Conversion Congress and Exposition (ECCE)* (2022), pp. 1–6.

28. Y. Yuan, X. Tang, W. Zhou, W. Pan, X. Li, H.T. Zhang, H. Ding, and J. Goncalves, Data driven discovery of cyber physical systems, *Nature Communications* 10 (2019), p. 4894.
29. S.M. Grigorescu, B. Trasnea, T.T. Cocias, and G. Macesanu, A survey of deep learning techniques for autonomous driving. *Journal of Field Robotics* 37 (2019), pp. 362–386, Available at https://api.semanticscholar.org/CorpusID:204744017.
30. E. Glaessgen and D. Stargel, *The digital twin paradigm for future NASA and US Air Force vehicles* (2012).
31. T. Hong and S. Fan, Probabilistic electric load forecasting: A tutorial review. *International Journal of Forecasting* 32 (2016), pp. 914–938.
32. X. Du, Z. Lang, M. Liu, and J. Wu, Regression analysis and prediction of monthly wind and solar power generation in China, *Energy Reports* 12 (2024), pp. 1385–1402, Available at www.sciencedirect.com/science/article/pii/S2352484724004487.
33. F. Tao, H. Zhang, A. Liu, and A.Y.C. Nee, Digital twin in industry: State-of-the-art. *IEEE Transactions on Industrial Informatics* 15 (2019), pp. 2405–2415.
34. J. Lee, M. Azamfar, J. Singh, and S. Siahpour, Integration of digital twin and deep learning in cyber-physical systems: Towards smart manufacturing. *IET Collaborative Intelligent Manufacturing* 2 (2020), pp. 34–36, Available at https://ietresearch.online library.wiley.com/doi/abs/ 10.1049/iet-cim.2020.0009.
35. G.F. Riley and T.R. Henderson, The ns-3 network simulator, in *Modeling and Tools for Network Simulation*, K. Wehrle, M. Güneş, and J. Gross, eds., Springer, Berlin, Heidelberg, 2010, pp. 15–34, Available at https://doi.org/10.1007/978-3-642-12331-3_2.
36. U.L. Richard Sharpe Ed Warnicke, *Wireshark User's Guide*, Wireshark Foundation (2007).
37. R. Rajkumar, I. Lee, L. Sha, and J. Stankovic, Cyber-physical systems: The next computing revolution, in *Design Automation Conference*, (2010), pp. 731–736.
38. Y. Luo, Y. Xiao, L. Cheng, G. Peng, and D.D. Yao, Deep learning-based anomaly detection in cyber-physical systems: Progress and opportunities, *ACM Computing Surveys* 54 (2021), Available at https://doi.org/10.1145/3453155.
39. N. Abosata, S. Al-Rubaye, G. Inalhan, and C. Emmanouilidis, Internet of things for system integrity: A comprehensive survey on security, attacks and countermeasures for industrial applications, *Sensors* 21 (2021), Available at www.mdpi.com/1424-8220/21/11/3654.

5 Vulnerability Assessment and Reporting for Cyber-Physical Systems Using AI and ML

5.1 INTRODUCTION

Cyber-physical systems (CPSs) represent integration of computational algorithms and physical processes of systems [1, 2]. These systems are embedded with intelligence, enabling them to autonomously monitor and control physical processes across various sectors, including manufacturing, transportation, healthcare, and energy [3]. With the combination of multifaceted processes and techniques, CPSs can change conditions and optimize performance in ways traditionally inaccessible including real-time processing. However, the increasing complexity and connectivity of CPSs have introduced new security challenges [4, 5]. For example, vulnerabilities within these systems can be exploited, leading to potential cyber-physical threats that compromise not only data integrity but also physical safety [6]. It is especially risky when the reliance on CPSs for critical infrastructure and services make them a prime target for cyber attacks and physical threats. Therefore, vulnerability assessment is imperative in establishing the security, safety, and resilience of these systems.

Traditional security measures are often insufficient to address these threats' dynamic and complex nature. For example, signature-based detection, predefined rules, and periodic vulnerability assessments can have limitations in addressing dynamic and modern threats in CPSs. These approaches are known for being static, reactive, and often adapting to the rapidly evolving threat landscape in addressing threats [7–9]. On the other hand, artificial intelligence (AI) and machine learning (ML) have emerged as powerful tools to enhance vulnerability assessment and reporting in CPSs [10, 11]. AI/ML can improve the resilience of CPS against cyber attacks by automating the detection of security gaps and predicting potential threats. These advanced technologies can analyse large amounts of data to identify anomalies, predict threats, and adapt defences to evolving risks [12, 13]. AI can transform CPS security from reactive to proactive, safeguard system integrity and availability against advanced cyber attacks by reducing false positives and enabling self-restoring mechanisms [14–16].

This chapter explores the transformation of vulnerability assessment in CPSs through the integration of AI/ML. It explores various methodologies, challenges, and solutions developed to assure robustness and reliability of AI-driven assessments. We

DOI: 10.1201/9781003489986-7

also discuss the operational integration of these technologies and address industry-specific challenges and solutions.

5.2 THE IMPORTANCE OF VULNERABILITY ASSESSMENT IN CYBER-PHYSICAL SYSTEMS

A vulnerability assessment can systematically identify and evaluate security weaknesses within CPSs. This helps organizations to implement corrective actions to mitigate risks. The interconnectedness of CPSs can lead even minor vulnerabilities into severe disruptions, for example, the *Stuxnet attack* [8], which exploited vulnerabilities in industrial control systems to cause physical damage. These assessments are critical in addressing various risks, from software bugs and misconfigurations to insecure communication protocols and physical tampering. Traditional security methods often fail to keep pace with the dynamic and evolving threat landscape. By incorporating advanced tools and techniques, vulnerability assessments provide a comprehensive understanding of potential entry points for attackers. They help organizations prioritize risks, allocate resources effectively, and maintain operational integrity. As cyber threats continue to increase, the importance of proactive and continuous vulnerability assessment in CPSs needs cannot be overstated. This section discusses the importance of vulnerability assessment in CPSs and also explores the challenges in conducting such assessments.

5.2.1 Understanding Cyber-Physical Systems

CPSs embed several complex systems from computers and networks monitor to control physical processes with feedback loops where physical processes affect computations. This tight coupling between the computational and physical elements allows for systems that can interact with the physical world in real time [1, 2]. This can lead to advancement across various industries including critical infrastructures (e.g., manufacturing, transportation, healthcare, space etc.).

The integration of physical components (such as sensors and actuators) with computational algorithms and communication networks [17] is the core concept of CPSs. First, sensors collect data from the physical environment, depending on the application, it can measure environmental factors (e.g., temperature, humidity, or light intensity in smart homes), mechanical states (e.g., pressure and vibration in industrial machinery), chemical and positioning (e.g., velocity, acceleration, or global positioning system (GPS) data in autonomous vehicles), and more. These data are then processed by computational units where it is filtered, analysed, and advanced ML techniques are applied. These computations result in decisions that are sent to actuators which help execute physical actions, such as adjusting a valve, steering a vehicle, or activating a robotic arm [18]. The actuator's actions will alter the physical environment, which is continuously monitored by sensors, completing the feedback loop [19]. This real-time process allows CPSs to adapt dynamically to changing conditions, optimize operations, enhance safety and function autonomously with minimal human intervention [20]. Let us discuss some of the examples of CPSs:

- **Manufacturing (Industry 4.0):** In modern manufacturing, CPSs form the backbone of smart factories where machines are equipped with sensors and connected via networks to collect and share data [3]. This allows for real-time monitoring of production processes, predictive maintenance, and optimization of workflows. For instance, Siemens' Amberg Electronics Plant uses CPSs to achieve a production quality rate of over 99%, where products communicate with machines to guide their manufacturing processes [21].
- **Autonomous Vehicles:** Self-driving cars are advanced form of CPSs that rely on a multitude of sensors such as cameras, radar, and LiDAR to perceive the environment [22]. Computational algorithms process this sensor data to make driving decisions like steering, acceleration, and braking effectively. For example, Tesla's Autopilot system continuously collects data from its fleet to improve its driving algorithms, enhancing safety and efficiency [23].
- **Smart Grids:** In energy systems, CPSs enable the development of smart grids to manage electricity generation, distribution, and consumption more efficiently [24]. By integrating sensors and communication technologies, utilities can monitor energy flow in real time, balance supply and demand, and incorporate renewable energy sources. This leads to reduced energy waste and improved grid reliability [25].
- **Healthcare Systems:** Medical CPS include implantable devices like pacemakers and insulin pumps that monitor physiological parameters and deliver therapy accordingly [26]. Wearable health monitors collect data on heart rate, activity levels, and other health indicators, which can be analysed to provide personalized healthcare solutions [27]. For example, continuous glucose monitoring systems help diabetic patients manage their condition more effectively by providing real-time data on blood sugar levels.
- **Critical Infrastructure:** CPSs are integral in managing critical infrastructure such as water treatment plants, transportation systems, and nuclear facilities [6]. Supervisory control and data acquisition (SCADA) systems monitor and control these industrial processes to ensure safety and efficiency [28]. For instance, traffic management systems use CPSs to adjust traffic signals based on real-time traffic conditions, reducing congestion and improving road safety [29].

5.2.2 Challenges

The previous section discusses several benefits of CPS; while they offer significant benefits, they also present unique challenges. Therefore, this section explores the critical challenges in CPS development and highlights key considerations for their successful deployment and operations. Some of the concerns are

- **Complexity and Interoperability:** CPSs often involve heterogeneous components and systems that must work together [2]. Determining interoperability among different devices and standards can lead to technical challenges [30].

- **Security Risks:** The integration of cyber and physical components makes CPSs vulnerable to cyber attacks that can have physical consequences [4]. For example, the Stuxnet worm demonstrated how malware could cause physical damage to industrial equipment by targeting control systems [8].
- **Real-Time Requirements:** Many CPS applications are time-sensitive, requiring precise timing and synchronization [1]. Delays or timing errors can lead to system failures or safety hazards, especially in critical applications like autonomous vehicles or medical devices [2].
- **Regulatory and Ethical Issues:** CPSs in sectors like healthcare and transportation must comply with strict regulations to ensure safety and privacy [17]. Ethical considerations also arise concerning data collection, surveillance, and the potential displacement of human labour [27].

CPSs represent a transformative shift in how computational and physical processes interact. By bridging the gap between the digital and physical worlds, CPSs can enable innovative solutions to complex problems, driving progress in industries ranging from manufacturing to healthcare. However, realizing the full potential of CPSs requires addressing significant challenges related to complexity, security, and ethical considerations. As research and development continue, CPSs are poised to play an increasingly integral role in shaping the future of technology and society.

5.2.3 The Nature of Vulnerabilities in CPSs

The nature of vulnerabilities in CPSs is inherently complex, rising from their integration of physical components, computational algorithms, and communication networks. These vulnerabilities can be broadly categorized into (i) cyber, (ii) physical, and (iii) hybrid threats. *Cyber vulnerabilities* typically involve weaknesses in software, protocols, or network configurations. For example, unpatched software flaws, weak authentication mechanisms, and unsecured communication protocols like Modbus and DNP3 [31] are often exploited for unauthorized access or data manipulation. *Physical vulnerabilities* stem from risks like tampering, sabotage, or environmental wear and tear of hardware components [17]. For instance, medical devices in healthcare CPSs can be physically altered to cause malfunctions, putting patient safety at risk. *Hybrid vulnerabilities*, on the other hand, combine both, such as attacks like Stuxnet [8], where cyber exploitation of SCADA systems caused physical damage to industrial machinery.

CPSs are particularly susceptible to advanced persistent threats (APTs) that exploit their real-time operations, distributed architecture, and interconnectivity. These systems often operate in dynamic environments with legacy components, creating additional entry points for attackers. Compounding these risks is the challenge of maintaining data integrity and system availability while ensuring compliance with safety-critical standards. Addressing the multifaceted vulnerabilities in CPSs required a systematic approach that combines proactive threat detection, robust security measures, and continuous monitoring to safeguard both their cyber and physical domains [4]. There are several factors, including increased connectivity, heterogeneity, reliance on legacy systems, and inadequate security measures, that can be the point of acceleration for CPS vulnerabilities.

- **Increased Connectivity and Attack Surface:** The increase of integration of advanced systems such as internet of things (IoT) devices and the push towards greater interconnectivity in CPSs have expanded the attack surface significantly [32]. As CPS components become more interconnected, vulnerabilities in a component can potentially be exploited to gain access to others, leading to cascading failures [33]. For example, in smart grids, an attacker who compromises a single smart meter could potentially access the wider network, disrupting energy distribution [24, 34].
- **Heterogeneity and Complexity:** CPSs often consist of heterogeneous components sourced from different manufacturers, employing various communication protocols and standards [2]. This diversity can result in compatibility issues and security gaps that are difficult to manage [30]. The complexity of integrating disparate systems increases the likelihood of configuration errors and unpatched vulnerabilities. For instance, discrepancies between proprietary industrial control systems and standard IT security practices can leave control systems exposed [35].
- **Legacy Systems with Outdated Security:** Many CPSs, especially in critical infrastructure sectors like energy and transportation, utilize legacy systems that were not designed with modern cybersecurity threats in mind [36]. These systems often lack basic security features such as encryption and authentication mechanisms [31]. Upgrading or replacing legacy components can be cost-prohibitive and operationally challenging, leading organizations to continue using outdated technology that is vulnerable to attacks. The use of unpatched operating systems in industrial control systems is a notable example [6].
- **Inadequate Coding and Security Practices:** Software flaws and poor coding practices contribute significantly to CPS vulnerabilities [37]. Attackers can exploit buffer overflows, input validation errors, and other common coding mistakes to gain unauthorized access or cause system failures [38]. Further, a lack of secure development life cycles and insufficient testing can result in the deployment of insecure systems.
- **Physical Tampering and Insider Threats:** Unlike traditional IT systems, CPSs are often deployed in physical environments where they may be exposed to tampering [39]. Physical access can allow attackers to bypass cyber defences entirely. For example, unauthorized individuals could manipulate sensors or actuators to disrupt operations in a manufacturing plant [40]. Insider threats, where employees misuse their access privileges, also pose significant risks to CPS security.

5.2.4 Case Studies of CPS Vulnerabilities

In the previous section, we discussed various vulnerabilities that can impact CPSs. In this section, we look into several real-world cases of CPS attacks in critical infrastructure and discuss their point of vulnerabilities.

- **Stuxnet Worm:** *A cyber-physical attack on nuclear facilities* – The Stuxnet worm represents one of the most significant and sophisticated attacks on CPSs, specifically targeting Iran's nuclear facilities. Discovered in 2010, Stuxnet exploited vulnerabilities in SCADA systems, which were responsible for controlling centrifuges at the Natanz uranium enrichment plant [8]. The worm infiltrated the system through USB devices, bypassing air-gapped security measures. Once inside, Stuxnet caused centrifuges to spin at speeds beyond their operational limits, leading to physical damage and reduced enrichment efficiency. Simultaneously, the worm fed false data to monitoring systems, masking its presence and ensuring operators were unaware of the sabotage. This dual impact of cyber manipulation and physical destruction highlighted the potential for cyber attacks to cross the digital–physical boundary, causing tangible harm. The attack demonstrated the vulnerabilities of industrial control systems, even in high-security environments, and underscored the need for rigorous security protocols, including robust patch management, secure communication channels, and enhanced monitoring capabilities [8].
- **Ukrainian Power Grid Attack:** *Disrupting critical infrastructure* – In December 2015, one of the first confirmed cyber attacks on a power grid occurred in Ukraine, affecting three energy distribution companies and causing widespread power outages [41]. The attackers used a multi-stage strategy, beginning with phishing emails to gain access to the industrial control systems. Malware, including BlackEnergy, was deployed to compromise the systems, enabling remote access to switch off power to substations. The attackers further compounded the disruption by launching a coordinated denial of service (DoS) attack against call centres, preventing customers from reporting outages. This incident highlighted critical vulnerabilities in the energy sector's CPSs, particularly the lack of segmentation between IT and operational technology (OT) networks and the inadequate security of legacy systems. It emphasized the importance of cybersecurity measures such as advanced intrusion detection systems (IDS), network segmentation, and regular security audits to protect critical infrastructure from similar threats [41].
- **Maroochy Shire Sewage Spill:** *The impact of insider threats* – The Maroochy Shire sewage spill in Queensland, Australia, is a stark example of the damage that insider threats can inflict on CPSs. In 2000, a disgruntled contractor with intimate knowledge of the system's operations manipulated the control systems of a sewage treatment plant, causing over 800,000 litres of raw sewage to spill into local parks, rivers, and residential areas [40]. Using stolen equipment and knowledge of the SCADA system, the attacker disrupted the

plant's operations, overriding commands and disabling alarms. The incident demonstrated the unique vulnerabilities posed by insiders who possess access to critical systems and understand their architecture. It also highlighted the need for strict access controls, thorough employee background checks, and robust monitoring to detect unauthorized activity. Beyond physical and environmental damage, the incident eroded public trust in the safety and reliability of critical infrastructure, underscoring the broader consequences of CPS vulnerabilities [40].

5.3 THE TRANSFORMATION OF VULNERABILITY ASSESSMENT

With the advancement of AI/ML, the landscape of vulnerability assessment in CPSs has transformed [11]. Traditional vulnerability assessment methods, which often involve manual processes and static analysis tools, have struggled to keep pace with the increasing complexity and dynamic nature of modern CPSs [10]. AI/ML technologies offer advanced capabilities for automating vulnerability detection, predicting potential threats, and adapting to new attack vectors in real time.

Traditional vulnerability assessment techniques rely heavily on signature-based detection and rule-based systems [42]. While effective against known threats, these methods may not be adequate for detecting advanced attacks that do not match existing signatures [11]. For example, manual code reviews and static analysis tools can be time-consuming and may not scale effectively for large, complex CPS with numerous interconnected components [43]. Furthermore, traditional methods often operate reactively, addressing vulnerabilities only after they have been exploited [10]. AI/ML, on the other hand, introduce a proactive and dynamic approach to vulnerability assessment. ML algorithms can analyse vast amounts of data from CPSs to identify patterns and anomalies that may indicate security vulnerabilities [44]. Some of the techniques such as anomaly detection, clustering, and classification can enable the discovery of previously unknown threats [45]. Deep learning models, including convolution neural network (CNN) and recurrent neural network (RNN) can process complex and high-dimensional data, improving the detection of subtle or multi-stage attacks [46]. AI-driven tools can perform continuous monitoring and real-time analysis of CPSs, allowing for immediate detection and response to threats [47]. They can adapt to evolving attack patterns by learning from new data, enhancing their effectiveness over time [46].

5.3.1 KEY PROCESSES AND TOOLS IN TRADITIONAL VULNERABILITY ASSESSMENT

Traditional vulnerability assessment typically follows a structured sequence of steps to identify and address weaknesses:

- **Asset Identification and Prioritization:** The process begins by cataloguing all assets within the CPSs, such as sensors, actuators, software components, and communication networks. This step ensures a comprehensive understanding of the system's architecture. For example, in a smart grid, assets like SCADA systems, substations, and communication links are identified and prioritized based on their criticality to system operations [48].

- **Vulnerability Scanning:** Automated tools such as Nessus, Qualys, or OpenVAS are employed to scan the system for known vulnerabilities. These tools rely on extensive databases of Common Vulnerabilities and Exposures (CVE) to detect weaknesses. Static application security testing (SAST) is used to analyse source code for programming errors, while dynamic application security testing (DAST) evaluates running applications for runtime vulnerabilities. Example: A SCADA system might be scanned for outdated firmware versions or misconfigurations in communication protocols like Modbus.
- **Threat Modelling:** Traditional threat modelling frameworks such as spoofing, tampering, repudiation, information disclosure, denial of service, and elevation of privilege (STRIDE) are used to map potential attack vectors. Example: In a water treatment CPS, STRIDE might identify tampering threats on remote terminal units (RTUs) as a significant risk [49].
- **Risk Analysis:** Each vulnerability is assessed for its potential impact and likelihood using frameworks like the common vulnerability scoring system (CVSS). This helps prioritize remediation efforts. Example: A weak password on a programmable logic controller (PLC) in an industrial system might be assigned a high-risk score due to the critical role of the PLC in system operations [50].
- **Mitigation and Patch Management:** Identified vulnerabilities are mitigated through software updates, reconfigurations, or implementing additional security controls. Example: Applying a firmware update to address a buffer overflow vulnerability in a medical device CPS [49].
- **Compliance Audits:** Traditional assessments often align with industry standards such as ISO/IEC 27001, NIST SP 800-53, or sector-specific guidelines like IEC 62443 for industrial automation. Example: A compliance audit might verify encryption standards for data transmissions in a transportation CPS [51].

In traditional contexts, vulnerability assessment relies on established tools and methodologies designed to identify and mitigate security weaknesses within systems. These tools are typically rule-based, rely on predefined vulnerability databases, and are geared towards static and periodic assessments. While effective for known vulnerabilities, they are limited in addressing modern, dynamic threats such as zero-day vulnerabilities or sophisticated attack vectors. Some of the widely used traditional vulnerability assessment tools are given in Table 5.1.

5.3.2 Challenges in Traditional Vulnerability Assessment

There are several challenges in these static-type traditional vulnerability assessments. We list some of the key challenges:

- **Static and Periodic Nature:** Traditional methods operate on predefined schedules, leaving systems vulnerable between assessments. This limitation is particularly concerning in CPSs, where threats can emerge rapidly (Figure 5.1).

TABLE 5.1
Popular Vulnerability Assessment Tools

Tool	Scope	Key Features	Limitations	Reference
Nessus	Networks, endpoints	Automated scanning, CVSS-based risk scores	Limited to known vulnerabilities	[48]
Qualys	Networks, web apps, endpoints	Cloud-based scanning, compliance integration	Requires frequent database updates	
OpenVAS	Networks, systems	Open-source, customizable, NVT library	Limited real-time capabilities	[52]
Nikto	Web servers	Detects outdated software, insecure headers	Limited to web servers	[53]
Burp Suite	Web applications	SQL injection, XSS detection, intercepting proxy	Focused on web application vulnerabilities	[54]
MBSA (discontinued)	Windows systems	Simple interface, patch management	Windows-only, outdated tool	[55]
OWASP ZAP	Web applications	Passive/active scanning, plugin support	Web application focus, expertise required	[56]

- **Dependency on Known Vulnerabilities:** These methods rely on databases like CVE, which means they are ineffective against zero-day vulnerabilities [48].
- **Resource-Intensive:** Manual inspections and patch management require significant time, expertise, and financial resources, especially for large-scale CPSs.
- **Scalability Issues:** Traditional methods struggle with the scale and complexity of modern CPSs, where the number of interconnected components continues to grow exponentially [2, 49].

5.3.3 Introduction of AI and ML in Vulnerability Assessment

The growing complexity and scale of modern systems, particularly CPSs, have caused traditional vulnerability assessment methods insufficient to address evolving cyber threats. AI is revolutionizing this domain by offering adaptive, efficient, and predictive capabilities that surpass conventional approaches. With AI, vulnerability assessment can transition from static, manual processes to dynamic, automated systems capable of real-time monitoring and decision-making. This section explores the fundamental principles of AI and ML in vulnerability assessment, their applications, and the transformative impact they have on identifying and mitigating risks.

FIGURE 5.1 Vulnerability assessment structure.

AI/ML have revolutionized vulnerability assessment by introducing automation, predictive analytics, and adaptive learning capabilities [10, 11]. It enhances vulnerability assessment by automating labour-intensive tasks, identifying patterns in large datasets, and adapting dynamically to evolving threats. Unlike traditional methods, which are often reactive and constrained to known vulnerabilities, AI-enabled systems operate in real time, analysing vast amounts of data from CPS components to detect anomalies and potential threats. For example, tools like Darktrace employ AI to monitor network traffic, identify unusual patterns, and take corrective actions autonomously [51]. Similarly, predictive ML models analyse historical data to forecast potential vulnerabilities and prioritize mitigation efforts [50].

5.3.3.1 Benefits of AI and ML Integration

- **Enhanced Detection Capabilities:** AI algorithms can identify patterns and anomalies that indicate potential vulnerabilities [43].
- **Real-Time Analysis:** ML models can process data in real time, allowing for immediate detection and response to threats [57].
- **Scalability:** Automated tools can handle large-scale systems with numerous components without the limitations of manual assessments [44].
- **Predictive Analytics:** AI can predict potential vulnerabilities based on historical data and emerging threat patterns [45].
- **Continuous Learning:** ML models improve over time as they are exposed to more data, enhancing their accuracy and effectiveness [11].

5.4 AI-DRIVEN TOOLS FOR VULNERABILITY ASSESSMENT AND REPORTING

AI-based tools are designed to provide proactive, automated, and scalable solutions for vulnerability management. This section presents a list of leading tools in the domain.

- Darktrace is a widely used AI-driven cybersecurity platform that uses unsupervised learning to detect anomalies and vulnerabilities. It builds a baseline of "normal" behaviour for networks, systems, and devices, flagging deviations that may indicate threats or vulnerabilities. It is commonly used in large-scale CPS environments such as smart grids and industrial automation, where real-time monitoring is critical [58]. Darktrace is easy to use and it leverages AI models to detect unusual behaviour without relying on predefined rules. It uses the Anitgena module which is an autonomous response system for detecting and mitigating threats in real time without disrupting normal operations. It also provides detailed visualization for network traffic and threat maps and generates automated reports based on anomalies detected by the systems. The reports are updated in real time and use context analysis to provide recommendations tailored to the specific incident.
- Cortex XDR is a comprehensive solution that unifies endpoint, network, and application security. It excels in correlating data across domains to provide holistic visibility into an organization's security posture [59]. It collects data from endpoints, networks, and cloud environments, integrating it into a unified dashboard. Cross-domain data correlation allows elimination of blind spots creating comprehensive threat analysis. Cortex XDR utilizes both user and system behaviour to detect anomalies indicative of security vulnerabilities or attacks. For example, it can identify insider threats by detecting unusual file access or data transfer activities. Cortex XDR utilizes ML-based threat scoring scheme to prioritize vulnerabilities. These vulnerabilities are assigned risk scores based on their exploitability and impact. The prioritization helps security teams focus their efforts on addressing the most critical threats first.

  Cortex XDR supports automated workflows for threat containment and mitigation. For example, it can quarantine infected files, terminate malicious processes, or block suspicious IP addresses in real time. One of the key features is customisable Dashboards and Reporting, this platform offers flexible reporting options, including customizable dashboards that present data in a visually intuitive format. These reports can be tailored to meet the needs of technical teams and executive management [59].

- Tenable.io combines traditional vulnerability scanning with advanced AI-driven analytics, making it a powerful tool for proactive risk management. For example, it can integrate with Nessus, one of the most widely used vulnerability scanners. This integration enables Tenable.io to leverage Nessus's extensive vulnerability database while enhancing it with AI-driven insights. Its predictive prioritization feature stands out as a key innovation. It uses machine learning to predict the likelihood of vulnerabilities being exploited. By combining data from vulnerability databases with real-world threat intelligence, it assigns risk scores that help prioritize remediation efforts.

  Automated Compliance Reporting: Tenable.io offers pre-configured templates for compliance reporting, aligned with regulations such as GDPR, PCI DSS, and ISO 27001. These templates simplify the process of demonstrating regulatory adherence.

  Dynamic Dashboards: The platform provides dynamic dashboards that visualize vulnerability trends, high-risk areas, and remediation progress. This feature aids in tracking the effectiveness of security initiatives over time.

  Cloud-Native Architecture: As a cloud-native tool, Tenable.io is highly scalable and suitable for hybrid environments that span on-premises and cloud infrastructures [60].

- Microsoft Defender for endpoint focuses on securing endpoints through a combination of AI-powered threat detection, automated remediation, and integration with the broader Microsoft ecosystem. Defender for Endpoint uses AI to monitor endpoint activities for anomalies. For instance, it can detect unusual login attempts or unauthorized software installations, flagging them as potential security threats.

  Automated Investigation and Remediation: The tool automates the process of investigating detected threats and takes immediate action to neutralize them. For example, it can isolate a compromised device or remove malware without requiring manual intervention.

  Integration with Microsoft Threat Intelligence: By leveraging Microsoft's global threat intelligence network, Defender for Endpoint provides enriched insights into emerging threats. This integration ensures that the tool stays up-to-date with the latest attack patterns and vulnerabilities.

  Compliance and Risk Management: Defender for Endpoint includes features for compliance reporting, helping organizations meet regulatory requirements. Reports are generated automatically, highlighting gaps and suggesting remediation measures.

  Endpoint-Specific Insights: The platform offers detailed insights into the security status of individual endpoints, including real-time monitoring, threat timelines, and actionable recommendations. This granularity is particularly valuable for industries like healthcare and finance, where endpoint security is critical [61].

AI-driven tools go beyond static reporting by delivering actionable insights that guide security teams in mitigating risks effectively. These insights are often tailored to the organization's environment, taking into account contextual factors such as asset criticality and network topology. For instance, Darktrace not only highlights anomalies but also provides recommendations for containment and remediation through its Antigena module. Such context-aware reporting allows organizations to respond more effectively to threats while minimizing disruptions to normal operations. Integration with external threat intelligence is another hallmark of AI-driven reporting tools. By incorporating data from global threat feeds, these tools enrich reports with information about emerging vulnerabilities, known attack patterns, and adversary tactics. This integration enhances situational awareness, enabling organizations to stay ahead of evolving threats. For example, Qualys integrates real-time threat intelligence to provide enriched insights in its vulnerability management reports.

5.4.1 Benefits and Challenges of AI-Driven Tools

The automation and intelligence provided by AI-driven tools offer several advantages over traditional reporting systems. These tools significantly reduce the time and effort required to generate reports, ensuring that security teams can focus on mitigation rather than data processing. Moreover, their ability to operate in real time enables organizations to respond to vulnerabilities as they are identified, reducing the window of exposure to potential attacks.

Scalability is another key benefit. AI-driven tools are well-suited to handle the large datasets generated by modern CPSs, which often involve thousands of interconnected devices and complex network architectures. This scalability ensures that organisations can maintain a comprehensive understanding of their security posture, regardless of size or complexity.

Despite their capabilities, AI-driven reporting tools are not without challenges. One significant limitation is the dependency on high-quality data. Inaccurate or incomplete data can lead to misleading reports, reducing their effectiveness. Additionally, the complexity of integrating AI-driven tools into legacy systems can be a barrier for some organizations. The "black-box" nature of AI algorithms also raises concerns about transparency and trust, particularly when reports are used for compliance or legal purposes (Table 5.2).

Each AI-driven vulnerability assessment tool offers a unique set of features tailored to specific use cases and industries. Darktrace excels in autonomous response and real-time anomaly detection, making it ideal for dynamic environments like IoT and CPSs. Cortex XDR provides cross-domain visibility and automated incident response, addressing the needs of complex, hybrid infrastructures. Tenable.io focuses on predictive risk management and compliance, bridging traditional and AI-driven methods. Meanwhile, Microsoft Defender for Endpoint specializes in endpoint security, leveraging Microsoft's extensive threat intelligence network to deliver advanced protection. Together, these tools represent the forefront of cybersecurity innovation, addressing modern challenges with precision and scalability (Table 5.3).

TABLE 5.2
AI-Driven Vulnerability Assessment Tools

Tool	Key Features	Applications	Reference
Darktrace	Anomaly detection, autonomous response	Network security, smart grids	[58]
Cortex XDR	Behavioural analytics, ML-based threat scoring	Industrial control systems, healthcare	[59]
Tenable.io with predictive prioritization	Predictive analytics, risk scoring	Hybrid environments, enterprise networks	[60]
Microsoft Defender for endpoint	Automated remediation, AI-powered detection	Healthcare, manufacturing	[61]

TABLE 5.3
Comparative Summary: Traditional versus AI-Driven Approaches

Aspect	Traditional Approaches	AI-Driven Approaches	Reference(s)
Detection scope	Limited to known vulnerabilities	Includes zero-day threats and unknown patterns	[48, 51]
Speed	Periodic and slow	Real-time and dynamic	[51, 62]
Scalability	Limited to smaller systems	Handles large-scale, interconnected CPSs	[50]
Adaptability	Static	Learns and adapts to evolving threats	[63]
Manual effort	High dependency on human input	Automated with minimal human intervention	[49, 62]
Predictive capabilities	Absent	Strong predictive analytics	[51]

5.5 AI AND ML ALGORITHMS IN VULNERABILITY ASSESSMENT

AI/ML have advanced the field of vulnerability assessment by automating threat detection, predicting future vulnerabilities, and offering adaptive solutions to evolving risks. These algorithms enable organizations to analyse vast datasets, identify patterns, and proactively respond to threats with precision and speed. This section provides a comprehensive overview of AI algorithms employed in vulnerability assessment, detailing their mechanisms, applications, and advantages. This section discusses different types of learning algorithms that can be used for vulnerability assessments in CPSs. There are three broad ML algorithms: supervised, unsupervised, and reinforcement learning. We will look into each of them in the next sections in detail.

5.5.1 Supervised Learning Algorithms

Supervised learning is one of the most commonly used categories of ML, where models are trained on labelled datasets to predict or classify. In vulnerability assessment, supervised learning algorithms can excel in identifying known threats and categorizing them based on their severity. There are different types of supervised learning algorithms that are suitable for the task. For example, decision trees use a tree-like structure to make decisions based on specific conditions. Random forests enhance this approach by combining multiple decision trees to improve accuracy and robustness. These models are particularly effective in classifying vulnerabilities based on their criticality and predicting exploitability. Classifying vulnerabilities into categories such as critical, high, medium, or low severity. Random forests are also based on tree-like structures and often used for prioritizing vulnerabilities by predicting their exploitability [64, 65]. A random forest model trained on historical attack data can classify incoming threats as benign or malicious based on features like source internet protocol (IP), file type, or access patterns.

Support vector machine (SVM) is another supervised learning, they are widely used for binary or multi-class classification tasks. They work by finding the optimal hyperplane that separates data points into distinct classes [64].

SVMs are effective in distinguishing between normal and anomalous network traffic, making them suitable for intrusion detection systems. For example, detecting malware signatures by classifying network packets based on extracted features.

Advanced deep learning algorithms are based on neural networks and are employed for complex problems where datasets have high dimensionality [46]. They can be applicable for advanced vulnerability prediction in dynamic systems, such as classifying zero-day vulnerabilities based on similarities to known exploits. For example, identifying suspicious file downloads by analysing file properties and metadata.

5.5.2 Unsupervised Learning Algorithms

Unsupervised learning is particularly valuable in vulnerability assessment when labelled datasets are unavailable. These algorithms identify hidden patterns within data, making them especially effective for detecting unknown threats and anomalies. Clustering algorithms, such as k-means and DBSCAN, group data points based on their similarity without requiring predefined labels. This ability to cluster related data is critical for analysing network traffic and user activity. For instance, a k-means algorithm might categorize login attempts into normal and anomalous groups by analysing features like login times, geographic locations, and devices. Such clustering helps security teams identify potentially malicious activities, such as brute-force attacks or unauthorized access attempts [63].

Another powerful unsupervised learning technique is principal component analysis (PCA), which reduces the dimensionality of large datasets while preserving the most critical variables. PCA is particularly useful in analysing complex, high-dimensional system logs. By focusing on the principal components, PCA identifies unusual patterns that may indicate vulnerabilities. For example, it can highlight atypical system events or unusual combinations of factors that are otherwise obscured in raw logs, helping to pinpoint potential attack vectors [63]. Similarly, autoencoders, a type of neural network, are designed for anomaly detection. They work by reconstructing input data and measuring reconstruction errors. Large reconstruction errors often signal outliers, which may represent vulnerabilities or malicious activities. For instance, autoencoders can detect unusual data transfer patterns in a smart grid, such as unexpected spikes in network load that may indicate cyber intrusions [50].

5.5.3 Reinforcement Learning Algorithms

Reinforcement learning (RL) is a dynamic and adaptive approach to vulnerability assessment, where models learn optimal strategies by interacting with an environment and receiving feedback through rewards or penalties. This adaptive capability makes RL highly effective for managing complex and evolving cybersecurity challenges. One prominent algorithm is deep Q-learning, which combines RL with deep learning to handle environments with large state spaces. Deep Q-learning can simulate attack scenarios, allowing it to identify the most vulnerable points in a system and test various defence strategies. For example, a digital twin of a smart city infrastructure could use Q-learning to test the resilience of its security measures, enabling preemptive identification and remediation of vulnerabilities [66].

Another RL technique, policy gradient method, focuses on learning policies directly by optimizing expected rewards over actions. These methods are especially effective in developing adaptive intrusion prevention systems. By continuously learning from real-time threat assessments, a policy gradient model can dynamically configure firewalls or adjust access controls in response to detected attacks. For instance, if the system identifies a sudden spike in unauthorized access attempts, the RL model can immediately modify network security policies to counter the threat [66]. The ability to adapt defences based on real-time feedback makes RL a powerful tool in modern vulnerability assessment.

5.5.4 Deep Learning Algorithms

Deep learning, a subset of machine learning, excels at processing unstructured data such as images, videos, and sequential logs. It is particularly effective in vulnerability assessment for identifying intricate patterns and relationships within complex datasets. CNNs, for instance, are widely used for spatial data and image analysis. In cybersecurity, CNNs can analyse visualized network traffic patterns to detect unauthorized access or configuration changes. For example, a CNN model might identify abnormal traffic patterns in a network map, flagging potential distributed denial of service (DDoS) attacks [50].

RNNs are another deep learning approach, tailored for sequential data like time-series logs or network traffic flows. These networks are adept at recognizing temporal dependencies and detecting unusual activity over time. For instance, an RNN can monitor real-time system logs and detect sequences of events that suggest APTs, such as repeated failed login attempts followed by successful unauthorized access [50]. Generative adversarial networks (GANs) also offer a unique capability by generating synthetic data to simulate attack scenarios. GANs consist of two competing neural networks—a generator and a discriminator—that improve each other's performance through competition. For example, GANs can create synthetic malware samples to enhance the training of intrusion detection systems, making them more effective against emerging threats [66].

5.5.5 Natural Language Processing (NLP) Algorithms

Natural language processing (NLP) plays a pivotal role in vulnerability assessment by extracting actionable insights from unstructured text data such as security advisories, vulnerability descriptions, and system logs. Named entity recognition (NER) is a technique that identifies and categorizes entities within the text, such as software names, versions, and CVEs. For example, NER can parse security advisories to extract critical details about vulnerabilities, helping organizations prioritize their mitigation efforts [64].

Topic modelling algorithms, such as latent dirichlet allocation (LDA), uncover hidden themes in textual data, making them valuable for identifying trends in cybersecurity threats. For instance, analysing forums and social media discussions using topic modelling can reveal emerging attack patterns, enabling organizations to prepare for new threats [64]. Similarly, text classification models categorize documents or logs based on predefined labels. These models are particularly useful for automating the prioritization of security tickets. For example, vulnerability reports can be classified into categories like "critical", "high", "medium", or "low" severity based on their descriptions, streamlining the decision-making process and reducing response times [64]. Together, NLP techniques enhance the ability of security teams to manage and interpret vast amounts of unstructured data efficiently.

5.6 CONCLUSION

The integration of AI and ML into vulnerability assessment and reporting for CPSs marks a significant advancement in securing these critical infrastructures [11]. By leveraging advanced algorithms and continuous learning, organizations can proactively identify and mitigate vulnerabilities, ensuring the safety and reliability of systems that are foundational to modern society.

However, the implementation of these technologies comes with challenges that require careful consideration and strategic planning [36]. Addressing issues related to data quality, model accuracy, operational integration, and industry-specific obstacles is essential for the successful adoption of AI-driven vulnerability assessments.

As cyber threats continue to evolve, the role of AI and ML in cybersecurity will become increasingly important [43]. Ongoing research, collaboration, and innovation are necessary to stay ahead of potential risks and to harness the full potential of these technologies in safeguarding CPSs.

PRACTICE QUESTIONS

Exercise questions are designed to reinforce the concepts and techniques discussed in this chapter. Through a combination of conceptual, practical, and case studies, these exercises provide an opportunity for readers to deepen their understanding and apply their knowledge to real-world scenarios.

I. Conceptual Understanding
   1. Define vulnerability assessment in the context of CPSs. Explain how AI and ML enhance traditional vulnerability assessment methods.
   2. Compare and contrast supervised learning and unsupervised learning algorithms in vulnerability assessment. Provide examples of their application in CPSs.
   3. Explain the importance of reporting in vulnerability assessment. What are the key features of AI-driven reporting tools?

II. Practical Application
   1. Design an experiment to detect anomalies in IoT device traffic using unsupervised learning. Choose a specific algorithm (e.g., k-means, PCA) and describe how you would apply it to identify potential vulnerabilities.
   2. Imagine you are implementing reinforcement learning for a cybersecurity system in a smart city. Describe how deep Q-learning could be used to simulate and test defensive strategies against cyber attacks.

III. Case Studies
   1. Research a real-world CPS vulnerability (e.g., the Stuxnet attack or the Ukrainian power grid attack). Discuss how AI and ML algorithms could have been used to detect or prevent the vulnerability.
   2. In the context of healthcare CPSs, such as connected medical devices, propose a framework using AI and ML to assess and report vulnerabilities.

REFERENCES

1. E.A. Lee, *Cyber physical systems: Design challenges*, in *2008 11th IEEE International Symposium on Object and Component-Oriented Real-Time Distributed Computing (ISORC)*. 2008, pp. 363–369.
2. R. Rajkumar, I. Lee, L. Sha, and J. Stankovic, *Cyber-physical systems: The next computing revolution*, in *Design Automation Conference*. 2010, pp. 731–736.
3. L. Monostori, *Cyber-physical production systems: Roots, expectations and R&D challenges*, *Procedia CIRP* 17 (2014), pp. 9–13, Available at www.sciencedirect.com/science/article/pii/S2212827114003497, variety Management in Manufacturing.
4. A.R. Sadeghi, C. Wachsmann, and M. Waidner, *Security and privacy challenges in industrial internet of things*, in *2015 52nd ACM/EDAC/IEEE Design Automation Conference (DAC)*. 2015, pp. 1–6.
5. J.S. Mertoguno, G. Briskin, J.H. Li, and K. Kwak, *Lessons learned and future directions for security, resilience and artificial intelligence in cyber physical systems*, in *Network Security Empowered by Artificial Intelligence*, Y. Chen, J. Wu, P. Yu, and X. Wang, eds., Springer Nature Switzerland, Cham (2024), pp. 403–432, Available at https://doi.org/10.1007/978-3-031-53510-9_15.
6. A.A. Cárdenas, S. Amin, and S. Sastry, *Research challenges for the security of control systems*, in *Proceedings of the 3rd Conference on Hot Topics in Security*, USA. USENIX Association, HOTSEC'08, 2008.

7. H.C. Chang, C.Y. Lin, D.J. Liao, and T.M. Koo, *The Modbus protocol vulnerability test in industrial control systems*, in *2020 International Conference on Cyber-Enabled Distributed Computing and Knowledge Discovery (CyberC)*. 2020, pp. 375–378.
8. R. Langner, *Stuxnet: Dissecting a cyberwarfare weapon*, *IEEE Security & Privacy* 9 (2011), pp. 49–51.
9. S.Y. Diaba and M. Elmusrati, *Proposed algorithm for smart grid ddos detection based on deep learning*, *Neural Networks* 159 (2023), pp. 175–184, Available at www.sciencedirect.com/science/article/pii/S0893608022005032.
10. R. Sommer and V. Paxson, *Outside the closed world: on using machine learning for network intrusion detection*, in *2010 IEEE Symposium on Security and Privacy (SP)*, May, Los Alamitos, CA. IEEE Computer Society, 2010, pp. 305–316, Available at https://doi.ieeecomputersociety.org/10.1109/SP.2010.25.
11. A.L. Buczak and E. Guven, *A survey of data mining and machine learning methods for cyber security intrusion detection*, *IEEE Communications Surveys & Tutorials* 18 (2016), pp. 1153–1176.
12. A.S. Jin, L. Hogewood, S. Fries, J.H. Lambert, L. Fiondella, A. Strelzoff, J. Boone, K. Fleckner, and I. Linkov, *Resilience of cyber-physical systems: Role of ai, digital twins, and edge computing*, *IEEE Engineering Management Review* 50 (2022), pp. 195–203.
13. A. Khadka, S. Sthapit, G. Epiphaniou, and C. Maple, *Resilient machine learning: Advancement, barriers, and opportunities in the nuclear industry*, *ACM Computing Surveys* 56 (2024), Available at https://doi.org/10.1145/3648608.
14. I.J. Goodfellow, J. Pouget-Abadie, M. Mirza, B. Xu, D. Warde-Farley, S. Ozair, A. Courville, and Y. Bengio, *Generative adversarial nets,* in *Proceedings of the 27th International Conference on Neural Information Processing Systems – Volume 2*, MIT Press, Cambridge, MA, NIPS'14, (2014), pp. 2672–2680.
15. I.J. Goodfellow, *et al.*, *Explaining and harnessing adversarial examples*, in *International Conference on Learning Representations*. (2015).
16. D.O. Esan, P.A. Owolawi, and C. Tu, *Generative adversarial networks: Applications, challenges, and open issues*, in *Deep Learning*, M. Domínguez-Morales, J. Civit-Masot, L. Muñoz-Saavedra, and R. Damaševičius, eds., Chapter 2, IntechOpen, Rijeka, (2023), Available at https://doi.org/10.5772/intechopen.113098.
17. R. Baheti and H. Gill, *Cyber-physical systems*, 2011, Available at https://api.semanticscholar.org/CorpusID:272843506.
18. A. Khoumsi, *Sensor and actuator attacks of cyber-physical systems: A study based on supervisory control of discrete event systems*, in *2019 8th International Conference on Systems and Control (ICSC)*. (2019), pp. 176–182.
19. X. Guan, B. Yang, C. Chen, W. Dai, and Y. Wang, *A comprehensive overview of cyber-physical systems: From perspective of feedback system*, *IEEE/CAA Journal of Automatica Sinica* 3 (2016), pp. 1–14.
20. V.K. Annanth, M. Abinash, and L.B. Rao, *Intelligent manufacturing in the context of industry 4.0: A case study of siemens industry*, *Journal of Physics: Conference Series* 1969 (2021), p. 012019, Available at https://dx.doi.org/10.1088/1742-6596/1969/1/012019.
21. H. Lasi, P. Fettke, H.G. Kemper, T. Feld, and M. Hoffmann, *Industry 4.0*, *Business & Information Systems Engineering* 6 (2014), pp. 239–242.
22. J. Guanetti, Y. Kim, and F. Borrelli, *Control of connected and automated vehicles: State of the art and future challenges*, *Annual Reviews in Control* 45 (2018), pp. 18–40, Available at www.sciencedirect.com/science/article/pii/S1367578818300336.
23. S. Grigorescu, B. Trasnea, T. Cocias, and G. Macesanu, *A survey of deep learning techniques for autonomous driving*, *Journal of Field Robotics* 37 (2020), pp. 362–386.

24. X. Fang, S. Misra, G. Xue, and D. Yang, *Smart grid—The new and improved power grid: A survey, IEEE Communications Surveys & Tutorials* 14 (2012), pp. 944–980.
25. A.H. Mohsenian-Rad, V. Wong, J. Jatskevich, R. Schober, and A. Leon-Garcia, *Autonomous demand-side management based on game-theoretic energy consumption scheduling for the future smart grid, IEEE Transactions on Smart Grid* 1 (2011), pp. 320–331.
26. S. Gollakota, H. Hassanieh, B. Ransford, D. Katabi, and K. Fu, *They can hear your heartbeats: non-invasive security for implantable medical devices*, SIGCOMM Computer *Communication Reveiw* 41 (2011), pp. 2–13, Available at https://doi.org/10.1145/2043164.2018438.
27. U. Varshney, *Pervasive healthcare and wireless health monitoring*, *Mobile Networks and Applications* 12 (2007), pp. 113–127.
28. R.L. Krutz, *Securing SCADA Systems*, John Wiley & Sons, (2005).
29. E.A. Lee, *The past, present and future of cyber-physical systems: A focus on models*, *Sensors* 15 (2015), pp. 4837–4869.
30. V. Gunes, S. Peter, T. Givargis, and F. Vahid, *A survey on concepts, applications, and challenges in cyber-physical systems*, KSII Trans. *Internet Information Systems* 8 (2014), pp. 4242–4268, Available at https://api.semanticscholar.org/CorpusID:16515889.
31. M. Bashendy, A. Tantawy, and A. Erradi, *Intrusion response systems for cyber-physical systems: A comprehensive survey*, *Computers & Security* 124 (2023), p. 102984, Available at www.sciencedirect.com/science/article/pii/S0167404822003765.
32. L. Atzori, A. Iera, and G. Morabito, *The internet of things: A survey*, *Computer Networks* 54 (2010), pp. 2787–2805, Available at www.sciencedirect.com/science/article/pii/S1389128610001568.
33. S. Nazir, S. Patel, and D. Patel, *Assessing and augmenting scada cyber security: A survey of techniques*, *Computers & Security* 70 (2017), pp. 436– 454, Available at www.sciencedirect.com/science/article/pii/S0167404817301293.
34. H. Li, L. Lai, and R.C. Qiu, *A denial-of-service jamming game for remote state monitoring in smart grid,* in *2011 45th Annual Conference on Information Sciences and Systems*. (2011), pp. 1–6.
35. E.D. Knapp and J.T. Langill, *Industrial Network Security: Securing Critical Infrastructure Networks for Smart Grid, SCADA, and Other Industrial Control Systems*, Syngress, (2011).
36. Y. Ashibani and Q.H. Mahmoud, *Cyber physical systems security: Analysis, challenges and solutions*, *Computers & Security* 68 (2017), pp. 81–97, Available at www.sciencedirect.com/science/article/pii/S0167404817300809.
37. M. Yampolskiy, W.E. King, J. Gatlin, S. Belikovetsky, A. Brown, A. Skjellum, and Y. Elovici, *Security of additive manufacturing: Attack taxonomy and survey*, *Additive Manufacturing* 21 (2018), pp. 431– 457, Available at www.sciencedirect.com/science/article/pii/S221486041730502X.
38. D. Gollmann, *Computer Security*, 3rd ed., Wiley, (2011).
39. H. Sandberg, S. Amin, and K.H. Johansson, *Cyberphysical security* in *networked control systems: An introduction to the issue*, *IEEE Control Systems Magazine* 35 (2015), pp. 20–23.
40. J. Slay and M. Miller, *Lessons Learned from the Maroochy Water Breach,* in *International Conference on Critical Infrastructure Protection*. Springer, (2007), pp. 73–82.
41. D. Jin, Z. Li, C. Hannon, C. Chen, J. Wang, M. Shahidehpour, and C.W. Lee, *Toward a cyber resilient and secure microgrid using software-defined networking*, *IEEE Transactions on Smart Grid* 8 (2017), Available at www.osti.gov/biblio/1488574.

42. P. García-Teodoro, J. Díaz-Verdejo, G. Maciá-Fernández, and E. Vázquez, *Anomaly-based network intrusion detection: Techniques, systems and challenges*, *Computers & Security* 28 (2009), pp. 18–28.
43. Z. Chen, Q. Yan, H. Han, S. Wang, L. Peng, L. Wang, and B. Yang, *Machine learning based mobile malware detection using highly imbalanced network traffic*, *Information Sciences* 433-434 (2018), pp. 346–364, Available at www.sciencedirect.com/science/article/pii/S0020025517307077.
44. M. Ahmed, A. Naser Mahmood, and J. Hu, *A survey of network anomaly detection techniques*, *Journal of Network and Computer Application* 60 (2016), pp. 19–31, Available at https://doi.org/10.1016/j.jnca.2015.11.016.
45. R.A.R. Ashfaq, X-Z. Wang, J.Z. Huang, H. Abbas, Y-L. He, *Fuzziness based semi-supervised learning approach for intrusion detection system*, *Information Sciences* 378 (2017), pp. 484–497.
46. Y. LeCun, Y. Bengio, and G. Hinton, *Deep learning*, *Nature* 521 (2015), pp. 436–444.
47. H. Liu, B. Lang, M. Liu, and H. Yan, *Cnn and rnn based payload classification methods for attack detection*, *Knowledge-Based Systems* 163 (2019), pp. 332–341, Available at www.sciencedirect.com/science/article/pii/S0950705118304325.
48. D. Wendlandt, *Nessus* (2019). Available at www.cs.cmu.edu/~dwendlan/personal/nessus.html.
49. K. Coffey, R. Smith, L. Maglaras, and H. Janicke, *Vulnerability analysis of network scanning on SCADA systems*, *Security and Communication Networks* 2018 (2018), p. 5, Available at https://doi.org/10.1155/2018/3794603.
50. Y. Luo, Y. Xiao, L. Cheng, G. Peng, and D.D. Yao, *Deep learning-based anomaly detection in cyber-physical systems: Progress and opportunities*, *ACM Comput. Surv.* 54 (2021), Available at https://doi.org/10.1145/3453155.
51. S. Wan, J. Saxe, C. Gomes, S. Chennabasappa, A. Rath , K. Sun, and X. Wang, *Bridging the gap: A study of AI-based vulnerability management between industry and academia* (2024). Available at https://arxiv.org/abs/2405.02435.
52. Greenbone, *Openvas–openvas–open vulnerability assessment scanner* (2022). Available at www.openvas.org/.
53. H. Target, *Nikto web vulnerability scanner | hackertarget.com* (2019). Available at https://hackertarget.com/nikto-website-scanner/.
54. EC-Council, *Burp suite for penetration testing of web applications | ec-council* (2024). Available at www.eccouncil.org/cybersecurity-exchange/penetration-testing/burp-suite-penetration-testing-web-application/.
55. Vinaypamnani msft, *Guide to removing Microsoft baseline security analyzer (MBSA) – Windows security* (2023). Available at https://learn.microsoft.com/en-us/windows/security/operating-system-security/device-management/windows-security-configuration-framework/mbsa-removal-and-guidance.
56. V. Botla, *Using owasp zap to find web app security vulnerabilities – triad article* (2019). Available at www.triad.co.uk/news/owasp-zap/.
57. P. Sangkatsanee, N. Wattanapongsakorn, and C. Charnsripinyo, *Practical real-time intrusion detection using machine learning approaches*, *Computer Communications* 34 (2011), pp. 2227–2235, Available at www.sciencedirect.com/science/article/pii/S014036641100209X.
58. D.A.R. Center, *Darktrace | the AI journal* (2024). Available at https://aijourn.com/companies/darktrace/.
59. Paloalto, *Palo alto cortex XDR | partner & reseller* (2024). Available at www.nomios.co.uk/partners/palo-alto-networks/cortex-security-operations/cortex-xdr/.

60. T.H.M. Le, H. Chen, and M.A. Babar, *A survey on data-driven software vulnerability assessment and prioritization, ACM Computing Surveys* 55 (2022), Available at https://doi.org/10.1145/3529757.
61. Microsoft, *Microsoft defender for endpoint | Microsoft security*. Available at www.microsoft.com/en-gb/security/business/endpoint-security/microsoft-defender-endpoint.
62. L. Almuqren, M.S. Maashi, M. Alamgeer, H. Mohsen, M.A. Hamza, and A.A. Abdelmageed, *Explainable artificial intelligence enabled intrusion detection technique for secure cyber-physical systems, Applied Sciences* 13 (2023), Available at www.mdpi.com/2076-3417/13/5/3081.
63. C. Fei and J. Shen, *Machine learning for securing cyber–physical systems under cyber attacks: A survey, Franklin Open* 4 (2023), p. 100041, Available at www.sciencedirect.com/science/article/pii/S277318632300035X.
64. S. Sharma and K. Guleria, *Machine learning techniques for intelligent vulnerability detection in cyber-physical systems*, in *2022 International Conference on Data Analytics for Business and Industry (ICDABI)*. (2022), pp. 200–204.
65. Y. Jiang and Y. Atif, *Towards automatic discovery and assessment of vulnerability severity in cyber–physical systems, Array* 15 (2022), p. 100209, Available at www.sciencedirect.com/science/article/pii/S2590005622000558.
66. W. Luo and L. Xiao, *Reinforcement learning based vulnerability analysis of data injection attack for smart grids*, in *2021 40th Chinese Control Conference (CCC)*. (2021), pp. 6788–6792.

6 Machine Learning Techniques for Cyber-Physical Systems Security Analytics

6.1 INTRODUCTION

The internet of things (IoT) which is a part of cyber-physical systems (CPSs) has transformed numerous sectors, notably healthcare, vehicular networks, and industrial systems, by providing unprecedented connectivity and data-driven insights.

As CPSss play crucial role in intelligent manufacturing, transportation, healthcare, and several other smart cities applications, there are huge security issues that make the systems to be protected from security attacks by hackers or intruders.

Due to the presence of several connected devices in the CPSs, the security solutions developed using traditional techniques are not really addressing the severity of the problem. Hence, it is essential to apply intelligent solutions to monitor and predict such kinds of attacks before they occur in the system.

In healthcare, the integration of IoT devices, termed medical IoT (MIoT), has revolutionised patient care through real-time monitoring and personalised interventions. However, the vast amounts of sensitive health data transmitted and stored by these devices underscore the pressing need for robust security measures to safeguard against intrusions and privacy breaches.

Recent research in IoT security has seen a surge in interest, particularly in the development of intrusion detection systems (IDSs) powered by machine learning (ML) algorithms. These IDSs are crucial for identifying and mitigating potential security threats in real time. For instance, Lazzarini et al. [1] proposed the deep integrated stacking for IoT (DIS-IoT), demonstrating the potential of ML to enhance security measures through a stacking ensemble of deep learning (DL) models. Similarly, Diab and Rodriguez-Villegas [2] highlighted the practical applications of TinyML, specifically utilizing Naïve Bayes and support vector machines (SVM) algorithms in healthcare tasks.

Furthermore, security concerns extend beyond healthcare to various IoT domains, including vehicular ad hoc networks (VANETs) and industrial systems. ML algorithms have been deployed to address security vulnerabilities in these environments, employing techniques like feature selection and ensemble learning to bolster accuracy and efficiency. Additionally, privacy concerns in federated learning-based

DOI: 10.1201/9781003489986-8

IDS have been tackled through decentralized approaches, aiming to enhance data privacy and mitigate single points of failure.

6.2 LAYER-WISE ATTACK TYPES IN CPS ARCHITECTURE

According to Kim and Part [3], the attacks targeting the network layer are specifically denial of service (DoS) attacks [4, 5] and makes the integrity of the CPSs questionable. The physical layer is attacked by false data injection (FDI) attacks. These attacks will affect the application layer too and lead to non-stability of the entire system.

Figure 6.1 shows a typical CPS with layers. Here, each layer of the CPS architecture faces several attacks and they need to be analysed and appropriately mitigated. At the application layer, software attacks such as malware attacks, firmware attacks etc. will happen. In the case of network layer, the DoS kinds of attacks on IoT devices make huge losses in several real-time applications especially in the social infrastructure. At the physical layer, attacks refocusing on the controller, sensor, and any other physical components occur.

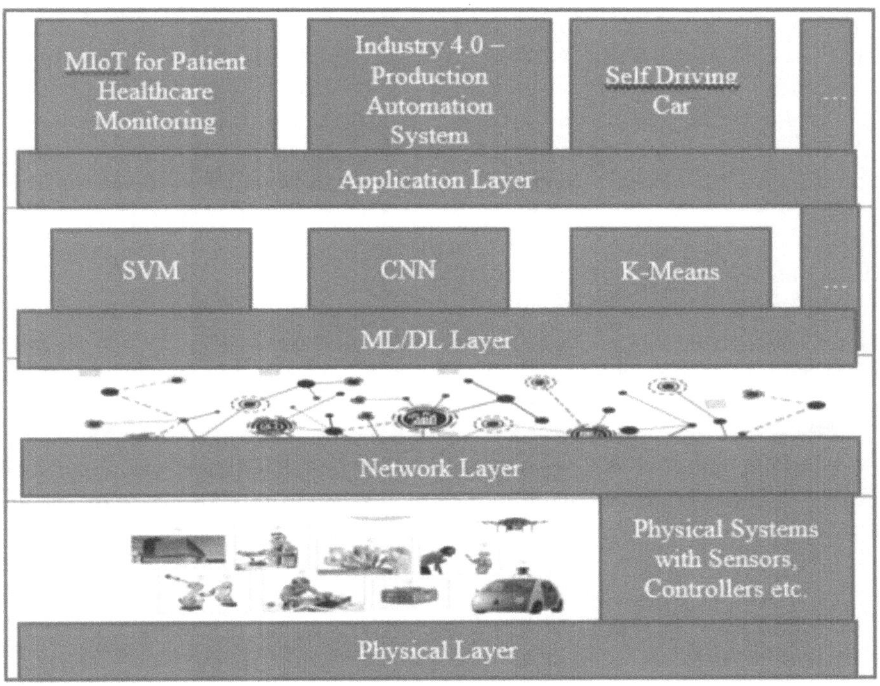

FIGURE 6.1 Layered architecture of smart CPSs that have application layer at the top that has several real-time applications, followed by ML/DL layer that has different ML/DL algorithms for prediction and decision-making, then the network layer, and finally at the bottom the physical layer that has physical systems with sensors, controllers etc.

6.2.1 Application Layer Attacks

The users who will access the application layer are generally facing issues such as flawed computations and services, erroneous commands executed in less resource systems, and this will make the entire system to fail [3].

6.2.2 ML/DL Layer Attacks

There will be a sporadic ML/DL component in the software portion of smart CPSs, and security risks will arise.

Papp et al. [6] divided the attack surface of embedded systems into layers. These layers are the application, firmware/OS, and hardware layers. Hardware backdoors, hardware Trojans, eavesdropping, fault injection, and hardware modification are some of the ways in which hardware layer assaults can happen, both during and after production.

Exploiting flaws or vulnerabilities in the underlying operating system (OS) or by installing malicious versions of the system's firmware or drivers through updates can carry out OS and firmware layer attacks. Application layer attacks take the form of assaults against the software applications or programmes that the embedded system operates.

Viruses, worms, software Trojans, and buffer overflow assaults are a few examples. As was previously said, CPSs like autonomous cars and drones are embedded systems, and these kinds of systems may contain ML algorithms as part of them. However, Papp et al.'s classification of the attack surface does not adequately represent ML security.

Although the hardware or sensors can directly affect it, the ML component is not a part of the CPS's hardware. Although it can be affected by both the OS and the firmware, it is not a component of either. Since the ML model is typically in charge of a specific system function, such as picture identification, some people would classify it as belonging to the application layer.

Nevertheless, an ML component is not a good match for a typical software programme. A software application or programme is generally understood to be a collection of instructions that are carried out logically in order to accomplish a certain purpose. A programmer wrote these instructions, and they remain the same unless the developer makes changes to the software. Because the software instructions are static, the application will always deliver a predefined output regardless of the input data it processes. This is not the case with ML algorithms, which are educated to provide an output rather than being coded.

The programme output may progressively alter over time when using an online learning technique like reinforcement learning. It is recommended that the ML layer be provided as a distinct layer that can be secured from assaults if the sorts of attacks are recognised, as ML does not fit well with the application layer.

6.2.3 Network Layer Attacks

6.2.3.1 DoS Attacks

The DoS attacks ill make the business operations to be affected by means of severe losses in terms of money and reputation. In such cases, the organisations have to take important measures to find and prevent such attacks by providing hardware and software protection mechanisms such as firewalls, traffic filtering systems, and IDSs. According to Xu et al. [7] and Hoque et al. [8], these attacks exploit the network protocols and also makes the system to be overloaded by irrelevant service requests multitude of times.

6.2.3.2 Replay Attacks

These types of attacks are generally by means of adversarial inputs to the system, and the way the system gives back the responses will be used by the attackers to infiltrate into the system. The replays from the system based on the historical data are transmitted from the sensors to infiltrate into the system. These kinds of attacks are generally unknown to the physical system layer which makes the system to be completely damaged. The attackers will make use of transmission channels of measurement part of the sensors and from control units.

6.2.4 Physical Layer Attacks

The physical layer is attacked through FDI attacks and they normally make it is highly challenging to identify. This leads to huge damage to the CPSs as they can perform attacks in two ways: random attacks –here the attacker changes the measurements from the sensors randomly; and targeted attacks – here the states of the system are specifically targeted and attacked.

6.2.5 Sample Layer-wise Attacks

Physical layer

- Hardware
- Hardware back doors
- Eavesdropping attacks through Trojans
- Injecting faults
- Modifying hardware parameters

Network layer

- DoS attacks
- Flooding attacks
- Spoofing attacks
- Replay attacks

Application layer

- The OS installed as the firmware attacks
  - Update the firmware and/or the driver software
  - Vulnerability attacks on OS itself

- Application
    - Virus code and worms-based attacks
    - Trojans targeting software application
    - Attacks such as buffer overflow
- ML model
    - ML model data corruption
    - Attacks using perturbation
    - Poisoning data

6.3 MACHINE LEARNING TECHNIQUES FOR SECURITY ANALYTICS

ML is a subset of artificial intelligence (AI) that focuses on the development of algorithms and statistical models that enable computers to learn and make predictions or decisions from data without being explicitly programmed. ML algorithms learn patterns and relationships from historical data and use them to make predictions or decisions on new, unseen data.

6.3.1 Key Concepts and Components of Machine Learning

- **Dataset:** Dataset is the foundation of ML. It can be structured (e.g., tabular data) or unstructured (e.g., text, images, audio). High-quality, relevant, and diverse data is essential for training accurate ML models.
- **Features:** Features are individual measurable properties or characteristics of the data. They represent the input variables used by ML algorithms to make predictions or decisions. Feature engineering involves selecting, transforming, and creating new features from raw data to improve model performance.
- **ML Models:** A model is a mathematical representation of the relationship between the input features and the output predictions or decisions. Models are trained using ML algorithms on historical data. Common types of models include linear regression, decision trees, SVMs, neural networks, and more.
- **Training:** Training is the process of fitting a model to the training data by adjusting its parameters or coefficients to minimise the difference between the predicted outputs and the actual target values. This process involves optimisation techniques such as gradient descent.
- **Evaluation:** Evaluation is the process of assessing the performance of a trained model on unseen data. Common evaluation metrics vary depending on the type of ML task (e.g., classification, regression) and can include accuracy, precision, recall, F1-score, mean squared error (MSE) etc.
- **Validation:** Validation is the process of tuning hyperparameters and selecting the best-performing model among different candidate models. It involves techniques such as cross-validation and grid search.
- **Testing:** Testing is the final step where the selected model is evaluated on a separate test dataset to assess its generalization performance. Testing helps ensure that the model performs well on unseen data and does not overfit to the training data.

- **Deployment:** Deployment involves deploying the trained model into production environments where it can make predictions or decisions on new, real-world data. This may involve integrating the model into software applications, web services, or IoT devices.

6.4.2 Feature Engineering

Feature engineering is crucial in extracting meaningful insights from IoT data. It involves creating new features or transforming existing ones to improve the performance of ML models or to reveal hidden patterns in the data.

Here are some common techniques for feature engineering with IoT data:

- **Timestamp Features:** Extract various features from timestamps such as hour of the day, day of the week, month, season, etc. These features can capture temporal patterns in the data.
- **Aggregation:** Aggregate sensor data over different time intervals (e.g., hourly, daily) to create summary statistics such as mean, median, minimum, maximum, standard deviation etc. This helps in capturing the overall trends and patterns in the data.
- **Windowing:** Create rolling windows of sensor data and compute statistics within each window. This can help capture short-term trends and patterns in the data.
- **Lag Features:** Create lag features by shifting the values of sensor data backward or forward in time. This can capture temporal dependencies and autocorrelation in the data.
- **Frequency Domain Features:** Convert time-domain sensor data into the frequency domain using techniques like Fourier transform or wavelet transform. Extract features such as dominant frequency, spectral entropy etc., which can capture frequency-based patterns in the data.
- **Seasonality and Trend:** Decompose time series data into seasonal, trend, and residual components using techniques like seasonal decomposition of time series (STL). Use these components as features to capture underlying patterns in the data.
- **Domain-Specific Features:** Incorporate domain knowledge to create features that are relevant to the specific IoT application. For example, in environmental monitoring, features related to weather conditions (e.g., temperature, humidity, wind speed) can be useful.
- **Dimensionality Reduction:** Apply dimensionality reduction techniques such as principal component analysis (PCA) or t-distributed stochastic neighbour embedding (t-SNE) to reduce the dimensionality of high-dimensional sensor data while preserving important information.
- **Feature Scaling:** Scale the features to a similar range (e.g., using Min-Max scaling or Z-score normalisation) to ensure that features with larger magnitudes do not dominate the model training process.
- **Feature Selection:** Use techniques like feature importance ranking, recursive feature elimination, or regularisation methods to select the most relevant features for model training and discard irrelevant or redundant features.

When engineering features for IoT data, it's important to consider the specific characteristics of the data, the requirements of the ML task, and the computational resources available for processing the data. Experimentation and domain expertise play a crucial role in determining which features are most informative for the given application.

Finding the best fit in different models involves selecting the model that best captures the underlying patterns in the data and generalizes well to unseen data. Here is a general process for finding the best fit in different models:

- **Define Evaluation Metric:** Choose an appropriate evaluation metric based on the nature of the problem you're solving. For example, for regression tasks, metrics like mean squared error (MSE) or root mean squared error (RMSE) are commonly used, while for classification tasks, metrics like accuracy, precision, recall, F1-score, or area under the ROC curve (AUC-ROC) may be used [9].
- **Split Data:** Split your dataset into training, validation, and test sets. The training set is used to train the models, the validation set is used to tune hyperparameters and select the best model, and the test set is used to evaluate the final model's performance.
- **Select Candidate Models:** Choose a set of candidate models that are suitable for your problem domain. This could include linear models (e.g., linear regression, logistic regression), tree-based models (e.g., decision trees, random forests, gradient boosting machines), SVMs, neural networks, etc.
- **Train Models:** Train each candidate model on the training data using default hyperparameters or initial parameter values.
- **Hyperparameter Tuning:** Use the validation set to tune the hyperparameters of each model. This can be done through techniques like grid search, random search, or Bayesian optimisation. Tune hyperparameters to optimise the chosen evaluation metric.
- **Evaluate Models:** Evaluate each model's performance on the validation set using the chosen evaluation metric. Select the model with the best performance as the final model.
- **Validate Model:** Validate the final model on the test set to ensure that it generalizes well to unseen data. Evaluate its performance using the chosen evaluation metric.
- **Compare Models:** Compare the performance of different models based on their evaluation metrics on the validation and test sets. Choose the model that performs the best overall.
- **Consider Complexity:** Consider the complexity of the models and the interpretability of their results. Sometimes simpler models may perform comparably to complex models and are easier to interpret.
- **Iterate:** Iterate on the process by trying different combinations of features, preprocessing techniques, and models to improve performance further.

By this process, we can systematically compare different models and select the one that best fits your data and problem domain. Remember that the choice of the best model may depend on various factors such as data characteristics, computational resources, interpretability requirements, and the specific goals of the project.

6.5 MACHINE LEARNING MODELS – INTRODUCTION

6.5.1 Logistic Regression

Several machine learning models have been applied in the recent past to identify security threats and attacks in the recent past [6, 10].
Despite its name, logistic regression is employed in binary categorisation. It models the predictor factors and outcome probability using a logit function (sigmoid function). The sigmoid function converts values to probabilities. Class labels can be provided to values above and below a threshold based on the desired result. Its formula is given in Equation (6.1).

$$f(x) = 1/\left(1 + \mathrm{e}^\wedge(-x)\right) \tag{6.1}$$

Equation (6.1) represents the function used to find the probability values to perform the classification.

Logistic regression can be used for multimodal classification tasks and is not limited to binary classification. The first step in logistic regression is similar to linear regression where each parameter has an associated weight. This is passed to the sigmoid function where the probability values are computed. During training, the parameter weights are adjusted and the decision point is also found. The decision point is a probability value where values obtained greater than the decision value belong to one class, while the rest belong to another class.

6.5.2 Decision Tree

Decision tree is a ML algorithm which is used for both classification and regression. It recursively splits the data based on a factor known as information gain. Information gain is computed using different methods such as entropy or gini index. This splitting creates a tree-like structure which makes the required decision. In a decision tree, the internal nodes represent the feature that contributes a decision and each outcome of the decision is represented by a branch emanating from that node. Each leaf node represents the outcome predicted from the internal decision node.

Equations (6.2) and (6.3) represent the functions used to find the parameter to perform split operation on. Equation (6.2) is used for calculating gini index which is computed by taking the sum of square probabilities of each instance of a parameter. Equation (6.3) is used in calculating the entropy values which measure the randomness. Similar to gini index, entropy takes the sum of logarithm of probability multiplied into the probability value of each instance of a parameter. These are used to find information gain.

$$\text{Gini index} = 1 - \Sigma(p^i) \tag{6.2}$$

$$\text{Entropy}(s) = -\Sigma(p_i \log_{2(p_i)}) \tag{6.3}$$

Information gain helps a model understand how useful one particular parameter is, the higher the information gain, the better it is to split using that parameter. Decision tree models run in every single scenario while trying to find the most ideal split locations.

6.5.3 Support Vector Machines

SVM is a supervised learning algorithm commonly used for classification and regression tasks. In radiology, SVM is particularly useful for image classification because it can effectively work with high-dimensional data.
The decision in SVM is taken based on the sign function.

$$f(x) = \sin(w.x + b) \tag{6.4}$$

In this context, w represents the weight vector, x is the input feature vector, b denotes the bias term, and the sign function is represented by sin().

6.5.4 *k*-Nearest Neighbours

k-Nearest neighbours (KNN) is a supervised learning algorithm that does not rely on a specific model structure and is often used for classification. KNN classifies a data point by locating the k nearest points in the feature space and then determining the data point's class based on the majority class of these neighbours.

KNN is effective for image classification, but it can be computationally intensive when working with large datasets, since it requires calculating distances between the query point and all training samples. The effectiveness of KNN is influenced by choices such as the value of k (number of neighbours to consider) and the distance metric used. Despite these challenges, KNN's straightforward approach and flexibility make it well-suited for radiology applications that involve datasets of different sizes and complexities.

A new image is classified by calculating the Euclidean distance. It classifies the image by selecting the least distance among all the distances.

$$f(x) = \sqrt{(x_1 - x_i)^2 + (y_1 - y_i)^2} \tag{6.5}$$

In this context, (x_1, y_1) refers to the coordinates of the new image, while (x_i, y_i) denotes the coordinates of the ith closest neighbour.

6.5.5 Random Forest

Random forest represents a supervised learning technique extensively utilized in radiology for both classification and regression tasks. Its operation entails the generation of multiple decision trees during the training phase, followed by the amalgamation of their outputs to ascertain the class mode (for classification) or mean prediction (for regression), renowned for its adaptability and resilience,

The algorithm's effectiveness stems from its ensemble learning approach, where each decision tree is trained using a random selection of the training data and features. This inherent randomness aids in curbing model variance and enhancing generalization performance. Moreover, random forest furnishes valuable insights into feature importance, enabling users to discern the significance of each feature towards the model's predictive efficacy. This attribute bolsters the interpretability of the model and facilitates informed decision-making within radiology applications.

6.5.6 XGBoost

Extreme gradient boosting (XGBoost) is a widely employed and effective ML algorithm specifically engineered for the purposes of classification and regression. It is an ensemble learning technique in which the predictions of numerous feeble learners are combined to construct a robust predictive model. Numerous ML tasks have been accomplished with the commendable efficiency, scalability, and predictive prowess of XGBoost model. Its objective function is given in Equation (6.6)

$$\text{Obj}(T) = \Sigma \text{l}(yi,\ y\_\text{pred},\ i) + \Sigma\Omega(f) \tag{6.6}$$

Equation (6.6) represents the objective function used to perform XGBoost classification. XGBoost objective function utilizes a loss value while computing the result where the loss value helps in improving the prediction results and boosts the performance of the model. This is the main principle of boosting. Boosting techniques make use of loss value from the previous iteration while training. Similar to bagging each model in boosting method receives different parts of the data. But the data received by the next model will involve the data which were not fit properly while training the previous model. This loss value can help every model learn from what the previous model was lacking and overcome the shortcomings to make robust and accurate predictions.

Table 6.1 summarizes the ML algorithms applied by researchers for security vulnerabilities prediction.

TABLE 6.1
Application of ML Algorithms in Security Vulnerabilities Prediction

Reference	Effective ML Algorithm	Accuracy
[3]	Random forest (RF)	97.70%
[4]	Support vector machines (SVMs)	98.52%
[5]	Random forest (RF)	98%
[6]	Random forest (RF)	97.76%
[8]	Polynomial SVM	95.38%
[9]	SVM	95.24%
[10]	Random forest (RF)	98.38%

6.6 RATIONALE OF SECURITY ANALYTICS USING ML TECHNIQUES

To be able to effectively recognize deviations that could indicate the occurrence of a DoS attack, an anomaly detection method for the detection of DoS attacks is implemented [9]. They are proactive in nature and detect threats that are not yet known or emerging and might not be recognized by signature-based methods.

This chapter intends to accomplish the following objectives through an anomaly detection method:

- **Early Detection:** The primary usefulness of the anomaly detection lies in its ability to spot non-typical or suspicious network behaviour as it happens. By training a model with historical data to establish a baseline of normal behaviour, any deviations from this norm can be promptly detected to achieve early warning of possible DoS attack.
- **Adaptability:** Anomaly detection mechanisms have an adaptive feature that enables them to deal with dynamically changing attack patterns and network dynamics. It is a learning algorithm whose threshold for what qualifies as anomalous behaviour keeps on updating itself as new data comes in thereby making sure that even against new-age vectors for attacks its detecting capabilities remain useful.
- **Contextual Analysis:** Anomaly detection allows for a comprehensive analysis of various network features, such as packet rates, bandwidth utilization, and system resource usage. By considering multiple dimensions of network behaviour, the method can provide a more accurate assessment of anomalies, minimizing false positives and enhancing the detection of DoS attacks.
- **Unknown Attack Detection:** On the other hand, rather than relying on known assault patterns as signature-based methods do, it (anomaly detection) can track novel attacks never seen before or zero-day attacks. By targeting deviations from the norm of what is expected, it can identify new attacking technologies or variations of old attacks that make this better at finding and countering more technical forms of DoS.
- **Non-stop Monitoring:** The anomaly detection system enables the real-time examination and decision-making in response to potential DoS attempts by monitoring network activities taking place. If those anomalies are greater than a given threshold, the system can generate alerts or trigger automated actions thus allowing instant investigation and mitigation steps.

This chapter provides novel approaches to improve overall DoS attack detection capability by utilizing an anomaly detection approach. This includes early warning systems for critical systems and networks, adaptability to emerging threats, situational awareness, and continuous monitoring. In conclusion, these measures will enable the chapter guarantee a more secure environment for targeted systems and networks; hence preventing any disruption caused by DoS attacks.

6.7 CASE STUDY – NOVEL FRAMEWORK FOR IDENTIFICATION OF DDOS ATTACKS IN IOT SYSTEMS

A DoS attack is whereby a computer overloads a server with many TCP and UDP packets thus causing denial of service [4]. In contrast, DDoS attacks [11, 12] happen when several systems work together to target one system with a DoS attack. In this case the network would be flooded with packets from multiple sources. DoS and distributed denial of service (DDoS) attacks [13] create significant risks to modern enterprises. The financial consequences that come along with successful DoS attack are quite dreadful as security surveys show that there is an average cost of around $20,000–$40,000 per hour for a DDoS attack. Even the biggest organizations can be brought down by this huge sum.

At this time, too many packets overload the network so much that it surpasses the server's capacity and prevents other devices and users on the network from being able to access the service; hence making targeted service unavailable during such attacks [5]. The purpose behind DoS attacks is to make individual machines or networks nonfunctional for their intended users.

6.7.1 Various Types of DoS Attacks

- **Buffer Overflows:** This is the most popular form of DoS attack. It involves flooding a network with so much traffic that it becomes unable to function.
- **Ping of Death (POD) or ICMP Flood:** When network devices which are misconfigured or unconfigured are used to send spoofed packets and ping every computer on a target network, this is called an ICMP flood attack which is also called PoD.
- **SYN Floods:** SYN flood attacks involve sending connection requests to a server and not completing the handshake process. Consequently, connection requests keep coming in until the system gets congested making it difficult for authorized users to connect.
- **Teardrop Attack:** In a teardrop DoS attack, fragmented IP data packets are sent by the attacker into a network. The network then tries to reassemble these fragments back into their original packets, and during this period the system crashes due to exhaustion from the process of reassembly which goes on and on without stopping. The main thrust of such an attack lies in manipulating specially designed fields which confuse the system thereby preventing proper reassembly from taking place.

DDoS attacks are detected on the Security System which is referred as SG (Security Guard) system by using various ML and deep learning (DL) techniques. The algorithm for detecting DDoS attacks is highly complicated and should be accurate, optimal, and performance oriented. Multiple ML/DL algorithms have been trained to detect DDoS attacks correctly.

The process begins with data extraction. Cyber attacks are classified and detected by using ML and DL algorithms. The algorithms applied for attacks detection are explored in the subsequent sections.

6.8 APPLICATION OF ML/DL ALGORITHMS IN DDOS ATTACKS DETECTION

6.8.1 Artificial Neural Networks (ANNs)

The workflow has four stages: initially, collect DDoS data from the database and pre-process it; then partition the dataset into training set of 70:30 ratio for testing set. ANN is a computational architecture with brain neural structure that has input, hidden, output layers. For every neural input set x in sample n, an initial value (x) and weight (w) are available. The most common one among neuron networks is backpropagation as well as feedforward. If we say that $x = [x_1, x_2, x_3 \ldots x_n]$ so $y(x) = [y_1, y_2, y_3 \ldots y_n]$, the calculations are given in Equations (6.7) and (6.8).

$$W_{\text{new}} = W_{\text{old}} - a\frac{\partial J}{\partial W} \tag{6.7}$$

$$f(x) = \frac{1}{1+e^{-(x)}} \tag{6.8}$$

6.8.2 Decision Tree

A decision tree methodology is employed in four stages: data preprocessing, dataset partitioning (70:30 ratio), decision tree construction inspired by neural architectures, and model evaluation. Decision trees offer interpretability and handle diverse data types efficiently. This approach capitalizes on decision trees' interpretability and ability to handle various data types effectively. By incorporating elements of neural networks, we aim to create a powerful and informative decision-making tool.

$$\text{Gini} = 1 - \sum_{i=1}^{c}(p_i)^2 \tag{6.9}$$

6.8.3 *k*-Modes Clustering

This algorithm groups similar data points into clusters according to the attributes that are categorical in nature. To find out the similar data points, distance metrics such as Euclidean distance, manhotten distance etc. are used. The working of this algorithm involves the identification of centroid which is the most frequent values within each cluster.

6.8.4 One-Class SVM in Outlier Detection

Outliers or anomalous data points can drastically affect the prediction of the ML model. Thus, it is necessary to identify and exclude any outliers in the dataset.

For outlier detection the one-class SVM algorithm has been employed. This algorithm constructs a hypersphere (or hyperplane in higher dimensions) to encapsulate the majority of data points, considering the rest as outliers. By defining a boundary around the normal data distribution, one-class SVM effectively identifies data points that deviate significantly from this norm.

On applying the one class SVM model, we observed a very low incidence of outliers. Only 6% (approximately) of the data points were identified as outliers. Due to the minute number of outliers in the dataset, the results obtained using the ML models can be considered accurate.

The performance of the ML algorithms is evaluated using several performance metrics: accuracy, precision, recall, and F1-score. A confusion matrix is used to determine how to measure such metrics [9].

It has four different cases:

True Positive (TP): It represents the number of correct prediction of the presence of attacks.

True Negative (TN): It is the metric that gives the number of correctly predicted packets which are not attacks.

False Positive (FP): This metric gives the number of wrongly predicted normal packets as attacks.

False Negative (FN): This metric shows the number of wrongly predicted normal packets which are actually attacks.

6.9 PERFORMANCE METRICS

Accuracy: The ratio of correct predictions to the total number of predictions.

$$\text{Accuracy} = \frac{\text{TP} + \text{TN}}{\text{TP} + \text{FP} + \text{TN} + \text{FN}} \tag{6.10}$$

Precision: The ratio of number of true positive cases to all the predicted positive cases.

$$\text{Precision} = \frac{\text{TP}}{\text{TP} + \text{FP}} \tag{6.11}$$

Recall: The ratio of number of true positive cases to the total positive cases.

$$\text{Recall} = \frac{\text{TP}}{\text{TP} + \text{FN}} \tag{6.12}$$

F1-Score: The ratio of number of true positive cases to the total positive cases.

$$\text{F1} = \frac{2 \times (\text{precision} \times \text{recall})}{(\text{precision} + \text{recall})} \tag{6.13}$$

The performance metrics evaluation is given in Table 6.2.

TABLE 6.2
Performance Metrics of ML Algorithms

Metrics/Model	Accuracy	Recall	Precision	F1-Score
ANN	96.94%	74.22%	94.67%	83.20%
Decision tree	99.99%	100%	99.98%	99.99%
K-modes	67.13%	56.13%	75.36%	64.36%

From Table 6.2, it is inferred that the decision tree classifier has given better results. In fact, the decision tree model offered a strong prediction accuracy, recall, precision, and F1-scores making it a good option in DDoS mitigation applications within IoT environment.

The ANN model is identified as the best model as it has higher accuracy and precision. However, there were problems with the *k*-modes model suggesting that IoT security demands customized algorithm selection. Later research should improve collaborative DDoS detection mechanisms first.

With help from ML advancements, we can solidify the security of IoT systems against new threats ensuring uninterrupted service to connected devices and users. To avoid financial risks caused by DDoS attacks, it's crucial that ML frameworks are constantly innovated and adapted to safeguarding IoT ecosystems.

6.10 RELATED WORK

Numerous research studies have focused on the identification of DDoS attacks. The scholarly literature extensively examines different techniques, including ML, for this purpose. Our research reveals that the utilization of ML in detecting DDoS attacks has been extensively researched and is crucial for organizations in their monitoring and resolution operations.

Dennis and Li [11] developed a system using a random forest and ML embedded algorithm with weighted voting. This system collects the statistics from the switches for every ten seconds and sends to the random forest classifier. If the model classifies that as an attack, then an attack mitigation process component will be invoked. This helps to prevent such attacks thereby reducing the breakdown of the entire network or the system.

Mishra et al. [14] have proposed ML integration to secure cloud-based operations by means of a mitigation process to protect from security attacks. The proposed framework is a multi-layer model that has ML model with a self-protection system that can detect and mitigate security attacks.

Doshi et al. [15] implemented various ML algorithms including random forest, *k*-nearest neighbours (KNN), and SVMs with linear kernel (LSVM) to distinguish normal IoT packets from DoS packets. Similarly, Khashab et al. [16] evaluated the ML models such as decision tree, logistic regression, naïve Bayes, KNN, Reward Model (RM), and SVM. Both researchers concluded that random forest to be the dominant algorithm.

Mohammed et al. [17] have employed a software-defined network (SDN) and they applied the proposed model by using realistic attacks.

Kyaw et al. [18] have proposed a method that compared SVM with other classifiers using different categories of attack types.

Ye et al. [19] applied a new SDN environment namely mininet, and a floodlight within a simulation platform constructed of six-tuple characteristic values applied to the switch flow table. They built a DDoS attack model by combining SVM classification algorithms.

Nurwarsito et al. [20] have analysed an SDN-architecture-based system to mitigate the DDoS attacks. They proposed an ML-based random forest algorithm that is able to classify packets as normal and attacks based on the packet-flow information.

A study [21] proposed an ML-based DoS attack detection framework for MQTT protocol to protect IoT systems from cyber threats. Further, the vulnerability of IoT due to DDoS attacks is also investigated in Reference [12] by highlighting the emergence of malware and botnets as significant threats.

Gaur and Kumar [22] proposed a hybrid methodology employing feature selection algorithms and ML classification algorithms to detect the DDoS attacks earlier in IoT devices. It demonstrated the higher feature reduction rate and higher accuracy using XGBoost.

Jia et al. [23] proposed an edge-centric IoT defense scheme called FlowGuard that detects DDoS attacks. It employed a DDoS attack detection algorithm based on patterns of traffic variations and also employs ML models. This algorithm achieves higher attack identification rate and classification accuracy.

A study [24] conducted by Javaid et al. focused on the use of blockchain for DDoS attack detection and mitigation. Another study conducted by Shah et al. [25]employed blockchain for DDoS attack detection and prevention. It also mentioned the benefits of the blockchain method for detection and prevention of attacks on IoT devices.

Zhang and Green [26] introduced an IoT DDoS defense algorithm focusing on lightweight detection mechanisms for working nodes, distinguishing between legitimate and malicious requests.

Other studies [27, 28–32] focused on DDoS prevention by using deep learning models, clustering strategies, graph structure features and hybrid deep learning algorithms.

Lazzarini et al. [1] proposed DIS-IoT, a novel approach employing a stacking ensemble of deep learning models for intrusion detection. While achieving improved accuracy, the computational requirements for such ensemble models remain a concern. Conversely, Diab and Rodriguez-Villegas [2] conducted a comprehensive review of embedded ML using microcontrollers in wearable healthcare systems, highlighting the practical applications of TinyML in various healthcare tasks. Despite its feasibility, challenges in benchmarking criteria for widespread TinyML support persist.

Furthermore, security concerns extend beyond healthcare to VANETs, as addressed by Gad et al. [33]. The researchers applied ML algorithms to the ToN-IoT dataset, employing techniques such as feature selection and SMOTE to enhance accuracy, with XGBoost yielding optimal results. Zeghida et al. [34] proposed an ensemble ML-based IDS for MQTT protocol in IoT environments, achieving improved efficiency through dataset conversion and classifier utilization.

Privacy concerns in federated learning-based IDS were tackled by Friha et al. [35], who proposed a decentralized approach to mitigate single points of failure. Similarly, Sharma et al. [36] proposed an anomaly-based IDS using a filter-based feature selection deep neural network (DNN) model, with generative adversarial networks (GANs) addressing class imbalance issues. The analysis by Prazeres et al. [37] explored a distributed IDS architecture leveraging fog computing to reduce overhead on central servers, with DNNs utilized for attack classification. Also, an ensemble learning-based IDS architecture was proposed to handle attacks in IoT devices [38], further a stacked ensemble learning-based IDS was proposed to handle attacks in IoT devices [39].

Hardware implementations for healthcare applications, such as seizure detection, were discussed by Torse et al. [40], focusing on edge computing to differentiate seizure signals based on characteristics. Kurapati [41] emphasized edge computing's role in intelligent healthcare management, highlighting security and privacy challenges compared with cloud computing. ML applications in healthcare wearable devices, including security and privacy concerns in edge computing environments, were reviewed by Sabry et al. [42]. The application of IDS at the edge of the IoT devices using TinyML was discussed with case studies in a literature [43].

Further, there are research works conducted on the application of TinyML in edge analytics for various IoT devices such as healthcare and industrial IoT [44–50].

Moreover, advancements in IDSs for industrial IoT networks were explored by Altunay and Albayrak [51], demonstrating superior performance with hybrid CNN+LSTM models. The automated IDS system for IoT networks proposed by Vishwakarma and Kesswani [52] combined different NIDS datasets to enhance attack detection. These studies collectively contribute to advancing IoT security practices, addressing challenges such as computational constraints, privacy concerns, and real-time processing requirements. A dependable federated learning for IoT intrusion detection against poisoning attacks has been proposed by the researchers to protect IoT devices in real-time applications [53]. Similarly several intrusion detection systems using deep learning techniques and co-design based attacks identification for IoT systems have been proposed in the recent past [3, 7, 8, 54–59].

6.11 SUMMARY

Our research on DDoS attack detection for IoT devices using ML techniques provided us with remarkable insights. In leveraging TinyML for processing at the network edge and preserving privacy through lightweight encryption, the presented results underscore a promising avenue for enhancing security in IoT environments. Across a range of ML classifiers, including KNN, random forest, and gradient boosting machine, notable accuracies have been achieved, signalling their efficacy in detecting network attacks. This not only demonstrates the feasibility of deploying ML solutions in resource-constrained edge devices but also highlights the robustness of these models in mitigating security threats.

Furthermore, the integration of lightweight encryption techniques ensures the privacy of sensitive data and network parameter IDs, aligning with regulatory requirements and safeguarding user confidentiality. By combining efficient edge processing with privacy-preserving measures, organizations can deploy scalable and secure IoT solutions that effectively detect and respond to threats in real time. This holistic approach not only enhances the resilience of IoT networks but also fosters trust among users by prioritizing their privacy and data security.

PRACTICE QUESTIONS

I. Short Answers – Remember/Understand
   1. Which sectors were transformed due to smart CPS?
   2. In healthcare applications with IoT, state the vulnerabilities caused by the attackers.
   3. What is the purpose of intrusion detection systems (IDSs) in smart CPS applications?
   4. List out the attacks related to the network layer of a CPS.
   5. Which attacks are related to the ML model if it is a component of CPS?
   6. Name some of the attacks that make the patient healthcare monitoring system to be breached.
   7. What is windowing?
   8. What do you mean by domain-specific features?
   9. What is hyperparameter tuning?
   10. Briefly write a note on contextual analysis.

II. Descriptive Questions – Apply/Analyse
   1. Analyse the various attacks happening at each layer in the CPS architecture.
   2. In a real-time image processing system to analyse the human activities, how the adversarial attacks happen? Explore the ways of such attacks happening in the layered architecture.
   3. Examine the various ways to improve the IoT security measures.
   4. You are involved in developing an automated home security system. Which are the criteria that you need to take into consideration to address adversarial attacks?
   5. Explore the key concepts and components of ML algorithms.
   6. Apply the feature engineering techniques to extract worthy insights from the dataset.
   7. Describe the general process guidelines to select a best model among different ML models.
   8. For a CPS placed in a product assembly line of an automobile industry that needs to take instant decision according to the current assembly-based image processing application embedded in it, analyse the appropriate ML algorithm that can be applied to classify whether the assembly done is correct or not.
   9. Elucidate the objectives to be derived through anomaly detection process.
   10. Elaborate the application of 1-class SVM applied for outlier detection with respect to a CPS that provides wrong predictions using the ML model.

III. Higher Order Thinking Questions – Analyse/Create/Evaluate
1. Showcase the various performance metrics associated with the evaluation of ML models.
2. Examine how ML or DL models to handle DoS and DDoS kinds of attacks efficiently?
3. Explore the application of edge analytics to find out the instant errors derived from the mode.
4. Analyse the working of security framework for MIoT-based wearable CPSs.
5. Write the algorithm applied for developing an edge analytics module to showcase the errors instantly.

REFERENCES

1. Lazzarini, Riccardo, Huaglory Tianfield, Prof Charissis. A stacking ensemble of deep learning models for IoT network intrusion detection. (2023).
2. Diab, Maha S., Esther Rodriguez-Villegas. Embedded machine learning using microcontrollers in wearable and ambulatory systems for health and care applications: a review. *IEEE Access* (2022).
3. Kim S, Park KJ. A survey on machine-learning based security design for cyber-physical systems. Appl. *Sci.* 11(12) (2021): 5458.
4. Guo S, Zhao X, Wang H, Xu N. Distributed consensus of heterogeneous switched nonlinear multiagent systems with input quantization and DoS attacks. *Appl. Math. Comput.*, 456 (Nov. 2023): 128127–128127, doi: https://doi.org/10.1016/j.amc.2023.128127.
5. Raghuvamsi Y, Teeparthi K. Detection and reconstruction of measurements against false data injection and DoS attacks in distribution system state estimation: a deep learning approach. *Measurement* 210 (Mar. 2023): 112565, doi: https://doi.org/10.1016/j.measurement.2023.112565.
6. Papp D, Ma Z, Buttyan L. Embedded systems security: threats, vulnerabilities, and attack taxonomy, in *Privacy, Security and Trust (PST)*, 2015 13th Annual Conference on, pp. 145–152.
7. Xu W, Ma K, Trappe W, Zhang Y. Jamming sensor networks: attack and defense strategies. *IEEE Netw.* 20(3) (2006): 41–47.
8. Hoque N, Kashyap H, Bhattacharyya DK. Real-time DDoS attack detection using FPGA. Comput. *Commun.* 110 (2017): 48–58.
9. Jayaswal V. Performance metrics: confusion matrix, precision, recall, and F1 score. *Medium*, Sep. 15, 2020. https://towardsdatascience.com/performance-metrics-confusion-matrix-precision-recall-and-f1-score-a8fe076a2262
10. Cheng Fei, Jun Shen. Machine learning for securing cyber–physical systems under cyber attacks: a survey, *Franklin Open*, 4 (September 2023): 100041.
11. Dennis JR, Li X. *Machine-Learning and Statistical Methods for DDoS Attack Detection and Defense System in Software Defined Networks.* Master's Thesis, College of Engineering and Science, Ryerson University, Toronto, ON, Canada, (2018).
12. Vishwakarma R, Jain AK. A survey of DDoS attacking techniques and defence mechanisms in the IoT network. *Telecommun. Syst.* 73 (2020): 3–25. https://doi.org/10.1007/s11235-019-00599-z
13. Alwabisi S, Ouni R, Saleem K. Using machine learning and software-defined networking to detect and mitigate DDoS attacks in fiber-optic networks. *Electronics*, 11(23) (2022): 4065. doi: 10.3390/electronics11234065.

14. Mishra S, Sharma SK, Alowaidi MA. Multilayer self-defense system to protect enterprise cloud. *Comput. Mater. Contin.* 66 (2021): 71–85.
15. Doshi R, Apthorpe N, Feamster N. Machine learning DDoS detection for consumer internet of things devices, in *2018 IEEE Security and Privacy Workshops (SPW)*, (2018). doi: 10.1109/spw.2018.00013.
16. Khashab F, Moubarak J, Feghali A, Bassil C. DDoS attack detection and mitigation in SDN using machine learning, in *Proceedings of the 2021 IEEE Conference on Network Softwarization: Accelerating Network Softwarization in the Cognitive Age (NetSoft)*, Tokyo, Japan, 28 June–2 July 2021, pp. 395–401.
17. Mohammed SS, Hussain R, Senko O, Bimaganbetov B, Lee JY, Hussain F, Kerrache CA, Barka E, Alam Bhuiyan MZ. A new machine learning-based collaborative DDoS mitigation mechanism in software-defined network, in *Proceedings of the 2018 14th International Conference on Wireless and Mobile Computing, Networking and Communications (WiMob)*, Limassol, Cyprus, 15–17 October 2018.
18. Kyaw AT, Zin Oo M, Khin CS. Machine-learning based DDOS attack classifier in software defined network, in *Proceedings of the 17th International Conference on Electrical Engineering/Electronics, Computer, Telecommunications and Information Technology (ECTI-CON)*, Phuket, Thailand, 24–27 June 2020, pp. 431–434.
19. Ye J, Cheng X, Zhu J, Feng L, Song L. A DDoS attack detection method based on SVM in software defined network. *Secur. Commun. Netw.* 2018 (2018): 9804061.
20. Nurwarsito H, Nadhif MF. DDoS attack early detection and mitigation system on SDN using random forest algorithm and Ryu framework, in *Proceedings of the 2021 8th International Conference on Computer and Communication Engineering (ICCCE)*, Kuala Lumpur, Malaysia, 22–23 June 2021, pp. 178–183.
21. Syed NF, Baig Z, Ibrahim A, Valli C. Denial of service attack detection through machine learning for the IoT. *J. Info. Telecomm.* 4(4) (2020): 482–503. https://doi.org/10.1080/24751839.2020.1767484
22. Gaur V, Kumar R. Analysis of machine learning classifiers for early detection of DDoS attacks on IoT devices. *Arab J Sci Eng* 47 (2022): 1353–1374. https://doi.org/10.1007/s13369-021-05947-3
23. Jia Y, Zhong F, Alrawais A, Gong B, Cheng X. FlowGuard: An intelligent edge defense mechanism against IoT DDoS attacks, in *IEEE Internet of Things Journal*, 7(10) (2020): pp. 9552–9562, doi: 10.1109/JIOT.2020.2993782.
24. Uzair Javaid, Ang Kiang Siang, Muhammad Naveed Aman, Biplab Sikdar. Mitigating loT device based DDoS attacks using blockchain, in *Proceedings of the 1st Workshop on Cryptocurrencies and Blockchains for Distributed Systems (CryBlock'18)*. Association for Computing Machinery, New York, NY, (2018), pp. 71–76. https://doi.org/10.1145/3211933.3211946
25. Shah Z, Ullah I, Li H, Levula A, Khurshid K. Blockchain based solutions to mitigate distributed denial of service (DDoS) attacks in the internet of things (IoT): A survey. *Sensors* 22(3) (2022): 1094. https://doi.org/10.3390/s22031094
26. Zhang C, Green R. Communication security in internet of thing: preventive measure and avoid DDoS attack over IoT network, in *Proceedings of the 18th Symposium on Communications & Networking* (2015, April), pp. 8–15.
27. Almaraz-Rivera JG, Perez-Diaz JA, Cantoral-Ceballos JA. Transport and Application layer DDoS attacks detection to IoT devices by using machine learning and deep learning models. *Sensors* 22(9) (2022): 3367. https://doi.org/10.3390/s22093367
28. Jing H, Wang J. *Detection of DDoS attack within industrial IoT devices based on clustering and graph structure features*. Security and Communication Networks, (2022).

29. Kumar S, Chandavarkar BR. DDOS prevention in IoT, in *2021 12th International Conference on Computing Communication and Networking Technologies (ICCCNT)*. IEEE (2021, July), pp. 1–6.
30. Efe A, Aksöz E, Hanecioğlu N, Yalman ŞN. Smart security of IoT against to DDoS attacks. *Int. J. Innov. Eng. Appl.* 2(2) (2018): 35–43.
31. Chen M, Tang X, Cheng J, Xiong N, Li J, Fan D. A DDoS attack defense method based on blockchain for IoTs devices, in *Artificial Intelligence and Security: 6th International Conference, ICAIS 2020*, Hohhot, China, July 17–20, 2020, Proceedings, Part II 6 (pp. 685–694). Springer, Singapore (2020).
32. Roopak M, Tian GY, Chambers J. An intrusion detection system against ddos attacks in iot networks, *in 2020 10th Annual Computing and Communication Workshop and Conference (CCWC)*. IEEE, (2020, January), pp. 0562–0567.
33. Gad, Abdallah R., Ahmed A. Nashat, Tamer M. Barkat. Intrusion detection system using machine learning for vehicular ad hoc networks based on ToN-IoT dataset. *IEEE Access* 9 (2021): 142206–142217.
34. Zeghida, Hayette, Mehdi Boulaiche, Ramdane Chikh. Securing MQTT protocol for IoT environment using IDS based on ensemble learning. *Int J. Inform. Security* (2023): 1–12.
35. Friha, Othmane, et al. 2DF-IDS: Decentralized and differentially private federated learning-based intrusion detection system for industrial IoT. *Comput. Security* 127 (2023): 103097.
36. Sharma, Bhawana, et al. Anomaly based network intrusion detection for IoT attacks using deep learning technique. *Comput. Elect. Eng.* 107 (2023): 108626.
37. Prazeres, Nuno, et al. Engineering the application of machine learning in an IDS based on IoT traffic flow. *Intell. Syst. Appl.* 17 (2023): 200189.
38. Alghamdi, Rubayyi, Martine Bellaiche. An ensemble deep learning based IDS for IoT using Lambda architecture. *Cybersecurity* 6.1 (2023): 5.
39. Cao, Yongzhong, et al. An intrusion detection system based on stacked ensemble learning for IoT network. *Comput. Elect. Eng.*
40. Torse, Dattaprasad A., et al. Hardware implementation of automated seizure detection system using EEG signals and edge computing. *2022 6th International Conference on Trends in Electronics and Informatics (ICOEI)*. IEEE, (2022).
41. Kurapati, Keerthana. Proactive and intelligent healthcare management using IoT. *2022 International Conference on Advances in Computing, Communication and Applied Informatics (ACCAI)*. IEEE, (2022).
42. Sabry, Farida, et al. Machine learning for healthcare wearable devices: the big picture. *J. Healthcare Eng.* 2022 (2022).
43. T'Jonck, Kristof, et al. Real-time activity tracking using TinyML to support elderly care. *2021 XXX International Scientific Conference Electronics (ET)*. IEEE, (2021).
44. Amin, Syed Umar, M. Shamim Hossain. Edge intelligence and internet of things in healthcare: a survey. *IEEE Access* 9 (2020): 45–59.
45. Ahmed, Khaled, Mohamed Hassan. tinyCare: A tinyML-based low-cost continuous blood pressure estimation on the extreme edge. *2022 IEEE 10th International Conference on Healthcare Informatics (ICHI)*. IEEE, (2022).
46. Gkogkidis, Anargyros, Vasileios Tsoukas, and Athanasios Kakarountas. A TinyML-based alcohol impairment detection system for vehicle accident prevention. *2022 7th South-East Europe Design Automation, Computer Engineering, Computer Networks and Social Media Conference (SEEDA-CECNSM)*. IEEE, (2022).
47. Mala, D. Jeya, and A. Pradeep Reynold. A novel machine learning approach in edge analytics with mathematical modeling for IoT test optimization. *Simulation*

and Analysis of Mathematical Methods in Real-Time Engineering Applications (2021): 317–344.

48. Sayeed, Md Abu, et al. eSeiz: an edge-device for accurate seizure detection for smart healthcare. *IEEE Transact. Consumer Electron.* 65.3 (2019): 379–387.
49. Olokodana, Ibrahim L., et al. EZcap: a novel wearable for real-time automated seizure detection from EEG signals. *IEEE Transact. Consumer Electron.* 67.2 (2021): 166–175.
50. Tabassum, Aliya, et al. Privacy-preserving distributed IDS using incremental learning for IoT health systems. *IEEE Access* 9 (2021): 14271–14283.
51. Altunay, Hakan Can, Zafer Albayrak. A hybrid CNN+ LSTM based intrusion detection system for industrial IoT networks. *Eng. Sci. Technol. Int. J.* 38 (2023): 101322.
52. Vishwakarma, Monika, Nishtha Kesswani. DIDS: A deep neural network based real-time intrusion detection system for IoT. *Decision Anal. J.* 5 (2022): 100142.
53. Yang, Run, et al. Dependable federated learning for IoT intrusion detection against poisoning attacks. *Comput. Security* 132 (2023): 103381.
54. Elsayed, Rania A., et al. Securing IoT and SDN systems using deep-learning based automatic intrusion detection. *Ain Shams Eng. J.* 14.10 (2023): 102211.
55. Syed, Naeem Firdous, Mengmeng Ge, Zubair Baig. Fog-cloud based intrusion detection system using recurrent neural networks and feature selection for IoT networks. *Comput. Netw.* 225 (2023): 109662.
56. Li S, Ahn CK, Xiang Z. Decentralized sampled-data control for cyber-physical systems subject to DoS attacks. *IEEE Syst. J.* 15(4) (2020): 5126–5134.
57. Xiaomeng L. *Research on Security Control of Cyber-Physical System under Multiple Attacks, Ph.D. Thesis*, Guangdong University of Technology (2021).
58. Zhao L, Li W, Li Y. Research on co-design of dual security control and communication for nonlinear CPS with actuator fault and FDI attacks. Meas. *Control.* 55(7–8) (2022): 767–782.
59. Hao J, Piechocki RJ, Kaleshi D, Chin WH, Fan Z. Sparse malicious false data injection attacks and defense mechanisms in smart grids. *IEEE Trans. Ind. Inform.* 11(5) (2015): 1–12.

Section III

Smart CPS Development for Biomedical and Healthcare Using Advanced AI

7 Smart AI-Based Biomedical Systems and Tools

7.1 INTRODUCTION: ARTIFICIAL INTELLIGENCE IN HEALTHCARE

Artificial intelligence (AI) is the simulation of human intelligence in machines, enabling them to perform tasks that typically require human cognition, such as learning from experience, problem-solving, and decision-making. Key techniques within AI include machine learning (ML), deep learning (DL), and neural networks (NN). These approaches allow systems to analyse large datasets, identify patterns, and generate insights with minimal human intervention.

John McCarthy coined the term "artificial intelligence" in 1956 during a workshop at Dartmouth College. Although progress in the field was initially slow, AI gained momentum in the late 1990s and early 2000s with logistics, data mining, and medical diagnosis applications. Over the past decade, AI development has accelerated significantly, fuelled by advancements in computational power and the availability of extensive datasets. Domains such as astronomy and meteorology have been early beneficiaries of AI's potential, leveraging vast amounts of structured and unstructured data.

Biomedicine has emerged as a particularly promising field for AI applications, owing to the wealth of "big data" projects and technological advancements. Clinical data from electronic health record (EHR) systems, genomics data from whole genome sequencing (WGS) studies, and digital imaging data from modalities like magnetic resonance imaging (MRI), ultrasound, and digital pathology systems represent just a few examples. AI thrives on diverse datasets, especially those that are both "wide" (many parameters) and "long" (large sample sizes), which help reduce issues like overfitting and improve algorithm robustness.

The medical applications of AI are vast and transformative. AI is now being utilized for diagnosis, prognosis, personalized treatment planning, surgical assistance, drug discovery, and other critical areas. For example, in 2023, a DL algorithm facilitated the discovery of a novel antibiotic called halicin. During the COVID-19 pandemic, AI-powered robots and telemedicine platforms enhanced safety for healthcare workers. In mental health, therapeutic chatbots have provided accessible support to individuals.

Major players in both academia and industry are investing heavily in AI-driven healthcare innovations. Renowned institutions like the Mayo Clinic and Cleveland

DOI: 10.1201/9781003489986-10

Clinic integrate data science and AI into their medical practices. Leading health-tech companies, including Philips, Siemens, and GE, are incorporating AI into imaging systems and other healthcare solutions. Biotech companies like BERG and pharmaceutical giants like Takeda are leveraging AI for biomarker discovery and drug development pipelines.

The involvement of "Big Tech" companies – Google, Apple, Facebook, Amazon, and Microsoft (collectively known as GAFAM) – is reshaping the healthcare landscape. These companies have developed AI-enabled products and services such as wearable health trackers, cloud-based healthcare data platforms, and tools for preventive health measures. For instance, Google's acquisition of Fitbit emphasizes its focus on health and lifestyle monitoring, while Facebook's Preventive Health tool connects users with resources and health checkup recommendations.

The integration of AI in healthcare promises unprecedented advancements in medical science, improving outcomes and efficiency across diagnostic, therapeutic, and operational domains. This chapter explains how AI is revolutionizing medicine, offering insights into its current applications and future potential.

7.2 RECONFIGURATION OF BIOMEDICAL SYSTEMS

7.2.1 Evolution of Biomedical Knowledge Systems

Digital technologies like AI, ML, and computational tools have driven significant changes in biomedical knowledge systems globally for decades. These transformations, studied by humanities and social sciences scholars, highlight shifts in how science, technology, and innovation (ST&I) are integrated into society.

7.2.2 Post-War Expansion of ST&I Policies

Since the post-war period of the 20th century, ST&I policies have focused on advancing the biotechnology and healthcare sectors. Initially centred in the United States and Europe, these efforts expanded in the 1990s and early 2000s to include regions like Southeast Asia. This decentralization of research and development (R&D) investments fostered the growth of local biotechnology industries, enabling emerging economies to build knowledge-based competencies.

7.2.3 Technological Advancements in the 2000s

The 2000s marked a surge in computational platforms for scientific research, leading to breakthroughs in genomics, biological engineering, and bio nanotechnology. Countries like China, South Korea, Singapore, and India became key players, supported by state-funded S&T policies and private R&D investments. These nations advanced biotechnology-related industries, including precision medicine and biopharmaceutical manufacturing.

7.2.4 Data-Driven Research in the 2010s

From the mid-2010s, biomedical research experienced another transformation driven by advancements in AI, big data analytics, and computing. This era saw a shift from labour-intensive R&D to automated systems that combined human expertise with machine-driven processes. Fields like bioengineering, systems biology, and digital health adopted AI to enhance hypothesis testing and accelerate scientific discovery.

7.2.5 Global Disparities in AI-Driven Research

Despite these advancements, the capabilities to develop complex AI-driven systems remain concentrated in a few global locations. Scholars emphasize the role of location-specific factors in fostering innovations, particularly in high-tech fields like healthcare, biotechnology, and information technology.

7.2.6 New Possibilities in Biomedical Research

AI's integration into scientific labs has revolutionized how biomedical knowledge is produced and applied. It influences problem design, hypothesis testing, and innovation in areas such as cancer vaccines, bioengineered devices, synthetic nanoparticle research, and molecular diagnostics. Today, multidisciplinary teams work in hybrid (physical–digital) environments, leveraging collaboration between scientists, engineers, automated systems, and computers to tackle complex research challenges.

7.3 ARTIFICIAL INTELLIGENCE IN BIOMEDICAL RESEARCH AND DEVELOPMENT

7.3.1 The Origins of AI in Biomedical Research

The integration of AI into biomedical research began in 1965 with the DENDRAL Project, a pioneering computer programme developed by scientists at Stanford University to identify chemical compounds. This marked the first step towards automating chemical experiments using AI, setting the stage for decades of innovation in laboratory processes.

7.3.2 Advancements in AI and Automation

Over time, the combination of ML, lab automation, and robotics has established data-intensive platforms as essential tools for discovering new compounds and materials. Advanced experimental systems, such as self-driving labs (SDLs) and materials acceleration platforms (MAPs), represent mature outcomes of this technological evolution. These systems are capable of screening thousands of chemical combinations using minimal starting reagents, identifying compounds with remarkable precision.

7.3.3 Impact on Biomedical Innovation

The introduction of SDLs and MAPs has significantly enhanced productivity and efficiency in biomedical research. These technologies enable rapid exploration of chemical and nanomaterial systems, allowing scientists to address complex biological challenges more effectively. Key areas of impact include

Drug Discovery Acceleration: AI reduces the time required to identify and develop potential therapeutic compounds.
New Materials Discovery: AI-driven systems streamline the identification of novel biomaterials for medical applications.
Nanomedicine: Advances in nanotechnology benefit from AI tools for designing and testing nanoparticles with clinical and therapeutic applications.

7.3.4 Convergence of Domains

A notable example of domain convergence is the use of AI in discovering nanomaterials for clinical purposes, developing AI-driven drug discovery tools, and leveraging generative AI to advance cancer research. This interdisciplinary approach exemplifies the transformative potential of AI in healthcare, addressing critical challenges in treatment development and precision medicine.

7.4 NANOMEDICINE

7.4.1 Challenges in Nanomedicine Development

The inherent complexity of nanomedicines has long posed challenges to their clinical success. However, recent advances in automated experimental systems have provided groundbreaking solutions. By employing AI, researchers can now design nanoparticles with precise properties, optimize drug delivery systems, and predict toxicity, significantly minimizing reliance on traditional trial-and-error methods. Automation has enabled the rapid synthesis and characterization of nanomaterials, accelerating the development of innovative drug carriers, imaging agents, and therapeutics.

7.4.2 Role of SDLs and MAPs in Nanomedicine

SDLs and MAPs have revolutionized the discovery and optimization of nanoscale materials for medical use. Utilizing high-throughput screening techniques and advanced data analytics, these platforms can simultaneously evaluate the properties and performance of thousands of materials. Research by Anselmo and Mitragrotri demonstrates significant advancements in nanoparticle studies over the past five years, highlighting the transformative role of AE systems in accelerating clinical trials of nanocarriers and therapeutic compounds.

7.4.3 Advancing Personalized Nanomedicine

The integration of AE technologies has brought the concept of personalized nanomedicine closer to reality. Personalized approaches aim to enhance treatment efficacy and minimize side-effects. Innovations, such as the NanoMAP system, address critical challenges in nanomedicine, including the stabilization of synthetic processes and the reproducibility of nanoscale experiments.

7.4.4 The Future of Nanomedicine

Automated experimental systems have propelled nanomedicine to the forefront of medical innovation. Researchers can now design, synthesize, and test nanomaterials with unprecedented speed and precision. These advancements promise more effective, personalized treatments, driving clinical innovation and delivering significant benefits to patients.

7.5 AI-ASSISTED DRUG DISCOVERY

7.5.1 Transforming Drug Discovery with AI

AI has revolutionized drug discovery by enabling the exploration of vast chemical spaces, leading to the identification of novel drug candidates. Some of these candidates have already progressed to clinical trials, marking a significant breakthrough in the pharmaceutical industry. The ability to efficiently pinpoint promising compounds represents a paradigm shift in R&D processes.

7.5.2 Accelerating Timelines and Reducing Costs

A recent report titled "AI-generated drugs will be available sooner than you think" highlighted the transformative impact of AI in medicine, particularly through automated experimental (AE) platforms. Historically, drug discovery was a lengthy and costly endeavour, often requiring over 12 years of effort with a failure rate exceeding 90% in clinical trials. Advances in AI and AE technologies have dramatically reduced timelines and costs, enhancing efficiency and accuracy. These developments are helping to bridge the so-called valley of death that often hinders preclinical and clinical innovation.

7.5.3 Integration of SDLs and MAPs

SDLs and MAPs have been at the forefront of AI-driven drug discovery. SDLs leverage robotics and high-throughput screening techniques to automate labour-intensive tasks, allowing researchers to concentrate on innovative and strategic aspects of drug development. Meanwhile, MAPs are instrumental in creating novel drug delivery systems and biomaterials, further advancing the field.

7.5.4 Precision and Patient Outcomes

AI-enabled drug discovery platforms offer the ability to precisely target diseased tissues, optimize drug release rates, and reduce side-effects. These capabilities improve patient outcomes and create opportunities for tailored treatments, particularly in areas like precision oncology.

7.5.5 Collaboration and Accessibility

The adoption of AI-assisted drug discovery has been bolstered by collaborations between pharmaceutical companies, AI start-ups, and academic institutions. These partnerships have lowered barriers to entry, enabling smaller companies and research groups to access cutting-edge technologies. This democratization of tools and resources is paving the way for more inclusive and widespread innovation in drug discovery and development.

7.6 PRECISION ONCOLOGY

7.6.1 AI in Cancer Detection and Treatment

Advancements in AI have revolutionized cancer treatment by improving drug delivery systems, detection methods, and therapeutic approaches. The integration of automated experimental (AE) systems with AI tools is enabling researchers to develop highly precise and personalized cancer therapies. AI can analyse vast datasets and design drug delivery systems tailored to individual patient profiles, offering treatments that are both effective as well as minimize side-effects.

7.6.2 Biomedical Engineering and Drug Delivery Systems

The convergence of biomedical engineering and biotechnology has resulted in innovative tools like microfluidics and lab-on-a-chip technologies. These systems mimic the biological environments of tumours, providing more accurate platforms for testing new drugs and materials. Microfluidic devices, for instance, allow precise manipulation of small fluid volumes, enabling researchers to quickly design and evaluate novel drug delivery systems. This approach reduces the reliance on animal models in preclinical trials and accelerates the discovery of effective cancer therapies.

7.6.3 Optimizing Drug Delivery with AE Systems

AE systems play a critical role in screening and optimizing materials for cancer treatment. They evaluate properties such as biocompatibility, drug release rates, and targeting precision, significantly reducing the time and cost of developing new therapies. These advancements help bring promising drug delivery systems to market faster, benefiting patients more quickly.

7.6.4 The Role of 3D Printing in Cancer Therapy

Three-dimensional (3D) printing has emerged as a powerful tool in creating highly customizable drug delivery vehicles, such as nanoparticles and liposomes. With precise control over size, shape, and surface properties, researchers can develop carriers that effectively target cancer cells while sparing healthy tissue. This level of precision is essential for enhancing treatment efficacy and minimizing side-effects.

7.6.5 Synergy of Precision Oncology Tools

AI-driven labs, microfluidics, 3D printing, and nanocarrier engineering are converging to advance cancer treatment. These tools work together to develop next-generation therapies tailored to the complex and heterogeneous nature of cancer. As precision oncology and AE systems continue to evolve, they offer hope for more targeted, less invasive treatments, improving outcomes for cancer patients worldwide.

7.7 MEDICAL IMAGE ANALYSIS FOR DISEASE DETECTION AND DIAGNOSIS

Medical image analysis is transforming healthcare by improving disease detection, diagnosis, and monitoring. Using AI and advanced computational techniques, medical professionals can now extract critical insights from imaging modalities like X-rays, MRIs, CT scans, and retinal images.

7.7.1 AI in Retinal Disease Detection

AI systems are significantly advancing the early detection and diagnosis of retinal diseases such as diabetic retinopathy (DR), retinopathy of prematurity (RoP), and age-related macular degeneration (AMD). By analysing images from optical coherence tomography (OCT) and fundus photography, AI can provide

- Early screening and grading of retinal diseases.
- Integration of multiple image sources for more accurate diagnoses.

While AI achieves high sensitivity and specificity, challenges remain, including data availability and the need for explainable models.

7.7.2 Brain Imaging and AI Applications

Brain imaging techniques like MRI, fMRI, PET, and EEG help diagnose neurological disorders. AI enhances this field by:

- Detecting and classifying brain tumours using DL models such as convolutional neural networks (CNNs).

- Optimizing AI model performance with advanced algorithms like sine–cosine fitness optimization.

AI-based approaches improve accuracy in identifying brain abnormalities while addressing challenges like data complexity and variability.

7.7.3 AI in Lung Imaging

Lung imaging is critical for detecting diseases like lung cancer and COVID-19. AI contributes by:

- Segmenting lung fields in chest X-rays using encoder–decoder architectures and DL techniques.
- Improving the analysis of lung CT scans with methods like synthetic data generation to enhance model training.

AI ensures earlier and more accurate disease detection, which is vital for improving patient outcomes.

7.7.4 Breast Cancer Imaging and Diagnosis

AI technologies are transforming breast cancer imaging by:

- Enhancing computer-aided detection (CAD) systems to differentiate between benign and malignant lesions using DL and radiomics.
- Improving model performance on small datasets through techniques like transfer learning and data augmentation.

AI also supports risk assessment, treatment planning, and prognosis prediction, ensuring more personalized care.

7.7.5 Liver Cancer Detection

AI aids in detecting liver cancer by automating tasks such as:

- Segmenting liver lesions using NN like ResNet and DenseNet.
- Quantifying tumour parameters such as volume and burden, speeding up image analysis and diagnosis.

AI-based methods provide accurate and efficient solutions for early cancer detection, crucial for better outcomes.

7.7.6 Diabetic Retinopathy (DR) Screening

AI is highly effective in detecting DR, a major cause of blindness worldwide. Benefits include

- High sensitivity and specificity in screening compared with human graders.
- Efficient analysis of retinal images, reducing the need for manual interpretation.
- AI systems enhance early diagnosis and reduce the burden on healthcare professionals.

7.7.7 AI in Age-Related Macular Degeneration (AMD)

AI tools assist in identifying and predicting the progression of AMD. Applications include

- Screening for early-stage AMD using retinal images.
- Providing detailed analyses to predict late-stage AMD development.

AI improves accuracy and supports timely intervention.

7.7.8 Optical Coherence Tomography (OCT) in Eye Diseases

AI-powered OCT is valuable in diagnosing and monitoring eye conditions like glaucoma, macular oedema, and AMD. Benefits include

- Improved sensitivity and specificity compared with traditional methods.
- Applications in telemedicine platforms to identify urgent and routine cases efficiently.

AI-assisted OCT enhances diagnosis and helps track disease progression.

AI is revolutionizing medical image analysis across various fields, improving early detection, diagnosis, and treatment planning. From brain tumours and breast cancer to retinal diseases and liver cancer, AI offers faster, more precise, and personalized solutions while addressing challenges such as data limitations and model interpretability. These advancements are shaping the future of healthcare, ensuring better outcomes for patients.

7.8 IMAGING AND MODELLING TECHNIQUES FOR SURGICAL PLANNING AND INTERVENTION

Advancements in imaging, 3D modelling, and AI are transforming surgical planning and intervention. These technologies enhance the understanding of complex anatomy, improve procedural accuracy, and support personalized treatments, ultimately benefiting both patients and healthcare providers.

7.8.1 AI-Enhanced Imaging for Surgical Planning

AI-powered algorithms process imaging data like CT and MRI scans to create precise, three-dimensional anatomical models. These advancements

- Enhance the visualization of patient-specific anatomy.
- Aid in planning complex surgeries with greater accuracy.
- Provide real-time insights during procedures.

For example, AI techniques such as the YOLOv7 algorithm and U-Net have been used to segment echocardiographic images into structures like LVendo, LVepi, and LA, assisting surgeons in understanding cardiac anatomy.

7.8.2 Printing for Surgical Applications

Combining 3D printing with imaging and modelling further revolutionizes surgical planning by:

- Creating physical anatomical models for surgeons to study before procedures.
- Enabling customization of surgical guides and implants tailored to patient-specific needs.

7.8.3 Surgical Guides for Precision Procedures

3D-printed surgical guides enhance precision in procedures like coronary artery bypass grafting (CABG). Key features include

- Custom designs based on patient anatomy.
- Biocompatible and sterile materials to meet surgical requirements.

7.8.4 Benefits of AI-Driven 3D Modelling and Printing

The integration of AI and 3D technologies offers numerous advantages, such as:

- Personalized surgical interventions.
- Reduced operative times and improved safety.
- Refined surgical strategies through detailed anatomical insights.

These innovations pave the way for a future of enhanced precision, efficiency, and customization in surgical practice, empowering healthcare professionals to deliver better patient care.

7.8.5 Image and Model Enhancement for Improved Analysis

Advancements in AI and ML have introduced transformative approaches to improving the analysis and interpretation of medical images. These techniques not only enhance decision-making and diagnostic accuracy but also provide more interpretable and visually enriched insights for clinical applications.

7.8.6 AI-Based Image Colorization for Enhanced Visualization

Adding chromatic information to medical images through AI-driven colorization techniques has shown promise in enhancing interpretation and analysis.

Key Technique:

Spatial Mask-Guided Colorization with GANs (SMCGAN)

- **Advantages:** Preserves structural details and improves visualization in targeted areas.
- **Performance:** Achieved an 8.48% improvement in the peak signal-to-noise ratio (PSNR) compared with other state-of-the-art GAN-based approaches using datasets like the Visible Human Project and NCI-ISBI prostate dataset.

Explainable AI (XAI) for Transparency in Decision Support
Explainable AI ensures transparency in complex models by making predictions interpretable.

- **NDG-CAM:** Combines semantic segmentation networks and Grad-CAM for nuclei detection in histopathology images.
- **Advantages:** Provides visual cues highlighting regions influencing predictions, enhancing both accuracy and interpretability.
- **Applications:** Improves disease diagnosis in digital pathology by segmenting nuclei regions effectively.

Open-Source COVID-19 CT Dataset
Includes automatic lung tissue classification for radiomics analysis, aiding in COVID-19 diagnosis and treatment planning.

- **Features:** Annotated regions like ground-glass opacities, consolidations, and normal lung tissue for improved insights.
- **Super-Resolution Techniques for Image Enhancement**

Super-resolution techniques focus on enhancing the resolution and quality of medical images:

GAN-Based Models:
Improved methods like squeeze-and-excitation blocks and enhanced deep super-resolution (EDSR) models recover fine details and improve visual quality.

- **Evaluation:** Compared against SRGAN, VDSR, and other state-of-the-art methods, achieving superior detail preservation.

Vision Transformers in Medical Imaging
Vision transformers offer new paradigms for handling medical images:

Applications in Super-Resolution:
Zhu et al. introduced a vision transformer with residual dense connections for single-image super-resolution.

Highlights: Utilizes local feature fusion and perceptual loss functions to enhance edges, textures, and organ details.

Applications in Segmentation:

Adaptation of SWIN transformers for medical image segmentation.

Architecture: A U-net-like design with hierarchical encoding and decoding processes.

Results: Demonstrates state-of-the-art segmentation performance on datasets like BraTS, LiTS, and KiTS.

7.8.7 Perceptual Loss Functions for Generative Models

Perceptual loss functions, derived from pretrained models, improve semantic and visual quality:

Focus: Optimize structural and high-level feature preservation while avoiding artifacts.

Applications: Generate high-quality reconstructions for super-resolution tasks.

7.8.8 Noise and Artifact Reduction in Medical Imaging

Addressing inherent challenges in medical imaging, such as noise and artifacts, remains a crucial research focus:

Goal: Develop robust methods to extract fine details beyond the limitations of imaging devices.

Potential: Unlock previously unattainable resolution and clarity for diagnosis and treatment.

These advancements collectively redefine the capabilities of medical image analysis, fostering more accurate diagnoses, improved visualization, and interpretable results, ultimately enhancing clinical workflows and patient care.

7.9 AI IN SURGICAL SETTINGS: TRANSFORMING THE OPERATING ROOM

In surgical environments, AI is significantly enhancing the capabilities of surgeons and improving surgical outcomes. By integrating AI into various aspects of surgery, from preoperative planning to postoperative recovery, AI provides valuable support that enables more precise, efficient, and personalized care. Here are the keyways AI is making an impact in surgical settings:

7.9.1 Preoperative Planning

AI plays a critical role in preoperative planning by analysing patient data, including medical imaging, to create highly detailed 3D models of a patient's anatomy. These models allow surgeons to plan the surgery with unmatched precision, identify potential challenges, and select the most optimal surgical approach.

Key Features:

- Analyses medical imaging data (e.g., CT scans, MRIs).
- Creates accurate 3D anatomical models of the patient.
- Helps surgeons identify key structures and potential complications ahead of time.
- Assists in determining the most effective surgical route or technique.

Impact:

- Reduces risks by providing a clear and detailed visualization of the surgical area.
- Enhances the precision of surgical interventions and improves outcomes.

7.9.2 Robot-Assisted Surgery

One of the most prominent uses of AI in surgery is robot-assisted surgery. These systems, such as the da Vinci Surgical System, integrate AI capabilities to assist surgeons during minimally invasive procedures, offering enhanced precision and control.

Key Features:

- Filters out hand tremors and refines the surgeon's movements.
- Allows for highly precise microsurgeries with small incisions.
- Provides real-time visual feedback with high-definition imaging.
- AI algorithms assist with decision-making and optimize movements during surgery.

Impact:

- Minimizes human error, ensuring greater accuracy.
- Reduces recovery time and surgical trauma for patients by enabling minimally invasive procedures.
- Expands the ability of surgeons to perform complex procedures with greater ease.

7.9.3 Image Recognition and Computer Vision

AI-powered image recognition and computer vision tools are increasingly being used in surgery to analyse medical images, such as X-rays, MRIs, and CT scans, to guide surgeons during procedures. These tools can highlight critical structures, such as blood vessels, tumours, or organs, ensuring that surgeons avoid damage to these vital areas.

Key Features:

- Automatically analyses images in real time to identify critical structures.
- Flags potential risks, such as tumours, abnormal growths, or bleeding vessels.
- Assists surgeons in navigating the surgical site with heightened awareness.

Impact:

- Reduces the likelihood of errors or missed abnormalities during surgery.
- Improves surgical outcomes by providing better visual guidance.
- Enhances safety by minimizing unintentional damage to surrounding tissues.

7.9.4 Intraoperative Decision Support

AI also plays a crucial role during surgery by providing real-time decision support. By analysing data from sensors and surgical instruments, AI systems offer recommendations based on the ongoing procedure, such as suggesting optimal incision sites or detecting complications as they arise.

Key Features:

- Monitors ongoing surgery through sensors and instruments.
- Provides real-time guidance and suggests corrections or improvements.
- Alerts surgeons to potential issues such as bleeding or equipment malfunctions.
- Enhances team collaboration by offering data-driven insights during the procedure.

Impact:

- Enhances decision-making during surgery, ensuring more accurate interventions.
- Reduces the risk of complications by identifying issues early.
- Supports the surgical team in making informed, evidence-based decisions during the procedure.

7.9.5 Postoperative Recovery and Monitoring

Post-surgery, AI continues to offer valuable assistance by monitoring patient recovery. It tracks vital signs, analyses data, and provides alerts if any complications arise, ensuring that the recovery process is as smooth and personalized as possible.

Key Features:

- Continuously monitors patient vitals, such as heart rate, blood pressure, and oxygen levels.
- Detects signs of postoperative complications, like infections or blood clots.

- Helps adjust recovery plans in real time based on the patient's specific needs and recovery progress.

Impact:

- Speeds up recovery by tailoring care to the individual patient.
- Detects potential issues early, reducing the risk of postoperative complications.
- Ensures ongoing care optimization to improve overall recovery outcomes.

AI in surgical settings is significantly transforming the landscape of surgery by providing enhanced precision, decision support, and personalized care (Figure 7.1). From preoperative planning to postoperative monitoring, AI helps ensure safer surgeries, better patient outcomes, and more efficient use of surgical resources. As AI technology continues to evolve, its role in surgery will expand, offering even more opportunities to improve healthcare delivery and patient care across the globe.

7.10 AI-INTEGRATED BIOMEDICAL TOOLS

In the realm of healthcare, AI is playing a transformative role in enhancing diagnostic accuracy, improving efficiency, and driving better patient outcomes. These AI-powered tools leverage advanced ML algorithms to assist healthcare professionals in detecting and diagnosing various medical conditions faster and more accurately than ever before. Mentioned further are five of the most prominent AI tools that are revolutionizing medical diagnosis across various specialties:

7.10.1 IBM Watson for Oncology

IBM Watson for Oncology (Figure 10.1) is a cutting-edge cognitive computing platform designed to assist oncologists in providing personalized treatment options for cancer patients. It utilizes a vast repository of medical records, research literature, and clinical guidelines to generate evidence-based treatment recommendations.

Key Features:

- Analyses patient medical history, tumour characteristics, and the latest research.
- Provides personalized, data-driven treatment options.
- Empowers oncologists with evidence-backed recommendations and confidence levels.

Challenges:

- Requires further validation of its algorithms.
- Faces concerns over biases in recommendations.
- Needs careful integration into clinical workflows, especially regarding data privacy and ethics.

7.10.2 ENDEX by Enlitic

ENDEX (Figure 10.2) is an AI-driven platform designed for medical image analysis, utilizing advanced DL algorithms to process and analyse X-rays, CT scans, MRIs, and ultrasounds with exceptional speed and accuracy. This tool aids healthcare-providers by offering valuable insights for earlier diagnosis and more effective treatment planning.

Key Features:

- AI models tailored to detect abnormalities in organs and tissues.
- Automates routine image analysis tasks to enhance consistency.
- Provides objective assessments for improved diagnostic accuracy.

User-Friendly Interface:

- Easily integrates into existing medical workflows.
- Allows healthcare professionals to upload images, receive AI-generated reports, and collaborate effectively.

7.10.3 IDx-DR

IDx-DR (Figure 10.3) is an FDA-approved autonomous AI system designed for diagnosing diabetic retinopathy (DR), a leading cause of blindness. By analysing retinal images captured by a fundus camera, IDx-DR detects early signs of DR, including microaneurysms, haemorrhages, and exudates.

Key Features:

- Operates autonomously, requiring no human interpretation.
- Provides fast and reliable diagnoses, even in underserved areas with limited access to specialists.
- Has demonstrated high sensitivity and specificity in detecting diabetic retinopathy.

Impact:

- Expands access to screening, enabling earlier detection and timely intervention to prevent vision loss.

7.10.4 Zebra Medical Vision's AI Solutions

Zebra Medical Vision is at the forefront of AI-driven medical imaging and diagnosis. The company offers a range of AI solutions across multiple medical specialties, including radiology, cardiology, and orthopaedics. Their AI tools are designed to analyse medical images such as X-rays, CT scans, and MRIs to detect a wide array of abnormalities.

Key Features:

- Offers solutions for detecting fractures, cardiovascular diseases, liver conditions, and breast cancer.
- Uses DL algorithms to provide highly accurate diagnostic insights.
- Validated through rigorous studies to ensure reliability in clinical settings.

Impact:

- Enhances early detection, enabling timely interventions that lead to better patient outcomes.

7.10.5 Arterys Cardio AI (Tempus Pixel Cardio)

Arterys Cardio AI is a pioneering platform that revolutionizes cardiovascular imaging. This AI-powered tool automates the analysis of cardiac MRI images, providing clinicians with precise, reproducible measurements for various cardiac parameters, such as cardiac function, blood flow, and tissue characterization.

Key Features:

- Uses DL algorithms to automate cardiac MRI analysis.
- Offers detailed insights into cardiac structure and function, even in complex cases.
- Facilitates real-time collaboration among healthcare professionals through its cloud-based platform.

Impact:

- Delivers accurate, objective assessments of cardiac health, improving diagnosis and treatment of cardiovascular conditions.

7.10.6 FDA Approved, AI-Powered Ultrasound Software

In a significant advancement for the medical field, the U.S. Food and Drug Administration (FDA) has approved the first AI-powered software designed to assist and guide medical professionals through the steps of a cardiac ultrasound exam, also known as an echocardiogram. This diagnostic tool is one of the most common methods used to detect and assess heart disease, a leading cause of death worldwide.

a. **AI as a Real-Time Co-Pilot during Echocardiograms**
   Developed by Caption Health, a San Francisco-based imaging company previously known as Bay Labs, the AI software acts as a real-time co-pilot during the echocardiogram process. It provides guidance on how to properly manoeuver the ultrasound probe and offers immediate feedback on the quality of the images being captured.

b. **Expanding Access to Cardiac Ultrasounds**
The FDA's approval of the AI software represents a major milestone in the field of medical diagnostics. It allows medical professionals who may not have specialized training in ultrasonography – such as registered nurses in family care clinics – to perform and interpret cardiac ultrasounds effectively.
c. **Enhancing Cardiac Care through AI**
According to Robert Ochs, the Deputy Director of the FDA's Office of In Vitro Diagnostics and Radiological Health, this marketing authorization will enable healthcare professionals with limited experience in performing ultrasounds to use this AI tool with confidence. This will enhance the accessibility and efficiency of cardiac care, ensuring that more patients receive timely and accurate assessments of heart health.
d. **Improving Diagnostic Accuracy in Primary Care Settings**
The development of this AI software is expected to have significant implications, particularly in primary care settings and rural areas where there may be a shortage of specialized cardiac technicians. The AI tool offers a streamlined diagnostic process, reducing wait times and improving patient outcomes through early detection of heart conditions.

The FDA's approval of the AI-powered cardiac ultrasound software is a breakthrough in integrating AI technology into medical diagnostics. This tool not only makes cardiac ultrasound exams more accessible for healthcare professionals with limited expertise but also improves patient care through early and accurate assessments of heart health.

7.11 REAL-TIME WEARABLE DEVICES: REVOLUTIONIZING BIOMEDICAL TOOLS

In recent years, wearable devices have become increasingly significant in healthcare, offering advanced solutions for monitoring vital signs and improving overall health outcomes. These innovative tools are reshaping healthcare delivery, providing greater accessibility, convenience, and proactive care for users. This chapter explores eight leading wearable devices in healthcare, highlighting their features and benefits.

1. **OrCam MyEye: Assistive Technology for the Visually Impaired**
Description: OrCam MyEye is a wearable device designed for individuals with visual impairments. Using AI and computer vision, it allows users to read printed text, recognize faces, and identify objects in real time.
Benefits:
   - Provides real-time assistance for reading and object recognition.
   - Enhances the independence and quality of life for visually impaired users.
   - Utilizes AI and computer vision technology to deliver real-time support.
2. **Owlet Smart Sock: Monitoring Infant Health**
Description: Owlet Smart Sock is a baby monitor that tracks an infant's heart rate and oxygen levels while they sleep. It offers real-time alerts to parents, ensuring optimal care for newborns by detecting potential health issues early.

Benefits:

- Monitors heart rate and oxygen levels during sleep.
- Sends real-time alerts to parents for peace of mind.
- Helps in early detection of potential health issues, ensuring timely intervention.

3. **Omron Heart Guide: Wearable Blood Pressure Monitor**
   Description: Omron HeartGuide is the world's first wearable blood pressure monitor in the form of a wristwatch. It enables users to monitor their blood pressure on the go, promoting proactive heart health management.
   Benefits:
   - Offers continuous blood pressure monitoring.
   - Compact and convenient wristwatch design.
   - Aids in the management of heart health, reducing the risk of cardiovascular diseases.
4. **TempTraq Wearable Temperature Monitor: Continuous Temperature Monitoring**
   Description: TempTraq is a discreet, wearable patch designed to monitor body temperature continuously. Ideal for children and patients with fever-related conditions, it provides caregivers with real-time alerts for timely intervention.
   Benefits:
   - Continuous body temperature monitoring for early detection of fever.
   - Discreet and comfortable patch design for ease of use.
   - Provides real-time alerts, helping caregivers respond quickly to changes in temperature.
5. **Lief Therapeutic Wearable: Managing Stress and Anxiety**
   Description: Lief Therapeutic Wearable uses biofeedback technology to help users manage stress and anxiety. By monitoring heart rate variability, the device guides users in regulating their breathing to promote mindfulness and emotional well-being.
   Benefits:
   - Monitors heart rate variability to assess stress levels.
   - Guides users in breathing regulation for improved emotional health.
   - Promotes mindfulness and helps manage anxiety, contributing to overall well-being.
6. **Oura Ring: Sleep and Activity Tracker**
   Description: Oura Ring is a sleek wearable that tracks sleep patterns, physical activity, and physiological signals, such as heart rate and body temperature. It provides personalized insights to optimize lifestyle choices and improve sleep habits.
   Benefits:
   - Monitors sleep patterns and daily activity levels.
   - Tracks heart rate and body temperature for a comprehensive health profile.
   - Provides personalized health insights to improve overall wellness.
7. **Bloomlife Smart Pregnancy Tracker: Monitoring Pregnancy Health**
   Description: Bloomlife Smart Pregnancy Tracker is designed specifically for expectant mothers. It tracks uterine activity and provides insights into

foetal movements, enabling mothers to monitor labour progression and assess pregnancy health.

Benefits:

- Monitors uterine activity and contractions to track labour progress.
- Offers insights into foetal movements for better understanding of pregnancy health.
- Enhances pregnancy monitoring and ensures a safe delivery process.

8. **Epson Moverio AR Smart Glasses: Enhancing Medical Procedures with Augmented Reality**

   Description: Epson Moverio AR Smart Glasses use augmented reality to overlay digital information onto the user's field of view. These glasses are designed to assist medical professionals by providing patient data and medical imaging in real time, improving procedure efficiency.

   Benefits:

   - Overlays digital information on the user's field of vision for improved decision-making.
   - Helps medical professionals access patient data and visualize medical imaging hands-free.
   - Enhances the efficiency and accuracy of healthcare procedures, contributing to better patient outcomes.

7.11.1 The Future of Wearable Health Tech

Wearable health devices are transforming healthcare by improving accessibility, enabling early detection, and providing personalized care. With advanced features like real-time monitoring and data analytics, these devices are paving the way for more proactive and efficient health management. As wearable technology continues to evolve, it promises even more innovative solutions to enhance the health and well-being of users around the world.

7.12 ARTIFICIAL INTELLIGENCE TO ADDRESS GLOBAL HEALTH ISSUES

AI holds immense potential for transforming healthcare, especially in emerging economies where health systems face unique challenges. These economies are often at the forefront of tackling neglected diseases, epidemics, and chronic health issues, and AI can play a pivotal role in addressing these concerns.

7.12.1 AI in Drug Discovery and Healthcare for Emerging Economies

Traditional drug discovery systems have been slow to address many global health challenges. In emerging economies, AI can provide a new paradigm for tackling health crises.

a. **Emerging Economies as Strategic Partners:**

   Countries such as India, Brazil, Taiwan, South Korea, and Indonesia possess substantial capabilities in vaccine R&D, public health, and planning. These nations are ideal partners in international AI consortia for areas like vaccine development for neglected diseases, precision oncology, and molecular diagnostics.

b. **Global Health Impact:**
AI's application in vaccine development and diagnostics can be especially important in addressing the social impacts of diseases like tropical infections and epidemics. AI-enabled tools can streamline the drug discovery process, identify biomarkers for diseases, and provide personalized treatments that address local health conditions.

c. **Science and Innovation Diplomacy (S&ID):**
The growing field of S&ID has been instrumental in fostering international collaborations that bridge the gap between science, technology, and foreign policy.

d. **S&ID and Emerging Economies:**
S&ID empowers emerging economies to engage in global scientific collaborations and policy-making, allowing them to influence international health innovation agendas.

e. **Institutional Support:**
Organizations like the Pan American Health Organization (PAHO), Global Alliance for Vaccine and Immunization (GAVI), and the Organization of American States' Inter-American Committee on Science and Technology (COMCyT) have facilitated such collaborations.

f. **Diplomatic Approaches to Health Innovation:**
S&ID can help harmonize the interests of various stakeholders – scientists, policymakers, and international organizations – towards advancing AI-driven health innovations for global health challenges.

7.13 ETHICAL AND REGULATORY LANDSCAPE FOR AI IN HEALTH

As AI technologies advance, emerging economies face significant challenges in developing ethical and regulatory frameworks that ensure responsible and inclusive use of AI in healthcare.

a. **Co-Producing Ethical Guidelines and Regulations:**
A collaborative approach to creating ethical guidelines for AI in biomedicine is crucial. This process can help ensure that AI interventions in healthcare are responsible, sustainable, and inclusive.

b. **Addressing Uncertainty:**
AI in healthcare – such as AI-assisted drug discovery and nanomedicine – can generate uncertainty about its benefits. Co-production of ethical and regulatory standards ensures that these technologies are deployed with the public's best interests in mind.

c. **Challenges in Governance:**
The lack of resources and knowledge infrastructures in many emerging economies can hinder their ability to fully engage with cutting-edge biotechnologies. Effective governance requires international collaboration to address these gaps.

d. **Diversity, Equity, Inclusion, and Trustworthiness (DEIT):**
AI in healthcare must prioritize diversity, equity, inclusion, and trustworthiness (DEIT) to ensure the technology benefits all global populations.
e. **Diversity in Research:**
Emerging economies bring invaluable expertise and insights based on their unique social and cultural contexts. Their inclusion can lead to more comprehensive research outcomes.
f. **Global Research Collaboration:**
Active participation from emerging economies in AI development is essential for global healthcare innovation. This includes capacity building, technology transfer, and equitable access to AI-driven tools.
g. **Trust in AI for Healthcare:**
For AI to be effectively integrated into healthcare systems, trust must be built among all stakeholders. Ethical and responsible use of AI technologies is key to fostering public trust, particularly in regions where there is historical scepticism about new biotechnologies.
h. **Strategic Advantage of DEIT:**
Promoting DEIT is not just an ethical imperative; it's a strategic advantage. Collaboration through international partnerships, funding mechanisms, and technology-sharing agreements can ensure that AI technologies are tailored to the needs of diverse populations and regions.

7.14 CONCLUSION: THE ROLE OF AI IN BIOMEDICAL RESEARCH AND HEALTH INNOVATION

7.14.1 The Promise of AI in Biomedical Research

The potential of AI in biomedical research and health innovation remains largely untapped. As AI technologies continue to evolve, we can anticipate significant breakthroughs in R&D as well as clinical practices. These advancements will ultimately lead to enhanced health outcomes across the globe, marking a new era in health-related innovations.

7.14.2 Opportunities for Emerging Economies in AI

AI in biomedical sciences presents a valuable opportunity for stakeholders from emerging economies to contribute to the global landscape. By co-producing knowledge and innovation in the field, these nations can address some of the most pressing health challenges. However, emerging economies often face challenges such as limited resources, funding gaps, and inadequate R&D infrastructure, which can hinder their participation in this rapidly advancing field.

7.14.3 Addressing Sociological Disparities in AI Development

A thorough sociological analysis is crucial to understanding the disparities in R&D capabilities between countries. Emerging economies frequently experience inconsistencies in access to resources, which can impede their involvement in global AI-driven biomedical research. Early intervention by policymakers and investors can help mitigate these issues, fostering an inclusive environment where stakeholders from the global south can actively participate in AI development.

7.14.4 Policy and Investment Priorities

For emerging economies to gain a foothold in AI and biomedical innovation, it is essential that policies and programmes are designed with a focus on addressing these disparities. By prioritizing equity in funding and resources, emerging nations can contribute meaningfully to solving global health issues. AI applications such as tackling high-prevalence diseases like cancer, accelerating biopharmaceutical development, and enhancing the quality of clinical study collaborations are just a few areas where these countries can play a pivotal role.

7.14.5 Enhancing Global Collaboration in AI for Health

Actively involving emerging economies in AI-driven biomedical research helps to build a more inclusive and equitable global science and technology landscape. By establishing decentralized AI infrastructures and initiatives, local restrictions can be overcome, fostering ongoing capabilities in these countries. This approach will also help to create a more culturally diverse innovation environment, which is crucial for the sustainable growth of AI in health.

7.14.6 The Importance of Inclusivity and Trust in AI Development

Promoting an inclusive, equitable, and trustworthy development of AI in health-related research is essential. By ensuring that emerging economies have a seat at the table, partnerships can be built that are grounded in trust and cooperation. This will increase the geographical representativeness of AI, facilitating its global diffusion and acceptance. As more nations contribute to AI in health R&D, the collective ability to address pressing global health issues will improve.

7.14.7 Strengthening AI-Driven Biomedical Knowledge Infrastructures

A social and political analysis of AI's implications in health innovation, particularly in biomedical research, is vital to strengthening the global infrastructure of AI. The focus should be on creating knowledge systems that are trustworthy, equitable, and accessible, allowing researchers worldwide to tackle health issues with public impact. Institutional preparedness in emerging countries is crucial for navigating the opportunities AI presents in biomedical research and health innovation.

7.15 SUMMARY: PREPARING FOR THE FUTURE

AI holds immense potential to address some of healthcare's most significant challenges, but realizing this vision remains a considerable hurdle. A primary barrier is the issue of data. While advanced technologies and ML algorithms are being developed, their effectiveness hinges on access to sufficient, diverse, and high-quality data. To fully harness AI's capabilities in healthcare, the industry must prioritize digitizing medical records, establishing standardized data infrastructure, and implementing robust systems to safeguard patient confidentiality and manage consent. Without these transformative efforts and collaborative action across the healthcare ecosystem, achieving AI's true promise to improve human health will remain an elusive goal. To fully realize the potential of AI in solving global health challenges, stakeholders must actively engage with emerging economies. By building equitable partnerships, addressing sociological disparities, and improving institutional readiness, AI can be leveraged to drive biomedical advancements that benefit all nations, ensuring a healthier future for people around the world.

PRACTICE QUESTIONS

1. What are the primary roles of AI in biomedical systems and tools?
2. Explain how AI enhances the accuracy and efficiency of biomedical tools in diagnostics and research.
3. How can AI be applied to analyse complex genomic and proteomic data in healthcare?
4. Discuss the role of AI in predictive modelling for disease prevention and management.
5. In what ways does AI improve medical imaging technologies, such as MRI or CT scans?
6. What emerging AI technologies are expected to shape the future of biomedical tools?
7. How can AI integration drive personalized medicine and tailored therapeutic interventions?
8. Describe the potential role of AI in real-time monitoring and decision-making in biomedical systems.
9. What ethical considerations should be addressed when implementing AI in biomedical tools?
10. Discuss the importance of interdisciplinary collaboration in the successful integration of AI into biomedical tools.

BIBLIOGRAPHY

Abdel-Hameed Al-Mistarehi M, Mijwil M.; Youssef Filali; Mariem Bounabi; Guma Ali; Mostafa Abotaleb. Artificial intelligence solutions for health 4.0: overcoming challenges and surveying applications. *MJAIH* 2023;2023:15–20.

Abolhasani M, Kumacheva E. The rise of self-driving labs in chemical and materials sciences. *Nat Synth*. 2023;2:483–92.

Abubaker Bagabir S, Ibrahim NK, Abubaker Bagabir H, Hashem Ateeq R. Covid-19 and artificial intelligence: genome sequencing, drug development and vaccine discovery. *J Infect Public Health*. 2022;15(2):289–96. https://doi.org/10.1016/j.jiph.2022.01.011.

Ahsan MM, Luna SA, Siddique Z. Machine-learning-based disease diagnosis: a comprehensive review. *Healthcare*. 2022;10(3):541.

Alami H, Rivard L, Lehoux P. Artificial intelligence in health care: laying the Foundation for responsible, sustainable, and inclusive innovation in low- and middle-income countries. *Global Health*. 2020;16:52. https://doi.org/10.1186/s12992-020-00584-1.

Alfaras M, Soriano MC, Ortín S. A fast machine learning model for ECG-based Heartbeat classification and arrhythmia detection. *Front Phys*. 2019;7. https://doi.org/10.3389/fphy.2019.00103.

Amarasingham R, Patzer RE, Huesch M, Nguyen NQ, Xie B. Implementing electronic health care predictive analytics: considerations and challenges. *Health Aff (Millwood)*. 2014;33(7):1148–54. https://doi.org/10.1377/hlth aff.2014.0352.

Amazon. *Healthcare & Life Sciences*. 2021. Available online: https://aws.amazon.com/health/

Ansari MS, Alok AK, Jain D, et al. Predictive model based on health data analysis for risk of readmission in disease-specific cohorts. *Perspect Health Inf Manag*. 2021;18(Spring):1j. Published 2021 Mar 15.

Artificial Intelligence and Machine Learning in Software as a Medical Device. *US Food and Drug Administration*. Released 2021. FDA website: www.fda.gov/medicaldevices/software-medical-device-samd/artificial-intelligence-and-machine-learningsoftware-medical-device.

Beam YuK, Kohane AL. Artificial intelligence in healthcare. *Nat Biomed Eng*. 2018;2(10):719–31. https://doi.org/10.1038/s41551-018-0305-z.

Becker J, Decker JA, Römmele C, Kahn M, Messmann H, Wehler M, et al. Artificial intelligence-based detection of pneumonia in chest radiographs. *Diagnostics*. 2022;12(6):1465. https://doi.org/10.3390/diag nostics12061465.

Berlyand Y, Raja AS, Dorner SC, Prabhakar AM, Sonis JD, Gottumukkala RV, et al. How artificial intelligence could transform emergency department operations. *Am J Emerg Med*. 2018;36(8):1515–7. https://doi.org/10.1016/j.ajem.2018.01.017.

Blasiak A, Truong A, Jeit W, Tan L, Kumar KS, Tan SB, et al. PRECISE CURATE. AI: a prospective feasibility trial to dynamically modulate personalized chemotherapy dose with artificial intelligence. *J Clin Oncol*. 2022;40(16suppl):1574–4. https://doi.org/10.1200/JCO.2022.40.16_suppl.1574.

Buch VH, Ahmed I, Maruthappu M. Artificial intelligence in medicine: current trends and future possibilities. *Br J Gen Pract*. 2018;68(668):143–4. https://doi.org/10.3399/bjgp18X695213.

Crossnohere NL, Elsaid M, Paskett J, Bose-Brill S, Bridges JFP. Guidelines for artificial intelligence in medicine: *literature review and content analysis of frameworks. J Med Internet Res*. 2022;24(8):e36823. https://doi.org/10.2196/36823.

da Silva RGL, Blasimme A. From lab to society: fostering clinical translation of molecular systems engineering. *Bioeng Translational Med*. 2023;e10564.

Davenport T, Kalakota R. The potential for artificial intelligence in Healthcare. *Future Healthc J*. 2019;6(2):94–8.

Davenport T, Kalakota R. The potential for artificial intelligence in healthcare. *Future Healthc J*. 2019;6:94–8. 10.7861/futurehosp.6-2-94

Dhamnani S, Singal D, Sinha R, Mohandoss T, Dash M. RAPID: rapid and precise interpretable decision sets. *2019 IEEE Int Conf Big Data (Big Data) IEEE*. 2019:1292–301.

Edelenbos J, van Buuren A, van Schie N. Co-producing knowledge: joint knowledge production between experts, bureaucrats and stakeholders in Dutch water management projects. *Environ Sci Policy*. 2011;14:675–84.

Esteva A, Kuprel B, Novoa RA, Ko J, Swetter SM, Blau HM, et al. Dermatologist-level classification of skin cancer with deep neural networks. *Nature*. 2017;542(7639):115–8.

Evens R, Kaitin K. The evolution of biotechnology and its impact on health care. *Health Aff (Millwood)*. 2015;34(2):210–9.

Feng YA, Chen CY, Chen TT, Kuo PH, Hsu YH, Yang HI, Chen WJ, Su MW, Chu HW, Shen CY, Ge T, Huang H, Lin YF. Taiwan Biobank: a rich biomedical research database of the Taiwanese population. *Cell Genom*. 2022;2(11):100197.

Ferguson AL, Brown KA. Data-driven design and autonomous experimentation in soft and biological materials engineering. *Annu Rev Chem Biomol Eng*. 2022;13:25–44.

Gandhi SO, Sabik L. Emergency department visit classification using the NYU algorithm. *Am J Manag Care*. 2014;20(4):315–20.

Gerke S, Minssen T, Cohen G. Ethical and legal challenges of artificial intelligencedriven healthcare. *Artif Intell Healthc*. 2020;295–336. https://doi.org/10.

Glikson E, Woolley AW. Human trust in artificial intelligence: *review of empirical research. Acad Manage Ann.* 2020;14:627–60. 10.5465/ annals.2018.0057

Google. *Google Cloud for Healthcare and Life Sciences*. 2021. Available online: https://cloud.google.com/solutions/healthcare-life-sciences

Graham S, Depp C, Lee EE, Nebeker C, Tu X, Kim HC, et al. Artificial intelligence for mental health and mental illnesses: an overview. *Curr Psychiatry Rep*. 2019;21(11):116. https://doi.org/10.1007/s11920-019-1094-0.

Guedj M, Swindle J, Hamon A, Hubert S, Desvaux E, Laplume J, et al. Industrializing AI-powered drug discovery: lessons learned from the patrimony computing platform. *Expert Opin Drug Discov*. 2022;17(8):815–24. https://doi.org/10.1080/17460441.2022.2095368.

Haenssle HA, Fink C, Schneiderbauer R, Toberer F, Buhl T, Blum A, et al. Man against machine: diagnostic performance of a deep learning convolutional neural network for dermoscopic melanoma recognition in comparison to 58 dermatologists. *Ann Oncol*. 2018;29(8):1836–42. https://doi.org/10.1093/ann onc/mdy166.

Han SS, Park I, Eun Chang S, Lim W, Kim MS, Park GH, et al. Augmented intelligence dermatology: deep neural networks empower medical professionals in diagnosing skin cancer and predicting treatment options for 134 skin disorders. *J Invest Dermatol*. 2020;140(9):1753–61. https://doi.org/10.1016/j.jid.2020.01.019.

Haug CJ, Drazen JM. Artificial intelligence and machine learning in clinical medicine, 2023. *N Engl J Med*. 2023;388(13):1201–8. https://doi.org/10.1056/NEJMra2302 038.

Healthineers S. Artificial intelligence in healthcare. 2021. Available online: www.siemens-healthineers.com/digital-health-solutions/artificial-intelligence-in-healthcare

https://gradstudies.musc.edu/about/blog/2024/01/ai-in-biomedical-research

Hulsen T, Obbink H, Van der Linden W, et al. 958 Integrating large datasets for the Movember Global Action Plan on active surveillance for low risk prostate cancer. *Eur Urol Suppl*. 2016;15:e958.10.1016/S1569-9056(16)60959-4

Hulsen T. Challenges and solutions for big data in personalized healthcare. In: Moustafa A, editor. *Big Data in Psychiatry*, Neurology, and Personalized Healthcare. 1 ed.: Elsevier; 2021.

Jiang F, Jiang Y, Zhi H, Dong Y, Li H, Ma S, et al. Artificial intelligence in healthcare: past, present and future. *Stroke Vasc Neurol*. 2017;2(4):230–43. https://doi.org/10.1136/svn-2017-000101.

Johnson KB, Wei WQ, Weeraratne D, Frisse ME, Misulis K, Rhee K, et al. Precision medicine, AI, and the future of personalized health care. *Clin Transl Sci*. 2021;14(1):86–93. https://doi.org/10.1111/cts.12884.

Kim H-E, Kim HH, Han B-K, Kim KH, Han K, Nam H, et al. Changes in cancer detection and false-positive recall in mammography using Artificial Intelligence: a retrospective, Multireader Study. *Lancet Digit Health*. 2020;2(3).

Li LR, Du B, Liu HQ, Chen C. Artificial intelligence for personalized medicine in thyroid cancer: current status and future perspectives. *Front Oncol*. 2021;10:604051. https://doi.org/10.3389/fonc.2020.604051.

Li S, Zhao R, Zou H. Artificial intelligence for diabetic retinopathy. *Chin Med J (Engl)*. 2021;135(3):253–60. https://doi.org/10.1097/CM9.00000 0000 0001 816

Lucas GM, Gratch J, King A, Morency LP. It's only a computer: virtual humans increase willingness to disclose. *Comput Human Behav*. 201;7:94–100.

Madabhushi A, Lee G. Image analysis and machine learning in digital pathology: challenges and opportunities. *Med Image Anal*. 2016;33:170–5. 10.1016/j.media.2016.06.037

Matheny ME, Whicher D, Thadaney Israni S. Artificial intelligence in health care: a report from the National Academy of Medicine. *JAMA*. 2020;323(6):509–10. https://doi.org/10.1001/jama.2019.21579.

McCorduck P, Cfe C. *Machines who think: a personal inquiry into the history and prospects of Artificial Intelligence*. AK Peters; 2004.

McKinney SM, Sieniek M, Godbole V, Godwin J, Antropova N, Ashrafian H, et al. International evaluation of an AI system for breast cancer screening. *Nature*. 2020;577(7788):89–94.

Microsoft. *Microsoft Cloud for Healthcare*. 2021. Available online: www.micros oft.com/en-us/industry/health/microsoft-cloud-for-healthcare

Mijwil MM, Aggarwal K. A diagnostic testing for people with appendicitis using machine learning techniques. *Multimed Tools Appl*. 2022;81(5):7011–23. https://doi.org/10.1007/s11042-022-11939-8.

Myszczynska MA, Ojamies PN, Lacoste AM, Neil D, Saffari A, Mead R, et al. Applications of machine learning to diagnosis and treatment of neurodegenerative Diseases. *Nat Reviews Neurol*. 2020;16(8):440–56.

Negnevitsky M. *Artificial intelligence: a guide to intelligent systems*. Pearson education; 2005.

Panch T, Szolovits P, Atun R. Artificial intelligence, machine learning and health systems. *J Global Health*. 2018;8(2). https://doi.org/ 10.7189/ jogh.08.020 303.

Peiffer-Smadja N, Dellière S, Rodriguez C, Birgand G, Lescure FX, Fourati S, et al. Machine learning in the clinical microbiology laboratory: has the time come for routine practice? *Clin Microbiol Infect*. 2020;26(10):1300–9. https://doi.org/10.1016/j.cmi.2020.02.006.

Predictive Analytics in Healthcare | Reveal. www.revealbi.io/blog/predictive-analytics-in-hea lthcare.

Pudjihartono N, Fadason T, Kempa-Liehr AW, O'Sullivan JM. A review of feature selection methods for machine learning-based disease risk prediction. *Front Bioinform*. 2022;2:927312. https://doi.org/10.3389/fbinf.2022.927312.Published 2022 Jun 27.

Radanliev P, De Roure D, Ani U, Carvalho G. The ethics of shared COVID-19 risks: an epistemological framework for ethical health technology assessment of risk in vaccine supply chain infrastructures. *Health Technol (Berl)*. 2021;11(5):1083–91.

Raghunath S, Pfeifer JM, Ulloa-Cerna AE, Nemani A, Carbonati T, Jing L, et al. Deep neural networks can predict new-onset atrial fibrillation from the 12-lead ECG and help identify those at risk of atrial fibrillation–related stroke. *Circulation*. 2021;143(13):1287–98. https://doi.org/ 10.1161/ cir cula tionaha.120.047 829

Reardon J, Lee SS, Goering S, Fullerton SM, Cho MK, Panofsky A, Hammonds EM. Trustworthiness matters: building equitable and ethical science. *Cell*. 2023;186(5):894–8. https://doi.org/ 10.1016/ j.cell.2023.01.008.

Regulatory Science Strategy to 2025. *European Medicines Agency*. Released 2020. www.ema.europa.eu/en/about-us/how-we-work/regulatory-science-strategy#regulatory-science-strategy-to-2025-section.

Ross MK, Wei W, Ohno-Machado L. "Big data" and the electronic health record. *Yearb Med Inform.* 2014;9:97–104. 10.15265/ IY-2014-0003

Russell SJ. *Artificial intelligence a modern approach.* Pearson Education, Inc.; 2010.

Sheu YH, Magdamo C, Miller M, Das S, Blacker D, Smoller JW. AI-assisted prediction of differential response to antidepressant classes using electronic health records. *NPJ Digit Med.* 2023;6:73. https://doi.org/10.1038/s41746-023-00817-8.

Smith KP, Kirby JE. Image analysis and artificial intelligence in infectious disease diagnostics. *Clin Microbiol Infect.* 2020;26(10):1318–23. https://doi.org/10.1016/j.cmi.2020.03.012.

Soler MG. Science diplomacy in Latin America and the Caribbean: Current landscape, challenges, and future perspectives. *Front Res Metr Anal.* 2021;6:670001. doi:10.3389/ frma.2021.670001

Sorlie T, Perou CM, Tibshirani R, Aas T, Geisler S, Johnsen H, et al. Gene expression patterns of breast carcinomas distinguish tumor subclasses with clinical implications. *Proc Natl Acad Sci.* 2001;98:10869–74. https://doi.org/10.1073/pnas.191367 098.

Subramanian M, Wojtusciszyn A, Favre L, Boughorbel S, Shan J, Letaief KB, et al. Precision medicine in the era of artificial intelligence: implications in chronic disease management. *J Transl Med.* 2020;18(1):472. https://doi.org/10.1186/s12967-020-02658-5.

Suleimenov IE, Vitulyova YS, Bakirov AS, Gabrielyan OA. Artificial Intelligence: what is it? *Proc 2020 6th Int Conf Comput Technol Appl.* 2020;22–5.

Takashima K, Morrison M, Minari J. Reflection on the enactment and impact of safety laws for regenerative medicine in Japan. *Stem Cell Rep.* 2021;16(6):1425–34. https://doi.org/ 10.1016/j.stemcr.2021.04.017.

Tolstikov V, Moser AJ, Sarangarajan R, et al. Current status of metabolomic biomarker discovery: impact of study design and demographic characteristics. *Metabolites* 2020;10:224.10.3390/metabo10060224

Topol EJ. High-performance medicine: the convergence of human and artificial intelligence. *Nat Med.* 2019;25(1):44–56.

Tran TTV, Surya Wibowo A, Tayara H, Chong KT. Artificial intelligence in drug toxicity prediction: recent advances, challenges, and future perspectives. *J Chem Inf Model.* 2023;63(9):2628–43. https://doi.org/10.1021/acs.jcim.3c00 200.

Undru TR, Uday U, Lakshmi JT, et al. Integrating artificial intelligence for clinical and laboratory diagnosis – a review. *Maedica (Bucur).* 2022;17(2):420–6. doi: 10.26574/ maedica.2022.17.2.420. PMID: 36032592.

Vollmer S, Mateen BA, Bohner G, Király FJ, Ghani R, Jonsson P, et al. Machine learning and artificial intelligence research for patient benefit: 20 critical questions on transparency, replicability, ethics, and effectiveness. *BMJ.* 2020;368:l6927. https://doi.org/10.1136/ bmj.l6927.

Wang H, Avillach P. Diagnostic classification and prognostic prediction using common genetic variants in autism spectrum disorder: genotype-based deep learning. *JMIR Med Inf.* 2021;9(4). https://doi.org/10.2196/24754.

Widen E, Raben TG, Lello L, Hsu SDH. Machine learning prediction of biomarkers from SNPs and of disease risk from biomarkers in the UK Biobank. *Genes (Basel).* 2021;12(7):991. https:// doi.org/10.3390/genes12070991. Published 2021 Jun 29.

Yersal O. Biological subtypes of breast cancer: prognostic and therapeutic implications. *World J Clin Oncol.* 2014;5(3):412–24. https://doi.org/10.5306/wjco.v5.i3.412.

Yuan B, Li J. The policy effect of the General Data Protection Regulation (GDPR) on the digital public health sector in the European Union: an empirical investigation. *Int J Environ Res Public Health.* 2019;16(6):1070. https://doi.org/10.3390/ije rph16061 070.

Zhang H, Chen Y, Li F. Predicting anticancer drug response with deep learning constrained by signaling pathways. *Front Bioinform.* 2021;1:639349. https://doi.org/10.3389/fbinf.2021.639349.

Zhang JY. Lost in translation? Accountability and governance of clinical stem cell research in China. *Regen Med.* 2017;12(6):647–56. https://doi.org/10.2217/rme-2017-0035. Epub 2017 Oct 4.

8 Smart Nutrition Analytics Using Machine Learning Algorithms

8.1 INTRODUCTION

Nutritionally mediated chronic diseases are highly prevalent and originate from complex, multifactorial interactions, requiring sophisticated, diverse data for effective analysis. Historically, research in this area has relied on focused, mechanistic methodologies that often fall short in capturing the full complexity of nutrition–disease interactions. However, recent technological and computational advancements have enabled researchers to leverage high-dimensional data, offering deeper insights into these diseases and addressing broader, complex research questions [1].

Key themes in nutrition science, including obesity, omics sciences, and the microbiome, alongside traditional areas like epidemiology, are increasingly benefiting from these high-dimensional data approaches. In response to the intricacies of modern data, emerging areas within nutrition research, such as precision nutrition (PN) and data-driven disease modelling, demand advanced analytical tools capable of interpreting complex datasets. Consequently, artificial intelligence (AI) and, more specifically, machine learning (ML) have become essential for addressing these demands. By enabling the integration and analysis of large-scale, complex datasets, AI and ML offer promising solutions to advance our understanding of nutritionally mediated chronic diseases.

8.1.1 Nutritionally Mediated Chronic Diseases: Challenges and Advanced Analytical Approaches

The prevalence of nutritionally mediated chronic diseases is rising globally, posing significant public health challenges. These conditions often result from complex, multifactorial interactions between diet, genetics, lifestyle, and environment. To understand and address these diseases, researchers need to analyse a wide range of interconnected data types [2]. Traditionally, nutritional science has relied on focused, mechanistic approaches to investigate these interactions, but such methods may oversimplify the intricate relationships between nutrition and chronic disease.

DOI: 10.1201/9781003489986-11

8.1.2 Limitations of Traditional Approaches

Conventional research methods in nutrition have often focused on isolated variables and linear relationships, making it challenging to capture the complex interplay of factors that influence health outcomes. While these methods have provided valuable insights, they may lack the depth required to address the multifaceted nature of chronic diseases, which involve diverse biochemical, genetic, and environmental factors.

8.1.3 Advances in High-Dimensional Data and Computational Technologies

With recent advancements in technology and computational power, researchers are now equipped to manage high-dimensional data, which is crucial for analysing the complexities inherent in chronic diseases. High-dimensional data approaches, which analyse numerous variables simultaneously, offer a more comprehensive perspective on health and disease [3]. Areas of nutrition science – such as obesity research, omics technologies (genomics, proteomics, metabolomics), and microbiome studies – are seeing substantial progress through these methods. Furthermore, traditional disciplines like epidemiology are also benefiting from these new analytical capabilities [1, 2].

8.1.4 Emerging Trends in Nutrition Research: Precision Nutrition and Data-Driven Disease Modelling

The complexity of data generated in modern nutrition science has given rise to innovative research areas, such as PN and data-driven disease modelling. PN aims to tailor dietary recommendations based on individual characteristics, including genetics, microbiome composition, and lifestyle factors. Data-driven disease modelling, meanwhile, uses large datasets to predict disease risk and progression, helping to create more personalized and effective interventions [4].

8.1.5 The Role of Artificial Intelligence and Machine Learning in Nutrition Research

To analyse complex, high-dimensional data, the field increasingly relies on advanced computational methods, particularly AI and ML. These technologies provide powerful tools for modelling and interpreting complex datasets, allowing researchers to uncover patterns and relationships that may be invisible to traditional methods. By facilitating a deeper understanding of the intricate interactions between nutrition and health, AI and ML are helping to transform nutrition science, making it possible to approach chronic diseases in a more holistic and precise manner [5].

This chapter provides an overview of how AI and ML are advancing our ability to analyse complex data in nutrition science, particularly in understanding and addressing nutritionally mediated chronic diseases. The following sections will explore specific applications of these technologies in areas such as precision nutrition, omics, and microbiome research, highlighting their potential to revolutionize the field and improve public health outcomes.

8.2 MACHINE LEARNING: AN OVERVIEW

8.2.1 Machine Learning as a Subdivision of Artificial Intelligence

ML is a specialized branch of AI that employs data-driven algorithms to accomplish tasks by recognizing and learning from patterns in data, rather than relying on explicit programming instructions. ML models are developed by defining a clear objective – such as predicting a numerical value – then recursively evaluating and optimizing the model's performance through experimentation [6]. Although AI currently cannot fully replicate the complexity of human cognition, it excels in certain aspects of learning, particularly in its speed, capacity to manage high-dimensional data, and ability to identify abstract patterns in data. These characteristics make ML particularly advantageous compared with traditional statistical and domain-specific techniques, giving it practical utility in a variety of complex applications, as will be discussed throughout this chapter.

8.2.2 Types of Machine Learning

ML includes four main types, each with unique approaches to learning, specific algorithms, and a range of applications which is given in Figure 8.1.

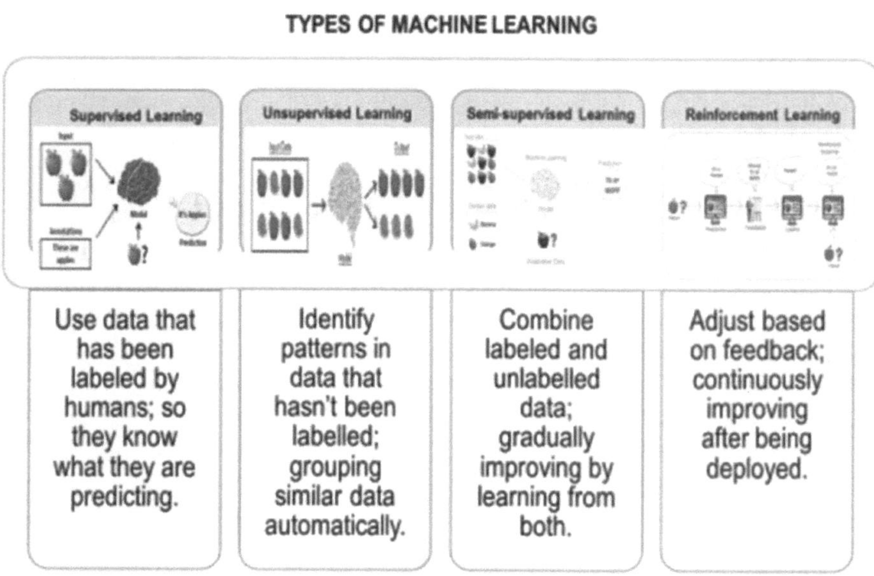

FIGURE 8.1 Types of machine learning.

8.2.2.1 Supervised Learning

In supervised learning, the data includes labels – values or categories for which the algorithm is trying to make predictions. Since these labels are known, it's easy to check if the algorithm's predictions are accurate. For instance, supervised learning is commonly used to build models predicting health outcomes like cardiovascular disease (CVD), type 2 diabetes (T2D), or nutrient levels in the blood. However, because each data point requires a label, human involvement is higher, which can increase costs and processing time [7].

8.2.2.1.1 Common Supervised Machine Learning Algorithms

8.2.2.1.1.1 Linear Regression

Linear regression is used to predict continuous values by establishing a linear relationship between input features and the target variable (see Figure 8.2).

8.2.2.1.1.2 Logistic Regression

Logistic regression is used for binary or categorical outcomes. It models the relationship between input features and the probability of a specific outcome using a logistic function.

Decision trees create a hierarchical structure to make decisions by splitting data based on feature values, useful for classification and regression [8].

8.2.2.1.1.3 Random Forests

Random forests combine multiple decision trees to enhance prediction accuracy. They reduce overfitting and improve robustness by averaging predictions from several trees [7, 8].

8.2.2.1.1.4 Support Vector Machines (SVM)

SVMs are used for classification and regression by creating decision boundaries that maximize the margin between classes, leading to accurate predictions [8].

8.2.2.1.1.5 Naive Bayes

Naive Bayes is a probabilistic classifier based on Bayes' theorem, assuming that features are independent. It's typically used for classification tasks [8].

8.2.2.1.1.6 k-Nearest Neighbours (KNN)

KNN classifies new data based on its proximity to labelled data points, assigning a class label based on the majority vote of its *k*-nearest neighbours [9].

8.2.2.1.1.7 Neural Networks

Neural networks are inspired by the human brain's structure, with layers of interconnected nodes that learn complex patterns and relationships in data [9].

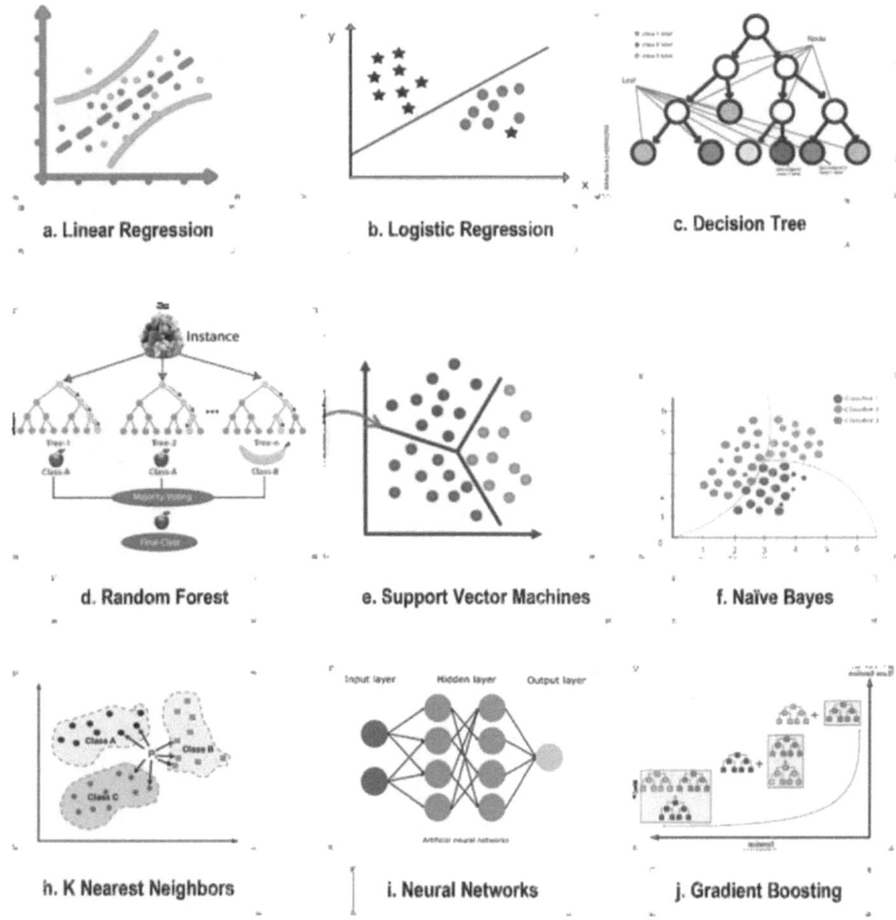

FIGURE 8.2 Common supervised machine learning algorithms.

8.2.2.1.1.8 Gradient Boosting

Gradient boosting methods, like XGBoost, sequentially build models that correct the errors of previous ones, achieving high accuracy for classification and regression tasks [9].

8.2.2.1.2 Advantages and Limitations of Supervised Machine Learning

Advantages:

i. Accurate Predictions
Supervised ML can deliver highly accurate predictions or classifications when trained on a diverse and representative dataset. It learns from labelled examples, making it effective in making reliable forecasts.

ii. Versatility
This approach can be applied across a wide range of domains, including healthcare, finance, and marketing. Its adaptability makes it useful for solving various problems in different industries.
iii. Interpretable Results
Supervised learning models often produce interpretable results, enabling users to understand how the model arrived at its predictions. This transparency is valuable in decision-making processes.

Limitations:

i. Dependency on Labelled Data
Supervised learning requires a large amount of labelled data to train the model, which can be time-consuming and costly to collect, especially for complex problems or specialized domains.
ii. Limited Generalization
Models trained on specific datasets may not generalize well to new, unseen data that differs significantly from the training set. This can lead to reduced performance when applied to real-world scenarios.
iii. Overfitting
If a model becomes too complex or is trained on insufficient data, it may overfit to the training set. This means it memorizes the data rather than learning the underlying patterns, resulting in poor performance on new data.

8.2.2.2 Unsupervised Learning

In unsupervised learning, the data lacks labels. Here, algorithms are designed to detect patterns and group similar data without prior instructions. This type of learning often involves clustering, dimensionality reduction, and anomaly detection. Since it doesn't need labelled data, unsupervised learning reduces the need for human intervention. It is often used for tasks like phenotyping in PN, where individuals are grouped based on similar characteristics. Unsupervised learning can also be a preprocessing step for supervised learning tasks, improving accuracy by organizing data before prediction [9 , 10]. For example, grouping individuals by certain traits (a process called "phenotyping") can improve predictions for variables like BMI, as shown in a study by Ramyaa et al. It is also valuable for hypothesis generation since it may reveal new groupings or insights in the data that researchers hadn't anticipated.

8.2.2.2.1 Types of Unsupervised Learning

Unsupervised learning refers to ML techniques that allow algorithms to identify patterns, structures, or relationships in data without prior knowledge of the outputs or labels. It is commonly used for grouping, reducing dimensions, and discovering hidden patterns within data [11]. Given further are some of the most common types of unsupervised learning (see Figure 8.3).

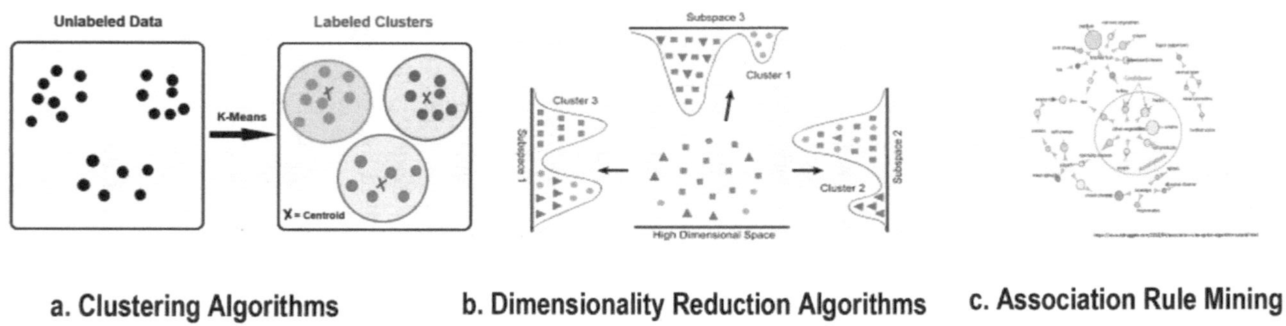

FIGURE 8.3 Types of unsupervised learning.

8.2.2.2.1.1 Clustering Algorithms

Clustering is the process of grouping similar data points together based on their inherent characteristics, without predefined labels [10]. The primary goal is to identify natural groupings or structures within the dataset.

- ***k*-Means Clustering:** This algorithm partitions data into a predefined number of clusters. It minimizes the sum of squared distances between the data points and their respective cluster centres (centroids). *k*-Means iteratively assigns points to clusters and updates centroids until convergence is reached.
- **Hierarchical Clustering:** Hierarchical clustering builds a tree-like structure (dendrogram) of clusters. It can either start with individual data points and merge them into larger clusters (agglomerative) or start with all data points in single cluster and recursively split them (divisive) based on similarity.
- **Density-Based Spatial Clustering of Applications with Noise (DBSCAN):** DBSCAN groups data points that are closely packed together, marking regions with higher density as clusters, and separating sparse regions as noise. This algorithm does not require the number of clusters to be specified beforehand, and it is effective at handling outliers.

8.2.2.2.1.2 Dimensionality Reduction Algorithms

Dimensionality reduction techniques aim to reduce the number of features or variables in the dataset while retaining as much relevant information as possible. These techniques are particularly useful for visualizing high-dimensional data or improving the efficiency of ML models [11].

- **Principal Component Analysis (PCA)**: PCA transforms the original set of variables into a smaller set of uncorrelated components, ordered by the amount of variance they explain in the data. The first few components capture the most significant information, enabling a reduction in dimensionality while retaining key characteristics of the data.
- ***t*-Distributed Stochastic Neighbour Embedding (t-SNE):** *t*-SNE is a technique for visualizing high-dimensional data in lower dimensions (typically 2D or 3D). It focuses on preserving local relationships in the data, making it effective for visualizing clusters or patterns in complex datasets.

8.2.2.2.1.3 Association Rule Mining

Association rule mining is a method used to identify interesting relationships or patterns among items in large datasets, particularly in transactional data. It is commonly applied in market basket analysis to discover associations between products purchased together.

- **Apriori Algorithm:** The Apriori algorithm is one of the most widely used techniques for association rule mining. It works by identifying frequent item sets in the dataset and then generating association rules based on these frequent item sets. The algorithm employs a "bottom-up" approach to iteratively find larger item sets that meet a minimum support threshold.

8.2.2.2.2 Advantages of Unsupervised Learning

Unsupervised learning offers several key advantages, especially when dealing with unstructured data or situations where labelling is impractical:

8.2.2.2.2.1 Use of Unlabelled Data

Unsupervised learning is particularly valuable as it does not require labelled data. This allows it to uncover hidden patterns, structures, or relationships within the data that may not be apparent initially. By revealing these underlying connections, unsupervised learning provides valuable insights that can guide further analysis or decision-making.

8.2.2.2.2.2 Scalability

Unsupervised learning algorithms are capable of handling large datasets efficiently. They do not rely on manually labelled data, making them more scalable compared with supervised learning, especially in cases where labelling data would be time-consuming or costly. This scalability makes unsupervised learning highly suitable for big data applications.

8.2.2.2.2.3 Anomaly Detection

Unsupervised learning is effective at detecting anomalies or outliers within datasets. This ability is particularly useful in areas like fraud detection, network security, or identifying rare events. By learning the normal patterns of data, unsupervised learning algorithms can flag data points that deviate significantly, indicating potential anomalies.

8.2.2.2.2.4 Data Preprocessing

Unsupervised learning techniques, such as dimensionality reduction, are helpful for preprocessing data. They reduce noise, eliminate irrelevant features, and improve the efficiency of subsequent tasks, particularly in supervised learning models. These techniques enhance the quality of the data before further analysis or predictive modelling.

8.2.2.2.3 Limitations of Unsupervised Learning

While unsupervised learning offers numerous benefits, there are several challenges and limitations that need to be considered:

8.2.2.2.3.1 Lack of Ground Truth

One of the main challenges of unsupervised learning is the absence of labelled data, which means there is no definitive measure of correctness or accuracy. Without ground truth, evaluating the quality or reliability of the results becomes subjective and depends heavily on domain knowledge and interpretation [9].

8.2.2.2.3.2 Interpretability

The results produced by unsupervised learning algorithms, such as clusters or patterns, often lack explicit labels or easy-to-understand explanations. Interpreting the meaning or significance of these patterns can be complex and subjective, requiring expertise to make sense of the findings in a meaningful way [10].

8.2.2.2.3.3 Overfitting and Model Selection

Unsupervised learning models are susceptible to overfitting, where the model captures noise or irrelevant patterns in the data rather than the underlying structure. Selecting the optimal model or parameters for a given dataset can be challenging, especially in the absence of a labelled validation set for comparison [11].

8.2.2.2.3.4 Limited Guidance

Unlike supervised learning, where the model receives explicit feedback to refine its learning, unsupervised learning lacks this type of guidance. As a result, the algorithm may discover patterns that are irrelevant or noisy, making it difficult to ensure that the results are meaningful or useful without human intervention and domain expertise.

8.2.2.3 Semi-Supervised Learning

Semi-supervised learning is a blend of supervised and unsupervised learning, where only some data points are labelled, while most are not. This approach strikes a balance: it improves prediction accuracy and generalizability by using a few labels while avoiding the high costs and time associated with labelling all data. Semi-supervised learning has been applied in genetics research to study how specific genes impact disease outcomes, where known genes (labelled data) are limited. Another example of semi-supervised learning is constrained clustering, where specific rules are enforced, such as keeping certain data points together or apart. This is particularly helpful in biological data, ensuring that clustering respects plausible groupings, like maintaining separate clusters for different biological sexes when relevant [11, 12].

8.2.2.3.1 Why Use Semi-Supervised Learning?

The appeal of semi-supervised learning stems from the abundance of unlabelled data and the challenges associated with labelling large datasets accurately. By utilizing a combination of both labelled and unlabelled data, this approach enhances model training efficiency and effectiveness. It provides a balance between the high-cost, resource-intensive process of labelling data and the availability of vast amounts of unlabelled data, making it an attractive solution for many real-world applications [10].

8.2.2.3.2 Applications of Semi-Supervised Learning

Semi-supervised learning is particularly effective in fields where labelled data is limited or difficult to obtain. Some key areas where this method excels include the following:

8.2.2.3.2.1 Computer Vision

In image recognition and analysis, manually labelling large datasets of images can be costly and time-consuming. Semi-supervised learning enables the use of a small set of labelled images in combination with a larger pool of unlabelled images, improving model performance while reducing the labelling effort.

8.2.2.3.2.2 Natural Language Processing (NLP)

Similar to computer vision, in NLP tasks such as sentiment analysis or language translation, obtaining labelled text data can be expensive. Semi-supervised learning can leverage a few labelled texts and a large corpus of unlabelled data to enhance language models without extensive annotation.

8.2.2.3.2.3 Medical Imaging

Annotating medical images, such as X-rays or MRI scans, requires expert knowledge and is costly. In medical imaging, using semi-supervised learning with a combination of labelled and unlabelled data can help improve diagnostic models while minimizing the need for expert-labelled data.

Semi-supervised learning offers a practical solution in scenarios where labelled data is scarce or costly to obtain. By using both labelled and unlabelled data, this approach maximizes the utility of available resources, leading to improved model performance with reduced reliance on manually labelled data. Whether in fields like computer vision, NLP, or medical imaging, semi-supervised learning helps bridge the gap between the abundance of unlabelled data and the need for accurate, labelled information.

8.2.2.4 Reinforcement Learning

Reinforcement learning (RL) is used in dynamic environments where an algorithm is rewarded or penalized based on its actions, enabling it to improve by maximizing rewards and minimizing penalties. RL allows an algorithm to become highly skilled at a task without explicit instructions. One well-known example is AlphaGo Zero, a system that learned to play the board game Go at a superhuman level by practicing against itself. RL is complex and not usually applied to simpler tasks like classification or regression [11]. Instead, it excels in scenarios that involve varied, complex data, like recommender systems and mobile fitness apps that personalize recommendations based on a user's activity [13]. The process of developing semi-supervised learning model is shown in Figure 8.4. The reinforcement learning model is shown in Figure 8.5.

8.2.2.4.1 Key Components of Reinforcement Learning

Agent: The intelligent entity that interacts with the environment.
Environment: The external system with which the agent interacts.
Action: The decisions made by the agent to transition between states in the environment.
Reward: Feedback given to the agent based on its actions; rewards encourage actions, while punishments discourage them.

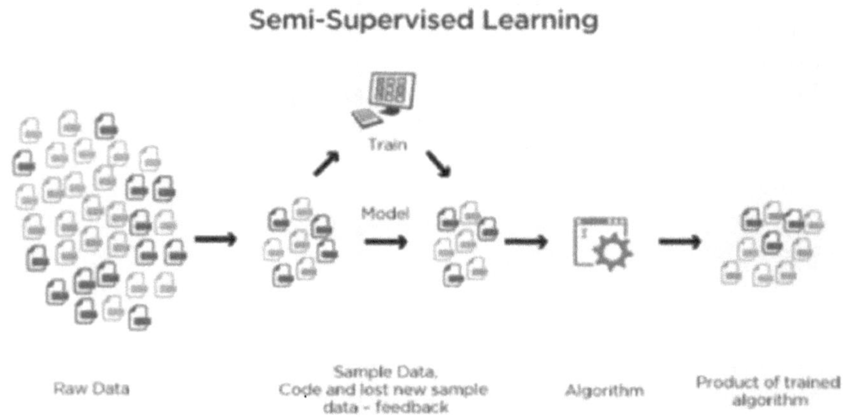

FIGURE 8.4 Process of developing a semi-supervised learning model.

FIGURE 8.5 Reinforcement learning.

8.2.2.4.2 Core Concepts in Reinforcement Learning

Markov Decision Processes (MDPs): A mathematical framework for modelling decision-making problems. MDPs help agents make sequential decisions where outcomes depend on current states and actions, assuming the future is independent of the past given the present state.

Value Functions: Estimate the expected return or utility of being in a particular state or taking a specific action, helping agents make decisions to maximize long-term rewards.

Policies: Strategies or rules that map states to actions. Policies guide an agent's decision-making, and they can be deterministic (one action per state) or stochastic (probabilistic actions).

Exploration versus Exploitation: The agent must balance exploring new actions to gather more information (exploration) and using known information to maximize rewards (exploitation).

8.2.2.4.3 Common Algorithms in Reinforcement Learning

Q-learning: A model-free algorithm that estimates the value of state–action pairs and learns an optimal policy through experience, without requiring knowledge of the environment's dynamics.

Deep Q-Networks (DQN): Combines Q-learning with deep neural networks to handle large, complex environments, enabling agents to make decisions in high-dimensional state spaces.

Policy Gradients: A method that optimizes the policy directly by adjusting the policy parameters based on the gradient of expected rewards.

Proximal Policy Optimization (PPO): A popular optimization algorithm that ensures stability and efficient learning by making small, controlled updates to the policy.

8.2.2.4.4 Challenges and Limitations of Reinforcement Learning

Sample Inefficiency: RL often requires a large number of interactions to learn optimal policies, which can be resource-intensive, especially in real-world applications.

Exploration–Exploitation Trade-off: Balancing exploration of new actions and exploiting known strategies is a challenge. Too much exploration can waste resources, while too little can lead to suboptimal solutions.

Reward Engineering: Designing appropriate reward functions is critical. Poorly designed rewards can lead to unintended behaviours or suboptimal policies.

Safety and Ethics: RL agents can exhibit harmful behaviours if not properly constrained. Ensuring safety and ethical considerations is a growing concern in real-world applications.

8.3 LEVERAGING ML AND TRADITIONAL METHODS: A BALANCED APPROACH

While this section highlights cases in which ML offers advantages over traditional methods, it is essential for researchers to view each approach as complementary tools rather than mutually exclusive options. By understanding the strengths and limitations of various methodologies, researchers can effectively combine ML and conventional techniques to achieve synergistic and more insightful results [14, 15]. A deep, thoughtful consideration of the research questions and the specific problem at hand is critical in selecting the most appropriate tools.

Researchers are encouraged to experiment with multiple techniques whenever feasible, ultimately choosing the solutions best suited to their unique data and research objectives. This approach of tailoring analytical methods to the research problem is elaborated upon in Section 8.7.

8.3.1 Machine Learning versus Traditional Statistical Methods

8.3.1.1 Choosing Between Machine Learning and Traditional Methods

In cases where inference is the primary goal, interpretability is crucial, and the dataset is low-dimensional with established features, traditional statistical techniques – such as regression methods – may be adequate [16, 17]. Researchers often select these familiar approaches, though in many scenarios ML methods may provide a more effective solution. ML is particularly suited to high-dimensional data and tasks focused on predictive performance [18, 19]. Unlike traditional methods, ML does not require predetermined variable selection; instead, it can explore numerous variables and automatically identify those most relevant through iterative experimentation [19]. This exploratory capacity of ML often uncovers novel predictive features, sometimes serendipitously [20].

8.3.1.2 Limitations of Traditional Statistical Approaches

Traditional statistical methods assume that data are generated by a stochastic model, which imposes constraints that can become unrealistic as data complexity increases, particularly in fields like nutrition and health sciences [21–26]. For example, in studies predicting mortality with epidemiologic datasets, Namkung [27] found that the nonlinear capabilities of sophisticated ML techniques led to superior predictive performance compared with traditional methods. Cammarota et al. [28] demonstrated the limitations of traditional statistics by integrating metabolic, nutritional, and lifestyle data to create a handheld device emulator for precision medicine in type 2 diabetes (T2D) prediction, a task that proved too computationally intensive for traditional methods [29–35].

ML approaches, such as ensemble predictors, consistently outperform linear classifiers or regressors in medical applications, even though ensemble methods are generally considered less interpretable. While these "black box" models lack transparency in how they generate outputs, their improved accuracy may more faithfully represent underlying data generation processes than traditional "white box" models [29, 36–40]. This benefit is particularly relevant as data complexity grows in research fields like obesity, omics, and microbiota studies, where traditional statistical methods may oversimplify or inadequately capture essential information.

8.4 ML AS A SUPPLEMENT TO DOMAIN-SPECIFIC TECHNIQUES

ML offers complementary strengths to domain-specific methods. For instance, Lecroy et al. [20] employed ML to generate risk scores for T2D based on microbiome data, achieving better predictive accuracy than conventional diabetes risk factors while identifying 11 new microbial taxa associated with T2D risk. Similarly, Tap et al. [22] used ML (Lasso) to differentiate microbiota signatures in patients with irritable bowel syndrome (IBS) and healthy controls, a distinct traditional ecological method that failed to detect [31].

In activity tracking, wearable devices generate unstructured movement data suited to ML [32]. ML classifiers improve the accuracy of activity type and calorie expenditure prediction, outperforming domain-specific cut points by reducing misclassification rates, enhancing generalizability, and allowing for nuanced grading of movement quality [33–36]. In cardiovascular disease (CVD) research, ML risk prediction models incorporate a range of biomarkers, outperforming traditional risk scores and identifying novel biomarkers in the process. Additionally, ML presents a cost-effective, non-invasive alternative to certain domain-specific methods, such as those used for diagnosing non-alcoholic fatty liver disease (NAFLD) and cancer [28, 37–41].

8.5 ML: PRACTICAL ADVANTAGES

8.5.1 Efficiency and Adaptability

ML offers practical advantages in its adaptability and reduced need for explicit programming. Since ML algorithms learn directly from data, they save significant time and effort otherwise spent on programming instructions [42, 43]. This adaptability means that the same algorithm can be applied to various datasets and tasks [44]. Furthermore, ML algorithms can process multiple data types, including structured (e.g., tabular) and unstructured (e.g., image-based) data. Some ML algorithms, like neural networks and *k*-nearest neighbours (kNN), are versatile enough to handle different problems and data types, enhancing their usability across research applications [32, 45–48].

8.5.2 Reducing Experimental Costs

By predicting outcomes based on existing data, ML algorithms reduce the need for costly and time-intensive experimental validation. For instance, Sorino et al. [49] demonstrated that using ML algorithms to analyse noninvasive data could eliminate 81.9% of unnecessary ultrasound scans in NAFLD diagnosis, saving resources and reducing wait times for patients [41].

8.5.3 Accessibility of ML Tools

The tools required for ML experiments are minimal. Apart from a computer and a virtual environment, the main requirement is access to a dataset. With the growing availability of public datasets and initiatives promoting data sharing, researchers can now conduct meaningful analyses without needing extensive funding, specialized equipment, or original data collection [48]. This accessibility empowers researchers worldwide to independently pursue scientific progress and generate valuable insights by leveraging both their own and shared datasets.

8.6 MACHINE LEARNING IN NUTRITION RESEARCH

The integration of ML into nutrition research offers powerful, practical advantages that traditional methods may not provide. Throughout this chapter, we will explore how ML's capabilities can address the increasing complexity of data in nutrition science, from activity tracking and disease modelling to biomarker discovery and predictive analytics [50]. By understanding when and how to employ ML alongside traditional techniques, researchers can enhance the precision, scope, and depth of their findings, driving forward our understanding of nutrition and health.

8.6.1 Precision Nutrition and Machine Learning

8.6.1.1 Overview of Precision Nutrition (PN)

PN is an approach to nutrition that tailors dietary advice based on individual differences, aiming to improve health outcomes more effectively than generic guidelines (see Figure 8.6). It is based on the idea that each person has unique nutritional needs due to factors like genetics, environment, and lifestyle, which are not captured by standard dietary recommendations.

The use of data to personalize nutrition advice has led to increased interest in ML, which is being applied to analyse large and complex datasets in PN [52–54]. For example, the National Institutes of Health (NIH) has committed $170 million to developing ML algorithms for PN over the next five years.

8.6.1.2 Application of ML in Precision Nutrition

ML models, such as random forest (RF) and gradient-boosting algorithms, are used to analyse a variety of data (plasma, microbiome, personal traits, diet, etc.) to predict individual health responses. For instance, Zou et al. [51] used an RF model to predict postprandial glucose response (PPGR) to specific foods with high accuracy, showing significant individual differences in glucose response to the same foods. This supports the idea that personalized nutrition can improve health compared with generic guidelines.

Similarly, Zou et al. [51] predicted not only glucose levels but also triglycerides and insulin (C-peptide) after meals, demonstrating that specific factors can be targeted to modify metabolic health.

8.6.2 Addressing Obesity and Overweight with PN

8.6.2.1 Obesity and Personalized Diets

Obesity is another area where PN is making a significant impact. Research by Ramyaa et al. [52] showed that women with different phenotypes (genetic and metabolic profiles) responded differently to macronutrients and exercise. This suggests that personalized diets and exercise plans would be more effective than one-size-fits-all approaches.

Zellerbach and Ruiz [53] attempted to predict overeating based on diet composition, but they found that their model didn't perform well. However, with better data quality (e.g., including variables like stress, sleep, and alcohol consumption), such models could potentially be more successful.

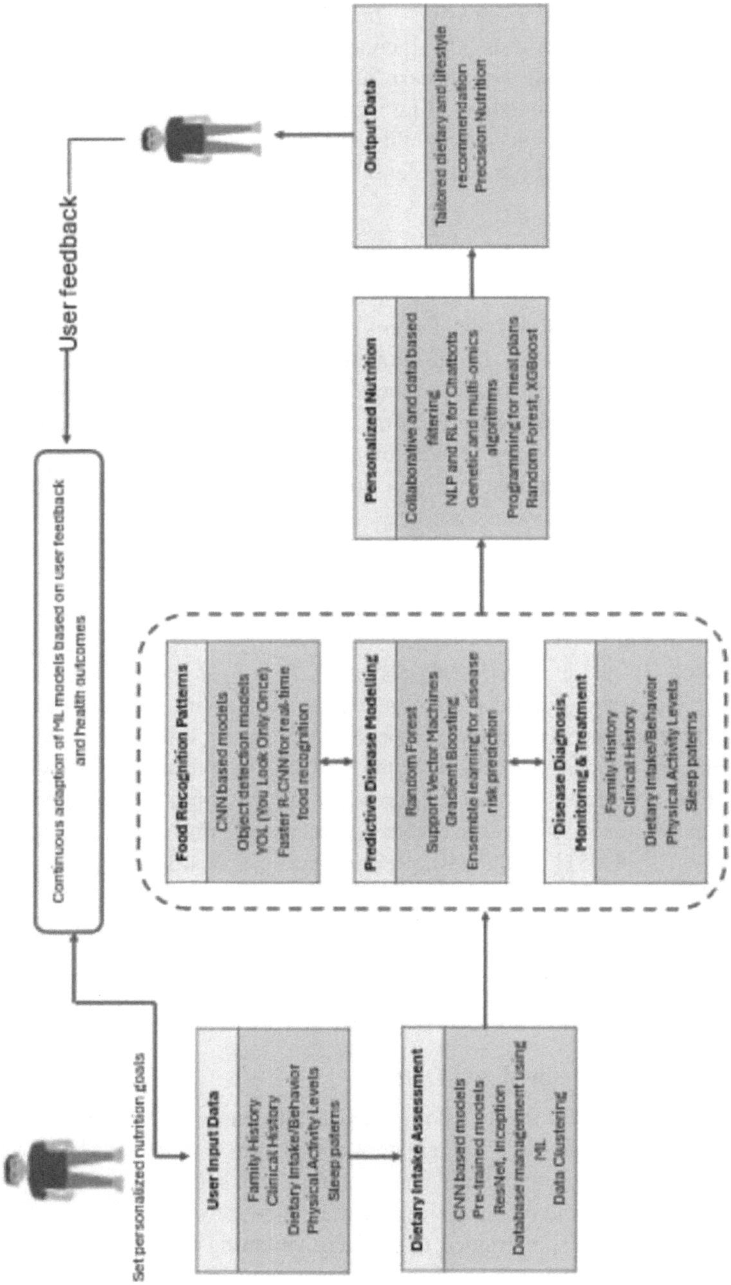

FIGURE 8.6 ML algorithms used in precision nutrition analytics.

8.6.3 The Role of the Gut Microbiota in PN

8.6.3.1 Gut Microbiota and Health

Research has also shown that gut microbiota (the collection of microbes in the digestive system) plays a key role in health conditions, including obesity. Studies like Wang et al. have used gut flora data to predict obesity, linking microbiota composition with individual health outcomes [33, 38]. By targeting the microbiome, PN could offer tailored interventions that improve metabolic health, particularly in individuals who may be predisposed to obesity or related conditions.

8.6.4 Malnutrition and Machine Learning

8.6.4.1 ML for Malnutrition Screening

PN is also being used to address malnutrition. Current screening methods for malnutrition in hospitals often suffer from poor adherence and lack of consistency. However, ML techniques, such as *k*-means clustering, have been applied to hospital data to categorize patients based on nutritional status, achieving high accuracy in predicting levels of malnutrition [46, 54].

For example, Yin et al. [55] used ML to classify patients into well-nourished or malnourished groups based on their hospital records. Subramanian et al. [56] used RF to predict age from the microbiota profile, finding that malnourished children have less mature microbiota compared with healthy children of the same age.

8.6.5 Improving Dietary Data Collection with ML

8.6.5.1 ML in Dietary Assessment

Accurately tracking dietary intake is a challenge, as traditional methods often lead to incorrect estimates. ML is being used to improve dietary data collection through more convenient and accurate methods. For example, ML can assist with analysing food intake using image recognition, wearable devices (like smartwatches), and audio-based tools [57–59].

Natural language processing (NLP) is also being used to automate the analysis of food diaries. NLP can process written or spoken text to convert it into data that can be used for personalized nutrition advice [57]. As this technology improves, it could provide more efficient ways to track food intake and offer personalized recommendations, reducing the need for direct interaction with health professionals.

8.6.6 Physical Activity Tracking in PN

8.6.6.1 Tracking Physical Activity

Physical activity is an important factor in PN. ML models have been applied to activity tracking data, such as that from accelerometers, to classify activity patterns and estimate energy expenditure [60, 61]. These models can improve the accuracy of activity data, making it a more valuable feature for personalized nutrition plans.

8.6.7 Omics and Machine Learning in Precision Nutrition

8.6.7.1 What Is Omics?

Omics is a field that studies various biological components, each identified by the suffix "omics", such as genomics, transcriptomics, epigenomics, proteomics, and metabolomics. These omics components represent different layers of biological information, with genomics focusing on genes, transcriptomics on gene expression, epigenomics on genetic modifications influenced by the environment, proteomics on proteins, and metabolomics on metabolites produced within the body. Other emerging areas include microbiome and lipidomics [62, 63].

8.6.7.2 Omics and Precision Nutrition (PN)

In PN, omics data provides valuable insights into individual health. By integrating multiple omics layers (known as multi-omics), a more comprehensive, systems-level view of an individual's health can be created. This holistic approach can uncover deeper insights than analysing single data types in isolation.

8.6.8 The Role of Genetics and Other Omics Layers in PN

8.6.8.1 Genetics and Machine Learning

Genetic data has been used in combination with ML to predict complex health conditions like obesity and diabetes. However, while genetics provides essential information, it often fails to fully explain the wide variation in health outcomes due to the influence of environmental factors. Environmental factors like food, lifestyle, and toxins can modify gene expression, making genetics alone insufficient for understanding complex conditions [64, 65]. This limitation led to the development of other omics fields, such as epigenomics, transcriptomics, and proteomics.

8.6.8.2 Epigenomics, Transcriptomics, and Proteomics

Epigenomics studies changes in gene activity without altering the DNA sequence itself, often caused by environmental factors, including diet.

Transcriptomics focuses on the RNA transcripts produced by genes, which can be influenced by both genetic and environmental factors.

Proteomics examines the proteins produced by cells, which can also be influenced by environmental inputs such as nutrition.

These layers interact to create a complex system that helps explain health outcomes beyond what genetics alone can reveal [21, 45].

8.6.8.3 The Microbiome's Impact on PN

The microbiome, the collection of microbes in the gut, has emerged as a key player in both health and disease. It can significantly influence health outcomes, sometimes explaining a large portion of the variance in predictive health models [33, 38]. For PN, the microbiome is both a target and an input variable – meaning that changes to the microbiome can be used to predict health outcomes and can also be modified to improve health.

8.6.9 Individual Responses in Health Research

8.6.9.1 Individual Variability in Health Responses

Health research often aggregates data across populations, but individual responses can vary significantly. A classic example is a study by Gardner et al. [66], which showed that while participants following a low-carbohydrate or low-fat diet for 12 months had similar group averages for weight loss, individual outcomes varied widely. Some participants lost a significant amount of weight, while others either lost little or gained weight. This highlights the importance of considering individual variability in health research, which is a key principle of PN.

8.6.10 Metabolomics and Machine Learning in Precision Nutrition

8.6.10.1 What Is Metabolomics?

Metabolomics is the study of metabolites – small molecules that are the end products of cellular processes. It provides valuable information about the metabolic state of an organism and plays a crucial role in PN as both a predictor of health outcomes and a method for data collection [67]. Metabolomics allows the profiling of many metabolites from a small sample, and with modern technologies, thousands of metabolites can be analysed simultaneously [68]. However, this large amount of data can be noisy, which makes data preprocessing and analysis essential. ML is particularly suited for these tasks, helping to extract meaningful patterns from complex metabolomics data.

8.6.11 Applications of Metabolomics in Precision Nutrition

8.6.11.1 Metabolomics for Phenotyping and Metabotyping

Metabolomics is often used for phenotyping, which is the process of characterizing an individual's metabolic profile. Metabolites reflect the biochemical reactions happening in the body, providing a more accurate representation of an individual's health status compared with other omics data [68, 69]. In PN, the concept of "metabotyping" is used, where individuals are grouped based on shared metabolite patterns. This allows for personalized health recommendations.

8.6.11.2 Personalized Nutrition Through Metabotyping

In another study, O'Donovan et al. [70] used metabolites to identify healthy and unhealthy metabolic clusters in two cohorts. Participants with elevated cholesterol were given targeted advice to lower their cholesterol. This kind of approach, using metabolomics and ML, helps create personalized nutrition plans based on an individual's specific metabolic profile, rather than applying one-size-fits-all guidelines [71].

8.6.12 Understanding Obesity Through Metabolomics

8.6.12.1 Metabolically Healthy Obese Phenotype

One intriguing concept in metabolomics is the "metabolically healthy obese" (MHO) phenotype. About one-third of obese individuals show no signs of metabolic disturbances (such as high blood pressure, high blood sugar, or high cholesterol), while the rest exhibit varying degrees of metabolic dysfunction [71]. Research using ML techniques has found distinct metabolic signatures between these groups, such as elevated levels of branched-chain amino acids (BCAAs), lipids, and acylcarnitines in the metabolically unhealthy obese group [72]. These differences can help guide targeted treatments for individuals who are obese but metabolically healthy, aiming to prevent future health problems.

PCA is often used to reduce the complexity of metabolomics data, making it easier to identify differences in metabolites and other metabolic markers that distinguish between healthy and unhealthy phenotypes.

8.6.13 Microbiota and Metabolomics in Nutrition

8.6.13.1 Microbiota-Derived Metabolomics

The microbiota, or the collection of microbes in the gut, is increasingly recognized as a key factor in health and disease. Studies using metabolomics and ML have shown that microbiota-derived metabolites can offer insights into various health conditions [74]. For example, ML has been used to differentiate between healthy and unhealthy metabolite signatures after a red wine intervention, to distinguish individuals with food addiction, and to identify paediatric IBS [75].

8.6.14 Metabolomics as a Tool for Dietary Assessment

8.6.14.1 Biomarkers of Food Intake

Metabolomics can serve as an objective tool for dietary assessment by identifying biomarkers associated with specific foods or food groups. For instance, metabolites have been identified that correlate with the consumption of bread, coffee, citrus fruits, meat, fish, and dietary components like polyphenols and fermented foods [73–76]. This allows researchers to track food intake more accurately than traditional methods like food diaries.

8.6.14.2 Monitoring Dietary Adherence

Metabolomics can also be used to estimate adherence to dietary patterns. For example, Acar et al. used metabolomics to identify participants who were not adhering to a prescribed diet using partial least squares discriminant analysis [75]. Their research found that, on average, about 10% of participants deviated from their prescribed diets at any given time, which could have significant implications for nutrition intervention studies.

8.6.15 Machine Learning for Data Analysis in Metabolomics

8.6.15.1 Data Processing and Feature Selection

Metabolomics data is high-dimensional, meaning it contains a large number of variables. This can make it difficult to extract useful information without sophisticated data processing methods. ML techniques, particularly deep learning models like convolutional neural networks (CNNs), are well-suited for feature selection and pattern recognition in metabolomics data [70–76]. These models can help identify important features within the data that may be relevant for predicting health outcomes or understanding disease mechanisms.

8.6.15.2 Applications of ML in Metabolomics

ML is useful in multiple stages of metabolomics research:

Data Preprocessing: Cleaning and transforming raw metabolomics data to make it ready for analysis.

Clustering and Classification: Grouping individuals based on their metabolic profiles and identifying patterns that distinguish different health conditions or dietary habits.

Predictive Modelling: Using processed metabolomics data to predict disease risk, identify biomarkers, and assess the effectiveness of nutritional interventions.

8.7 A SIMPLE FRAMEWORK FOR APPLYING MACHINE LEARNING IN NUTRITION SCIENCE

Applying ML in nutrition research can bring valuable insights, but it requires careful consideration of the problem, data, and the tools available. Given further is a simplified framework to help researchers decide when and how to use ML effectively in their studies. The framework for application of ML in precision nutrition is shown in Figure 8.7.

8.7.1 Understanding the Problem and the Data when Applying ML

Before applying ML, it's important to understand the problem you're trying to solve and the data you have. If your goal is to predict an outcome (e.g., health conditions or dietary patterns) based on existing data, ML might be a good approach.

Data Complexity: ML works well when the data has many features (variables) and complex, nonlinear relationships between them. If you have a lot of data that might be hard to analyse manually, ML can help you spot patterns and make predictions.

Data Preprocessing: ML can be useful for cleaning and preparing data, such as reducing noise in metabolomics data or handling missing values. For example, missing data (where some values are missing) is common in nutrition research. ML can help fill in missing values (called "imputation") or decide whether to exclude incomplete data.

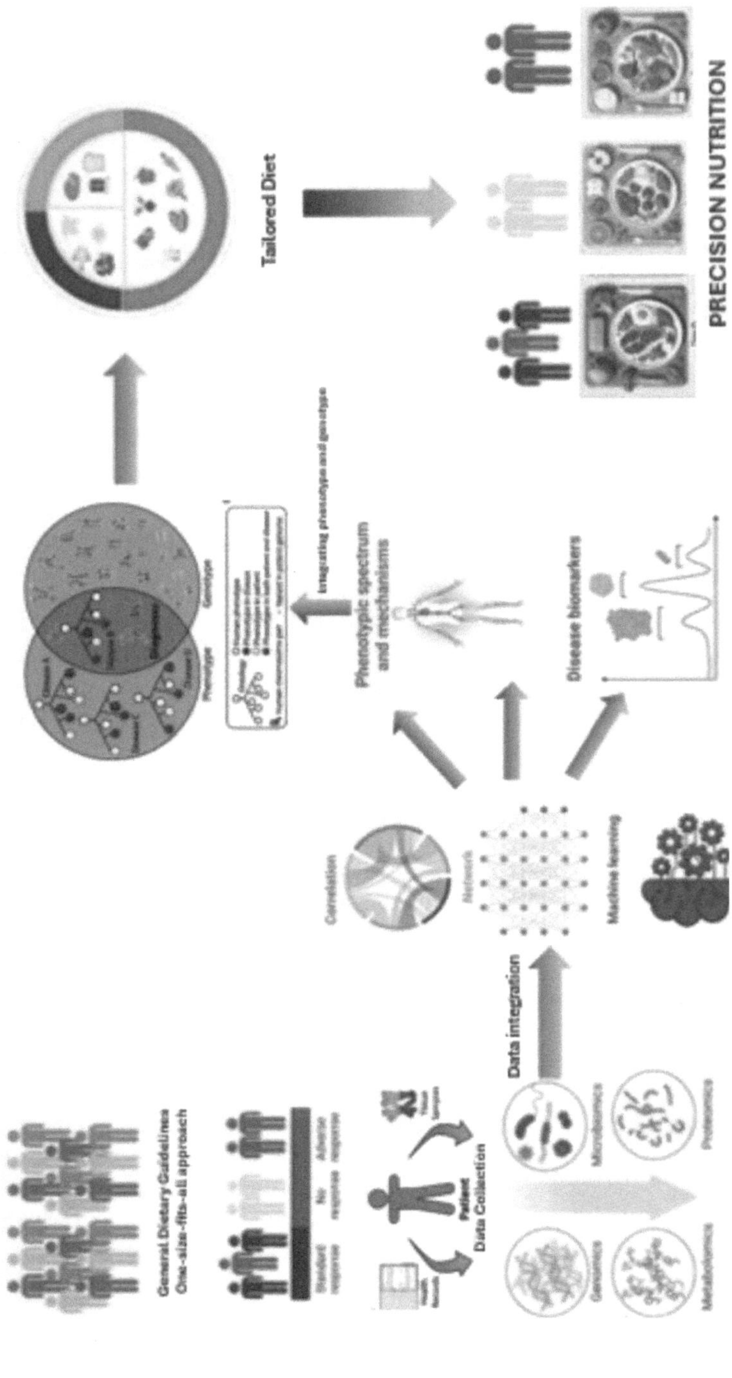

FIGURE 8.7 A framework for application of ML in precision nutrition.

8.7.2 Background Research and Existing Solutions

Before diving into ML, check if there are already existing methods that work well for your problem. If existing methods are already effective, using ML may only offer limited improvements. However, ML is particularly useful when:

Current Methods Are Limited: If traditional methods (like basic statistics) can't handle the complexity of your data, ML might help. For example, predicting personal postprandial glucose response (PPGR) can be improved with ML because traditional methods are less accurate.

Improving Efficiency: ML can be useful in situations where existing methods are inefficient or prone to human error. For example, using ML to assist healthcare professionals in identifying nutritional needs or diagnosing conditions like non-alcoholic fatty liver disease (NAFLD) can improve accuracy and save time.

8.7.3 Choosing the Right ML Solution

Once you've decided to use ML, you need to choose the best algorithm for your problem. The task you're working on and the type of data you have will guide this decision.

Task Type: Different ML algorithms are suited for different tasks. For example, if you're working on a classification task (e.g., predicting whether someone will develop a disease), decision trees or RF might be a good choice. If you're dealing with numerical predictions (like predicting weight change), linear regression could be appropriate.

Data Characteristics: Consider the type of data you're working with. If your data has nonlinear relationships, models like SVM with a nonlinear kernel or RF may perform better than linear models.

Dimensionality Reduction: In nutrition research, where there's often collinearity (overlap between variables), you might need techniques that reduce the complexity of your data, like PCA. This can help speed up computations and improve the quality of the analysis.

8.7.4 Testing Multiple Solutions

One of the strengths of ML is the ability to test multiple models and configurations. It's a good practice to try various ML algorithms and compare their performance.

Avoid Overfitting: When testing ML models, focus on their performance on test data (not just training data) to avoid overfitting (where the model performs well on training data but not on new data).

Hyperparameter Tuning: ML algorithms have parameters (called hyperparameters) that affect their performance. It's essential to fine-tune these parameters using techniques like grid search or random search to find the best settings.

Try Different Models: Even if you have a clear choice, it's worth testing other models. For example, while ensemble methods like RF typically outperform simpler models, there are cases where logistic regression could be better.

Stacking Models: Sometimes, combining multiple models (called stacking) can improve results. This approach merges the strengths of different algorithms, providing a potentially better solution.

8.7.5 Evaluating and Communicating the Results

Once you've tested your models, it's time to evaluate their performance and communicate the results.

Evaluation Metrics: Choose the right metrics for your task. For example, use area under the receiver operating characteristic (AUROC) curve for classification problems, or R^2 (coefficient of determination) for regression tasks.

Visualization: Visualizing results can help you understand model performance. You can use:

Plots: AUROC plots, R^2 plots, PCA score plots, etc.

Feature Importance: Tools like Shapley additive explanations (SHAP) or LIME can show which features are most important for a model's decision-making process.

XAI Techniques: If you're using complex models (e.g., RF or neural networks), explainable AI (XAI) methods like SHAP or partial dependence plots can help make the model's decision process more transparent.

8.8 LIMITATIONS OF MACHINE LEARNING IN NUTRITION RESEARCH

While ML holds great potential in nutrition research, it also has several limitations that researchers must be aware of. Here are the key challenges and drawbacks associated with the use of ML in this field.

8.8.1 Key Challenges Due to Non-rigorous Research Practice

8.8.1.1 Overoptimism and Rigorous Methodology

There is often an overestimation of the capabilities of ML due to non-rigorous research practices. For example:

Data Splitting Issues: If the dataset is small, simple training-test splits can lead to inconsistent results, making findings less reliable.

Lack of Cross-Validation: Nested cross-validation (CV), which is essential for testing model robustness, is rarely used in nutrition research, affecting the reliability of results.

These issues can result in seemingly impressive outcomes that may not be reproducible or generalizable.

8.8.1.2 Generalizability Concerns

ML models are typically tested on homogeneous datasets, often from affluent populations. As a result:

Performance in Diverse Populations: The performance of ML algorithms in different populations or settings (e.g., low-income or rural populations) is not guaranteed. This limits the generalizability of findings to broader, more diverse groups.

Data Collection Bias: ML models might be less effective when applied to data collected using different methods or from different geographic regions, reducing their overall utility.

For ML models to be widely useful in nutrition research, they must be adaptable and effective across diverse contexts and populations.

8.8.1.3 Flawed Feature Selection and Importance Estimation

A common issue in ML research is flawed feature selection, especially in algorithms like RF or XGBoost:

Correlated Features: These algorithms may downplay the importance of correlated features, while overestimating the significance of less relevant ones.

Inaccurate Importance Scores: Many studies report feature importance scores from ML algorithms without verifying them with additional techniques. This can lead to misleading conclusions about which features are most influential.

Such inaccuracies can distort interpretations and undermine the quality of the research.

8.8.2 Practical Challenges in Data Handling

While ML excels in transforming unstructured data into usable forms, this process is not without its challenges:

High Costs: Data collection, hardware, infrastructure, and skilled ML engineers can be costly. The financial and technical demands of ML applications in nutrition research can be prohibitive for some researchers.

Data Integration: Nutrition research often involves diverse data sources, collected using different tools and languages. Unifying this data into a coherent system for analysis can be difficult and time-consuming.

Complexity of Data Transformation: ML models can require sophisticated data preprocessing, which may not always be feasible for nutrition researchers who lack expertise in data science.

These practical issues can hinder the effective implementation of ML in nutrition studies, especially in resource-constrained settings.

8.9 OPPORTUNITIES, LIMITATIONS, AND CHALLENGES IN THE APPLICATION OF MACHINE LEARNING TO PRECISION NUTRITION (PN)

8.9.1 Opportunities in Machine Learning for Precision Nutrition (PN)

ML has emerged as a promising tool for addressing various questions in the field of PN. The coming years are expected to see a significant increase in studies applying ML to PN, fuelled by advancements in technology and computing power. Several institutions, including the NIH in the United States, are providing substantial financial support for the development of algorithms aimed at advancing PN research [76–81]. This growing investment and technological progress are likely to lead to more refined and accurate applications of ML in the field.

8.9.2 Integration of Broader Contexts into Precision Nutrition

Finally, PN tends to prioritize biomedical parameters, often overlooking the broader social, cultural, economic, and environmental factors that influence dietary choices. These factors play a crucial role in determining the acceptability and long-term adherence to dietary recommendations [75]. The integration of such diverse aspects into ML-based PN algorithms is still in its infancy. This challenge, known as the $p > n$ problem (where the number of parameters far exceeds the number of available data points), underscores the need for more comprehensive and better-quality data to allow ML to contribute effectively to a multidisciplinary approach to true PN.

8.10 FUTURE DIRECTIONS FOR RESEARCH AND DEVELOPMENT IN PN

To overcome these limitations, future research in ML and PN should focus on addressing the challenges of reproducibility, replicability, and data representativeness. More longitudinal studies with diverse populations are required to test the generalizability of current findings. Furthermore, the development of more standardized and universally accepted outcome measures is crucial to enable better comparisons across studies [80, 81]. Finally, a more holistic approach to PN should consider the integration of non-biomedical factors, such as cultural and environmental aspects, into the algorithms. These advancements will pave the way for ML to play a transformative role in personalized nutrition.

8.11 SUMMARY

The integration of ML into PN offers exciting opportunities for advancing personalized dietary recommendations. However, challenges such as limited reproducibility, lack of standard outcome measures, and the need for more comprehensive datasets must be addressed. Additionally, incorporating nonbiomedical factors into the decision-making processes is essential for making PN truly effective and applicable to a wide range of individuals. As these issues are resolved, ML is poised to make significant contributions to the field of PN.

PRACTICE QUESTIONS

1. What are the risks of overoptimism in ML research in nutrition?
2. How can improper data splitting affect the reliability of ML results?
3. Why is nested cross-validation (CV) important in ML, and how does its absence impact the findings?
4. What is meant by "generalizability" in the context of ML models in nutrition research?
5. Why is the performance of ML models in affluent societies not guaranteed to translate to diverse populations?
6. What factors limit the generalizability of ML findings from specific populations to broader, more diverse groups?
7. How do correlated features impact the importance ranking in ML algorithms like RF and XGBoost?
8. What is the potential consequence of reporting feature importance scores without verification from other methods?
9. Why is it essential to cross-check feature importance estimates when using ML in nutrition research?
10. What are some of the high costs associated with applying ML in nutrition research?
11. Why can data integration be challenging in nutrition studies, and how does this affect the use of ML?
12. How do unstructured data and complex data transformation processes pose challenges for nutrition researchers?
13. What are the main limitations of ML in nutrition research, and how can they be addressed?
14. Why is it important for ML models to be tested in diverse settings before they can be widely used in nutrition research?
15. How can further development and refinement of ML methods help reduce nutrition-related health problems?

REFERENCESS

1. Jorm LR. Commentary: towards machine learning-enabled epidemiology. *Int J Epidemiol* 2021;49(6):1770–3.
2. Breiman L. Statistical modeling: the two cultures (with comments and a rejoinder by the author) [Internet]. 2001;16(3):199–231. Available from: https://projecteuclid.org/journals/statistical-science/volume-16/issue-3/Statistical-Modeling–The-Two-Cultures-withcomments-and-a/10.1214/ss/1009213726.full
3. Song X, Mitnitski A, Cox J, Rockwood K. Comparison of machine learning techniques with classical statistical models in predicting health outcomes. *Stud Health Technol Inform* 2004;107:736–40.
4. Stolfi P, Castiglione F. Emulating complex simulations by machine learning methods. *BMC Bioinformatics* 2021;22(Suppl 14):483.
5. Sarker IH. Machine learning: algorithms, real-world applications and research directions. *SN Comput Sci* 2021;2(3):160.

6. Machine Learning in Nutrition Research. Downloaded from https://academic.oup.com/advances/advancearticle/doi/10.1093/advances/nmac103/6724380 by Wageningen University and Research – Library user on 01 Nov 2022.
7. Pavey TG, Gilson ND, Gomersall SR, Clark B, Trost SG. Field evaluation of a random forest activity classifier for wrist-worn accelerometer data. *J Sci Med Sport* 2017;20(1):75–80.
8. Basu S. Semi-supervised learning. In: Liu L, Özsu MT, editors. *Encyclopedia of database systems* [Internet]. Boston, MA: Springer; 2009. p. 2613–5. Available from: https://link.springer.com/referenceworkentry/10.1007/978-0-387-39940-9_609
9. Nguyen TP, Ho TB. Detecting disease genes based on semi-supervised learning and protein–protein interaction networks. *Artif Intell Med* 2012;54(1):63–71.
10. Davidson I. Clustering with constraints. In: Liu L, Özsu MT, editors. *Encyclopedia of database systems* [Internet]. Boston, MA: Springer; 2009 [cited 2022 Feb 28]. p. 393–6. Available from: https://link.springer.com/referenceworkentry/10.1007/978-0-387-39940-9_610
11. Silver D, Hubert T, Schrittwieser J, Antonoglou I, Lai M, Guez A, et al.A general reinforcement learning algorithm that masters chess, shogi, and go through self-play. *Science* 2018;362(6419):1140–4.
12. Mulani J, Heda S, Tumdi K, Patel J, Chhinkaniwala H, Patel J, et al. Deep reinforcement learning based personalized health recommendations. In: Dash S, Acharya B, Mittal M, Abraham A, Kelemen A, editors. *Deep learning techniques for biomedical andhealth informatics*. Cham, Switzerland: Springer; 2020. p. 231–55.
13. Zhou M, Mintz Y, Fukuoka Y, Goldberg K, Flowers E, Kaminsky P, et al. Personalizing mobile fitness apps using reinforcement learning [Internet]. *CEUR Workshop Proc* 2018;2068. Available from: www.ncbi.nlm.nih.gov/pmc/articles/PMC7220419/
14. Breiman L. Random forests. *Mach Learn* 2001;45(1):5–32.
15. Awad M, Khanna R, Awad M, Khanna R. Support vector regression. In: *Efficient learning machines*. Cham, Switzerland: Springer; 2015. p. 67–80.
16. Zhou M, Fukuoka Y, Goldberg K, Vittinghoff E, Aswani A. Applying machine learning to predict future adherence to physical activity programs. *BMC Med Inform Decis Mak* 2019;19(1):169.
17. Perakakis N, Yazdani A, Karniadakis GE, Mantzoros C. Omics, big data and machine learning as tools to propel understanding of biological mechanisms and to discover novel diagnostics and therapeutics. *Metabolism* 2018;87:A1.
18. Naimi AI, Balzer LB. Stacked generalization: an introduction to super learning. *Eur J Epidemiol* 2018;33(5):459.
19. Colmenarejo G. Machine learning models to predict childhood and adolescent obesity: a review. *Nutrients* 2020;12(8):2466.
20. Lecroy MN, Kim RS, Stevens J, Hanna DB, Isasi CR. Identifying key determinants of childhood obesity: a narrative review of machine learning studies. *Child Obes* 2021;17(3):153–9.
21. Reel PS, Reel S, Pearson E, Trucco E, Jefferson E. Using machine learning approaches for multi-omics data analysis: a review. *Biotechnol Adv* 2021;49:107739.
22. Tap J, Derrien M, Törnblom H, Brazeilles R, Cools-Portier S, Doré J, et al. Identification of an intestinal microbiota signature associated with severity of irritable bowel syndrome. *Gastroenterology* 2017;152(1):111–23.e8.
23. Li R, Li L, Xu Y, Yang J. Machine learning meets omics: applications and perspectives. *Brief Bioinform* 2022;23(1):bbab460.

24. Bzdok D, Altman N, Krzywinski M. Points of significance: statistics versus machine learning. *Nat Methods* 2018;15(4):233–4.
25. De Silva K, Jönsson D, Demmer RT. A combined strategy of feature selection and machine learning to identify predictors of prediabetes. *J Am Med Inform Assoc* 2020;27(3):396–406.
26. Wang X, Liu J, Ma L. Identification of gut flora based on robust support vector machine. *J Phys Conf Ser* 2022;2171(1):012066.
27. Namkung J. Machine learning methods for microbiome studies. *J Microbiol* 2020;58(3):206–16.
28. Cammarota G, Ianiro G, Ahern A, Carbone C, Temko A, Claesson MJ, et al. Gut microbiome, big data and machine learning to promote precision medicine for cancer. *Nat Rev Gastroenterol Hepatol* 2020;17(10):635–48.
29. Wiemken TL, Kelley RR. Machine learning in epidemiology and health outcomes research. *Annu Rev Public Health* 2020;41:21–36.
30. Wiens J, Shenoy ES. Machine learning for healthcare: on the verge of a major shift in healthcare epidemiology. *Clin Infect Dis* 2018;66(1):149–53.
31. Kirk D, Catal C, Tekinerdogan B. Precision nutrition: a systematic literature review. *Comput Biol Med* 2021;133:104365.
32. Goecks J, Jalili V, Heiser LM, Gray JW. How machine learning will transform biomedicine. *Cell* 2020;181(1):92–101.
33. Vilne B, ¸Kibilds J, Siksna I, Lazda I, Valci¸na O, Krumina A. Could artificial intelligence/machine learning and inclusion of diet-gut microbiome interactions improve disease risk prediction? Case study: coronary artery disease. *Front Microbiol* 2022;13:627892.
34. Chollet F. On the measure of intelligence [Internet]. 2019 Nov 5 [cited 2022 Jul 29]. Available from: https://arxiv.org/abs/1911.01547v2
35. Wang H, Ma C, Zhou L. A brief review of machine learning and its application. In: *2009 International Conference on Information Engineering and Computer Science*. New York, NY: IEEE; 2009. doi:10.1109/ICIECS.2009.5362936
36. Witten IH, Frank E, Hall MA, Pal CJ. *Data mining: practical machine learning tools and techniques*. Amsterdam, Netherlands: Elsevier; 2016.
37. Poss AM, Maschek JA, Cox JE, Hauner BJ, Hopkins PN, Hunt SC, et al. Machine learning reveals serum sphingolipids as cholesterolin dependent biomarkers of coronary artery disease. *J Clin Invest* 2020;130(3):1363.
38. Gou W, Ling CW, He Y, Jiang Z, Fu Y, Xu F, et al. Interpretable machine learning framework reveals robust gut microbiome features associated with type 2 diabetes. *Diabetes Care* 2021;44(2):358.
39. Dinh A, Miertschin S, Young A, Mohanty SD. A data-driven approach to predicting diabetes and cardiovascular disease with machine learning. *BMC Med Inform Decis Mak* 2019;19(1):211.
40. Canbay A, Kälsch J, Neumann U, Rau M, Hohenester S, Baba HA, et al. Non-invasive assessment of NAFLD as systemic disease—a machine learning perspective. *PLoS One* 2019;14(3):e0214436.
41. Khusial RD, Cioffi CE, Caltharp SA, Krasinskas AM, Alazraki A, Knight-Scott J, et al. Development of a plasma screening panel for pediatric nonalcoholic fatty liver disease using metabolomics. *Hepatol Commun* 2019;3(10):1311–21.
42. Ambale-Venkatesh B, Yang X, Wu CO, Liu K, Gregory Hundley W, McClelland R, et al. Cardiovascular event prediction by machine learning: the multi-ethnic study of atherosclerosis. *Circ Res* 2017;121(9):1092–101.

43. De Silva K, Lim S, Mousa A, Teede H, Forbes A, Demmer RT, et al. Nutritional markers of undiagnosed type 2 diabetes in adults: findings of a machine learning analysis with external validation and benchmarking. *PLoS One* 2021;16(5):e0250832.
44. Choi TY, Chang MY, Heo S, Jang JY. Explainable machine learning model to predict refeeding hypophosphatemia. *Clin Nutr ESPEN* 2021;45:213–9.
45. Khorraminezhad L, Leclercq M, Droit A, Bilodeau JF, Rudkowska I. Statistical and machine-learning analyses in nutritional genomics studies. *Nutrients* 2020;12(10):3140.
46. Ahmadi MN, Brookes D, Chowdhury A, Pavey T, Trost SG. Free-living evaluation of laboratory-based activity classifiers in preschoolers. *Med Sci Sports Exercise* 2020;52(5):1227–34.
47. Chowdhury AK, Tjondronegoro D, Chandran V, Trost SG. Ensemble methods for classification of physical activities from wrist accelerometry. *Med Sci Sports Exercise* 2017;49(9):1965–73.
48. Catal C, Akbulut A. Automatic energy expenditure measurement for health science. *Comput Methods Programs Biomed* 2018;157:31–7.
49. Sorino P, Campanella A, Bonfiglio C, Mirizzi A, Franco I, Bianco A, et al. Development and validation of a neural network for NAFLD diagnosis. *Sci Rep* 2021;11(1):20240.
50. Sorino P, Caruso MG, Misciagna G, Bonfiglio C, Campanella A, Mirizzi A, et al. Selecting the best machine learning algorithm to support the diagnosis of non-alcoholic fatty liver disease: a meta learner study. *PLoS One* 2020;15(10):e0240867.
51. Zou Q, Qu K, Luo Y, Yin D, Ju Y, Tang H. Predicting diabetes mellitus with machine learning techniques. *Front Genet* 2018;9:515.
52. Ramyaa R, Hosseini O, Krishnan GP, Krishnan S. Phenotyping women based on dietary macronutrients, physical activity, and body weight using machine learning tools. *Nutrients* 2019;11(7):1681.
53. Zellerbach, K, Ruiz, C. Machine learning to predict overeating from macronutrient composition, *2019 IEEE International Conference on Bioinformatics and Biomedicine (BIBM)*, San Diego, CA, USA, 2019, p. 1268–70, doi: 10.1109/BIBM47256.2019.8983166.
54. Ahmadi MN, Chowdhury A, Pavey T, Trost SG. Laboratory-based and free-living algorithms for energy expenditure estimation in preschool children: a free-living evaluation. *PLoS One* 2020;15(5):e0233229.
55. Yin Z, Sulieman LM, Malin BA. A systematic literature review of machine learning in online personal health data. *J Am Med Inform Assoc* 2019;26(6):561–76. https://doi.org/10.1093/jamia/ocz009.
56. Subramanian I, Verma S, Kumar S, Jere A, Anamika K. Multi-omics data integration, interpretation, and its application. *Bioinf Biol Insights* 2020;14:117793221989905.
57. Mezgec S, Eftimov T, Bucher T, Koroušic Seljak B. Mixed deep ´learning and natural language processing method for fake-food image recognition and standardization to help automated dietary assessment. *Public Health Nutr* 2019;22(7):1193–202.
58. van Erp M, Reynolds C, Maynard D, Starke A, Ibáñez Martín R, Andres F, et al. Using natural language processing and artificial intelligence to explore the nutrition and sustainability of recipes and food. *Front Artif Intell* 2021;3:115.
59. Bodnar LM, Cartus AR, Kirkpatrick SI, Himes KP, Kennedy EH, Simhan HN, et al. Machine learning as a strategy to account for dietary synergy: an illustration based on dietary intake and adverse pregnancy outcomes. *Am J Clin Nutr* 2020;111(6):1235–43.
60. Jones PJ, Catt M, Davies MJ, Edwardson CL, Mirkes EM, Khunti K, et al. Feature selection for unsupervised machine learning of accelerometer data physical activity clusters—a systematic review. *Gait Posture* 2021;90:120–8. Downloaded from https://academic.oup.com/advances/advance-66.

61. O'Driscoll R, Turicchi J, Hopkins M, Horgan GW, Finlayson G, Stubbs JR. Improving energy expenditure estimates from wearable devices: a machine learning approach. *J Sports Sci* 2020;38(13):1496–505.
62. Korduner J, Nilsson PM, Melander O, Gerl MJ, Engström G, Bachus E, et al. Proteomic and metabolomic characterization of metabolically healthy obesity: a descriptive study from a Swedish cohort. *J Obes* 2021;2021:6616983.
63. Wang DD, Hu FB. Precision nutrition for prevention and management of type 2 diabetes. *Lancet Diabetes Endocrinol* 2018;6:416–26.
64. Montañez CAC, Fergus P, Hussain A, Al-Jumeily D, Dorak MT, Abdullah R. Evaluation of phenotype classification methods for obesity using direct to consumer genetic data. In: *Lecture notes in computer science*. Berlin, Germany: Springer Verlag; 2017. p. 350–62.
65. Rodríguez-Pardo C, Segura A, Zamorano-León JJ, Martínez-Santos C, Martínez D, Collado-Yurrita L, et al. Decision tree learning to predict overweight/obesity based on body mass index and gene polymporphisms. *Gene* 2019;699:88–93.
66. Gardner CD, Kiazand A, Alhassan S, et al. Comparison of the Atkins, Zone, Ornish, and LEARN diets for change in weight and related risk factors among overweight premenopausal women: the A to Z Weight Loss Study: a randomized trial. *JAMA*. 2007;297:969–77. doi: 10.1001/jama.297.9.969. [Erratum, *JAMA* 2007;298: 178].
67. Fiehn O. Metabolomics—the link between genotypes and phenotypes. *Plant Mol Biol* 2002;48(1):155–71.
68. Acar E, Gürdeniz G, Khakimov B, Savorani F, Korndal SK, Larsen TM, et al. Biomarkers of individual foods, and separation of diets using untargeted LC-MS-based plasma metabolomics in a randomized controlled trial. *Mol Nutr Food Res* 2019;63(1):1800215.
69. Garcia-Aloy M, Llorach R, Urpi-Sarda M, Tulipani S, Salas-Salvadó J, Martínez-González MA, et al. Nutrimetabolomics fingerprinting to identify biomarkers of bread exposure in a free-living population from the PREDIMED study cohort. *Metabolomics* 2015;11(1):155–65.
70. O'Donovan, CB, Walsh, MC, Gibney, MJ, et al. (2016) Can metabotyping help deliver the promise of personalised nutrition? *Proc Nutr Soc* 75, 106–114.CrossRefGoogle ScholarPubMed
71. O'Sullivan A, Gibney MJ, Connor AO, Mion B, Kaluskar S, Cashman KD, et al. Biochemical and metabolomic phenotyping in the identification of a vitamin D responsive metabotype for markers of the metabolic syndrome. *Mol Nutr Food Res* 2011;55(5):679–90.
72. O'Donovan CB, Walsh MC, Woolhead C, Forster H, Celis-Morales C, Fallaize R, et al. Metabotyping for the development of tailored dietary advice solutions in a European population: the Food4Me study. *Br JNutr* 2017;118(8):561–9.
73. Cornelis MC, Erlund I, Michelotti GA, Herder C, Westerhuis JA, Tuomilehto J. Metabolomic response to coffee consumption: application to a three-stage clinical trial. *J Intern Med* 2018;283(6):544–57.
74. Heinzmann SS, Brown IJ, Chan Q, Bictash M, Dumas ME, Kochhar S, et al. Metabolic profiling strategy for discovery of nutritionalbiomarkers: proline betaine as a marker of citrus consumption. *Am J Clin Nutr* 2010;92(2):436–43.
75. Cheung W, Keski-Rahkonen P, Assi N, Ferrari P, Freisling H, Rinaldi S, et al. A metabolomic study of biomarkers of meat and fish intake. *Am J Clin Nutr* 2017;105(3):600–8.
76. Montanez CAC, Fergus P, Hussain A, Al-Jumeily D, Abdulaimma B, Hind J, et al. Machine learning approaches for the prediction of obesity using publicly available

genetic profiles. In: *Proceedings of the International Joint Conference on Neural Networks*. Piscataway, NJ: Institute of Electrical and Electronics Engineers Inc.; 2017. p. 2743–50.

77. López B, Torrent-Fontbona F, Viñas R, Fernández-Real JM. Single nucleotide polymorphism relevance learning with random forests for type 2 diabetes risk prediction. *Artif Intell Med* 2018;85:43–9.
78. Wang Y, Zhang L, Niu M, Li R, Tu R, Liu X, et al. Genetic risk score increased discriminant efficiency of predictive models for type 2 diabetes mellitus using machine learning: cohort study. *Front Public Heal* 2021;9:606711.
79. Celis-Morales C, Livingstone KM, Marsaux CFM, Macready AL, Fallaize R, O'Donovan CB, et al. Effect of personalized nutrition on health-related behaviour change: evidence from the Food4Me European randomized controlled trial. *Int J Epidemiol* 2017;46(2):578–88.
80. An J, Yoon SR, Lee JH, Kim H, Kim OY. Importance of adherence to personalized diet intervention in obesity related metabolic improvement in overweight and obese Korean adults. *Clin Nutr Res* 2019;8(3):171.
81. Letertre MPM, Giraudeau P, de Tullio P. Nuclear magnetic resonance spectroscopy in clinical metabolomics and personalized medicine: current challenges and perspectives. *Front Mol Biosci* 2021;8:698337.

9 Smart AI in Digital Twins for Healthcare

9.1 INTRODUCTION

Precision or personalized medicine is an emerging field that aims to better understand individual differences in patients with the same condition. This approach allows for customized screening, diagnostics, and treatment plans. It leverages biological databases, like the human genome sequence, along with biomarkers, to group patients for more targeted therapies. For years, technologies known as "omics" have been used to measure the activities of thousands of genes, proteins, and other molecular elements, generating complex data that helps link genetic information to physical traits. These genetic, microbial, proteomic, metabolic, clinical, and behavioural characteristics define a patient's health. Advanced computational tools can analyse these large datasets to address individual variations, resulting in more accurate clinical decisions and personalized treatments.

A promising application of precision medicine is the concept of a digital twin (DT), where a patient-specific virtual model is used to predict treatment outcomes and personalized care. The idea of DTs originates from the industry practice of creating virtual models of physical systems for planning, testing, risk assessment, and maintenance. Using technologies like cloud computing, 5G networks, and predictive software, these DTs interact with real-world data through sensors and devices. This setup enables real-time data exchange and analysis, allowing for the detection of patterns, prediction of issues, and optimization of performance. This concept has been widely used in various industries, including transportation, supply chain management, and manufacturing, where DTs help maintain machinery, predict failures, and improve operational efficiency.

In healthcare, a DT for a patient combines data from two main sources: deep phenotyping from medical records, genetic information, and imaging data, and real-world data from sensors and wearable devices. This data is processed and integrated to provide meaningful insights for decision-making and prediction. For example, in cardiology, DTs use statistical and mechanistic models to analyse data and predict patient outcomes. A patient's DT is selected based on similarities with a group of individuals in a population-based databank. The outcomes of virtual treatments applied to this twin are then used to refine the digital model and update the databank,

DOI: 10.1201/9781003489986-12

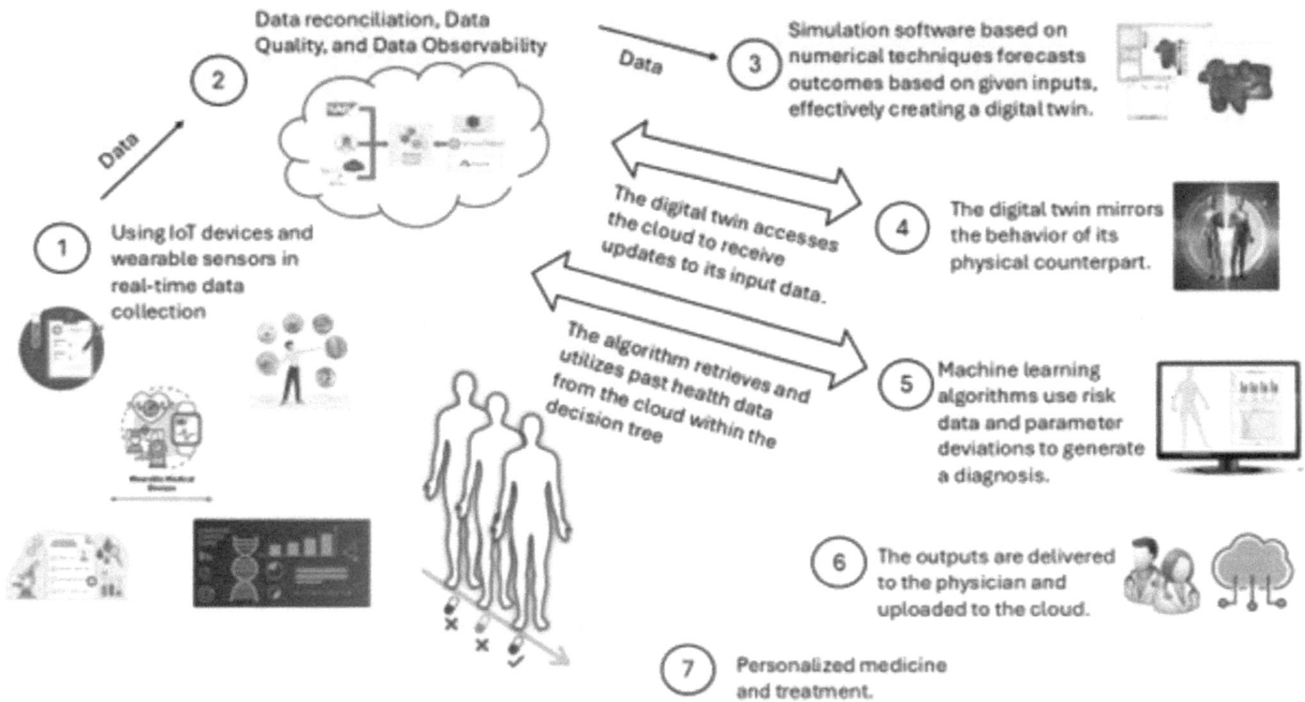

FIGURE 9.1 Core concepts of a digital twin model in precision medicine.

creating a dynamic feedback loop that enhances the accuracy and diversity of the dataset. This process draws on bioengineering and computer science to combine and analyse data from large patient groups.

DTs are transforming healthcare by enhancing industry practices and yielding tangible benefits for medical organizations and patient care. Notably, the global market for healthcare DTs was valued at approximately $16 billion in 2023 underscoring its substantial economic footprint. Industry experts project a strong annual growth trajectory for DT adoption, marking a remarkable milestone in healthcare innovation. Furthermore, studies indicate that around 66% of healthcare executives plan to increase investment in healthcare robotics within the next three years, reflecting confidence in DT technology's transformative impact on patient care and operational efficiency. This commitment underscores a lasting shift, signalling a foundational change in how technological advancements are shaping health management practices.

Given the potential of DTs in medical research, it is crucial to understand their capabilities fully. This chapter aims to explore the DT concept, technique, application, and research, summarize key concepts, identify contributing disciplines, and highlight emerging challenges in integrating DTs into precision medicine. It is shown in Figure 9.1.

9.2 WHAT IS A DIGITAL TWIN?

A DT is more than a static virtual representation of a physical system. It is a sophisticated, dynamic model that mirrors real-world systems in real time, enabling the study of their behaviour and offering predictive insights. By leveraging advanced simulations, machine learning (ML), and analytical reasoning, DTs facilitate data-driven decision-making, distinguishing them from simple static models or dummy replicas.

The concept of the "digital twin" was first introduced by Michael Grieves in 2005 within the context of product lifecycle management. Later, by 2010, NASA, along with John Vickers, broadened its scope, describing the DT as a virtual representation of a physical system. This definition emphasized three core components: the physical entity, its virtual model, and a live data connection between the two. This real-time linkage ensures that the virtual representation accurately reflects changes occurring in the physical system.

9.2.1 The Origin of the Digital Twin Concept

The DT concept traces its roots back to the 1960s when NASA adopted it for their space programme. Initially, DTs were used to simulate spacecraft systems, allowing the team to troubleshoot and resolve issues in real time as they arose. One notable application of the DT was during the Apollo 13 mission. When the spacecraft experienced a critical malfunction, the NASA team successfully used the DT to replicate onboard conditions and guide the astronauts and spacecraft safely back to Earth.

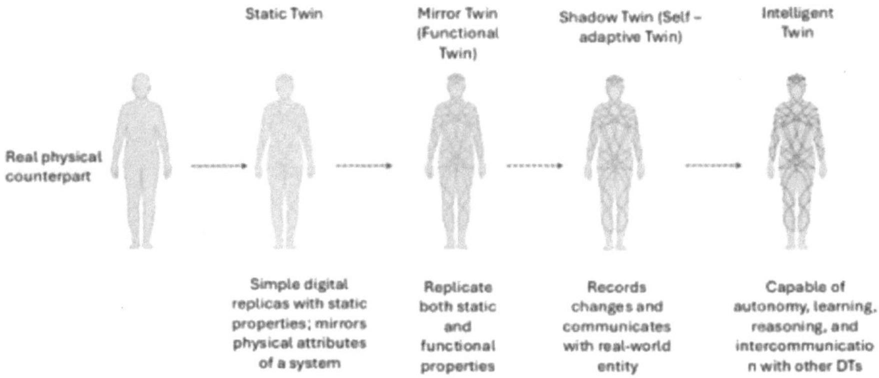

FIGURE 9.2 Evolution of digital twins.

9.2.2 Evolution of Digital Twins

Over time, the DT concept has evolved, with the literature describing several types of DTs (Figure 9.2):

Static Twin: Initially, DTs were simple digital replicas with static properties, mirroring only the physical attributes of a system.

Mirror Twin (Functional Twin): The next evolution introduced dynamic behaviour capabilities, enabling the twin to replicate both static and functional aspects of the physical system. Examples include DTs used in surgical planning and in silico clinical trial design.

Shadow Twin (Self-Adaptive Twin): This iteration incorporated real-time data acquisition and model updates, connected through a digital thread that records changes and communicates with the real-world entity. Shadow twins are used in medical device design, biomarker discovery, and biomanufacturing.

Intelligent Twin: The most advanced form, the intelligent DT, integrates artificial intelligence (AI) making it capable of autonomy, learning, reasoning, and intercommunication with other DTs. Intelligent twins dynamically exchange data with their physical counterparts, optimizing simulations and ML algorithms for decision-making and system optimization.

9.2.3 Characteristics of a Digital Twin

DTs possess several key characteristics that distinguish them as a powerful tool in the digital era (Figure 9.3):

1. **Connectivity**

   A DT is based on connectivity. It enables connection between the physical element and its digital counterpart. The sensors create the connectivity of physical products that obtain, integrate, and communicate data using various integration technologies.

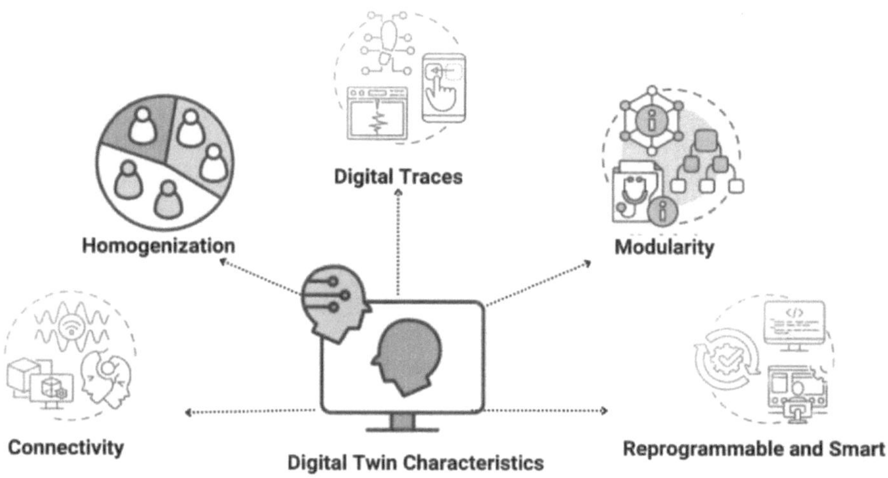

FIGURE 9.3 Digital twin characteristics.

2. **Homogenization**
   Homogenization has two components; first, spatially aligning data, and second to ensure that the quality of the data is compatible. DTs are both the consequence and enabler of homogenization of data. It allows decoupling of information from its physical form.
3. **Digital Traces**
   DT technologies leave digital traces. The trails are helpful to diagnose the source of the problem that occurred in case of machine malfunctions.
4. **Modularity**
   Modularity refers to the design and customization of products and production modules. The addition of modularity to functional models helps manufacturers gain the ability to tweak machines and models.
5. **Reprogrammable and Smart**
   DTs automatically enable re-programmability through sensors, AI techniques, and predictive analysis.

9.3 DT IN PRECISION HEALTH

We define a DT for precision health as a virtual model of an individual that enables dynamic simulations of potential treatment strategies, monitors and predicts health outcomes, and facilitates early interventions and prevention. This is achieved through multi-scale modelling that integrates various data sources, including clinical, genetic, molecular, environmental, and social factors.

DTs integrate diverse data sources, including electronic health records (EHRs), disease registries, and omics data – spanning proteomic, genomic, metabolomic, and bionic information – alongside demographic and lifestyle factors for each individual. The advent of new technologies powered by the internet of things (IoT) and artificial intelligence (AI), combined with the extensive availability of multidimensional data types (e.g., behavioural, biometric, cognitive, emotional, and psychological information) has significantly enhanced the exploration and application potential of DTs in healthcare. When fully leveraged, DTs could provide a transformative framework for future approaches to lifestyle, health, wellness, and chronic disease management.

In healthcare, DT functions across multiple stages, requiring sophisticated modelling techniques and tools for accurate model construction. Health IoT and mobile internet technologies play a crucial role in enabling real-time connectivity and data exchange between physical and virtual entities, supporting the dynamic interaction necessary for comprehensive DT applications.

A DT should possess five key attributes: it should be individualized, interconnected, interactive, informative, and impactful (referred to as the "5Is"). For a patient-specific DT, the model can offer personalized treatment strategies, leading to improved outcomes and reduced side-effects. Simultaneously, real-world patient data can be utilized to benchmark, validate, and refine the DT, enhancing its accuracy and effectiveness over time.

9.4 SMART AI AND ML TECHNIQUES IN DEVELOPING A DT MODEL

Understanding the differences and nuances among various virtual models is crucial, particularly regarding their data sources, applications, interaction methods, and visualization techniques. For instance, DTs focused on simulating specific organs, such as the living heart model, rely heavily on high-resolution imaging data. This approach differs significantly from disease-specific models in precision medicine, which are based on extensive molecular profiling and clinical datasets.

In medical research, constructing models forms the cornerstone of DT development, combining human anatomy with advanced digital technologies through techniques like image processing, digital data acquisition, and mathematical modelling. Researchers create intuitive three-dimensional (3D) representations by reconstructing two-dimensional (2D) cross-sectional images of the human body, aiding in the visualization of anatomical structures. A notable example is the Blue Brain Project, launched in 2005 as a collaboration between École Polytechnique Fédérale de Lausanne (EPFL) and IBM, aimed at modelling the neocortical column. Similarly, Schütt et al. developed a DT for the dynamic colon model, incorporating parameters like media viscosity and peristaltic wave speed, offering innovative insights into dissolution testing beyond traditional pharmacopoeial methods.

Modelling in this domain involves several methodologies, including mechanical modelling, mechanical network modelling, and statistical modelling. Mechanical models provide a framework for exploring theoretical concepts such as mass transfer within the human body and the electrophysiological phenomena linked to mechanics. These models can predict the emergent behaviour of biological systems in health and disease by integrating interactions among biological components.

Mechanical network models are typically formulated as coupled ordinary differential equations that describe the evolution of molecular concentrations over time in response to interactions and network stimuli. On the other hand, statistical models complement mechanical models by validating their outputs and offering mechanistic insights through inductive reasoning. Rigorous quantitative analysis is essential to validate the data generated from mechanical models. Sensors, including traditional and advanced spectroscopic types, are used in process analytical technology to capture and integrate real-time data or signals from interconnected bioreactors. Whole-cell projects focus on modelling specific cell types or organisms to enhance understanding of the molecular mechanisms underlying various diseases.

To improve real-time problem-solving, faster surrogate models can be developed using ML techniques. These models accelerate development, enhance quality, and allow modular approaches while addressing interactions between different subprocesses. However, challenges such as the need for further validation and achieving clinical interpretability continue to limit DT advancements. When applied to patients across different regions or medical centres, these models often require recalibration, and dataset quality must be carefully evaluated before use in clinical decision-making.

For successful clinical translation, creating synergies among models can enhance interpretability, verify adaptability, improve research outcomes, and expedite the integration of new technologies into clinical practice.

9.5 DIGITAL TWIN APPLICATIONS ACROSS SECTORS

Recent technological advancements – such as generative AI, cognitive computing, IoT, and sensor technologies – have paved the way for DTs to be utilized in various sectors, including aerospace, automotive, energy management, urban planning, construction, environmental management, climate modelling, and healthcare (Figure 9.4).

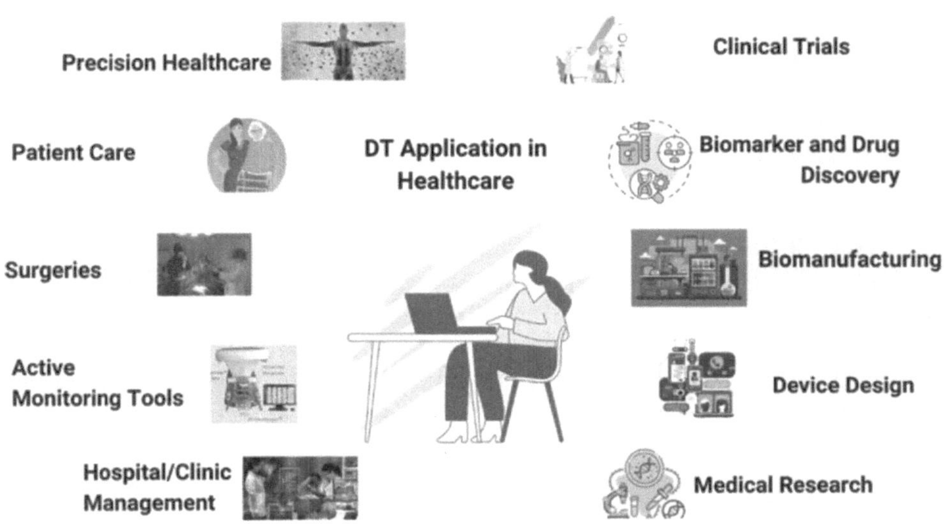

FIGURE 9.4 DT application in healthcare.

9.5.1 Digital Twins in Healthcare

The concept of DTs is increasingly being applied in areas such as precision medicine, cancer treatment, personalized training, and individual well-being. In healthcare, their applications extend to hospital operations, facility design, workflow optimization, decision-making, and tailored patient care. DTs can be customized to meet individual patient needs, serving as "well-being twins" to support health maintenance or as "illness twins" to enable personalized approaches to diagnosis, treatment planning, and survivorship care.

9.5.1.1 Types of DTs in Healthcare

In the context of healthcare, various types of DTs can be developed, as outlined in Table 9.1. These may focus on individual body systems or organs (such as the lungs or heart), specific bodily functions, finer components (like cellular or subcellular levels), or the entire human body. Additionally, DTs can be tailored for specific diseases or disorders. Composite DTs can integrate multiple DT types, and their accuracy depends on the incorporation of real-time, real-world data, including updates from clinical sensors and the autonomous capabilities of the DT.

A DT instance represents the physical twin throughout its lifespan, while a DT aggregate consists of multiple instances or other aggregates. The concept of a DT bank refers to a structured repository of these instances and aggregates, which can aid in matching clinical trials and advancing drug development. Conversely, a DT thread serves as a temporal data pipeline that tracks the relationships of data elements from birth to death over time (Table 9.1).

9.5.1.2 Integration of DT in a Healthcare Setting

The integration of DTs into healthcare presents numerous opportunities, such as:

- In silico clinical trial design
- Medical device design and testing
- Drug discovery and development
- Coordinated care and treatment planning
- Surgical planning and simulation
- Biomanufacturing process

9.5.1.3 In Silico Trials: A Game-Changer in Clinical Research

Traditional clinical trials are often prolonged and resource-intensive. A transformative alternative is emerging in the form of in silico trials (ISTs), which leverage digital simulations and advanced modelling to conduct virtual studies. With fewer than 10% of adult cancer patients participating in traditional trials and only a 5.1% success rate in oncology drug approvals, the inefficiencies are clear. Recruitment costs, particularly for rare diseases, further complicate the process, as patients may be unwilling to risk assignment to placebo or standard care groups. These issues frequently lead to underpowered studies or trial failures. ISTs address these challenges by creating virtual control and efficacy groups, optimizing recruitment strategies, and enabling more robust trial designs.

TABLE 9.1
Types of DTs in Healthcare Practices

Disease/Organ System	Mechanism of Action	Application	Source
Lung	Simulates blood flow and oxygen distribution	Forecasts ventilation needs	Lexma
Heart	Models the heart's structural and physiological functions	Personalization and optimization of cardiac devices Cardiac synchronization	Dassault, Medtronic, Boston Scientific, FDA; Siemens Healthineers23
	Simulates TAVR implantations using various aortic prostheses	Surgical preparation	Heart Navigator40
Spine	Forecasts vertebral fracture risk following stereotactic ablative radiotherapy	Enhanced radiation planning to minimize treatment-related side-effects	Ahmadian et al.
Alzheimer's disease	Projecting individual outcomes in neurological diseases	Digital twin (DT) simulation for managing clinical trial controls and, eventually, clinical interventions	Unlearnai
Mammary lesions	Image-driven virtual patients evaluating digital mammography versus tomosynthesis	Assess which imaging technique is most effective in identifying mammary lesions	VICTRE Trial
Pharyngeal cancer	Best treatment choice	Determine the most effective treatment strategy for pharyngeal cancer	Tardini et al.
Type 2 diabetes	Reversal of type 2 diabetes	Randomized controlled trial comparing precision treatment using digital twins to standard care	Cleveland Clinic Twin Health NCT05181449[88]
Mental health		General well-being	MindBank AI, IBM, Babylon, DigiTwin

DTs offer an advanced approach compared with external control arms that rely on historical data. DTs simulate individualized, model-based outcomes, providing more precise comparisons. For instance, companies, like Unlearn.AI, employ DTs in neurology trials for Alzheimer's and Parkinson's diseases. As regulatory agencies increasingly recognize in silico tools, their applications in medical device development and trial control arms are gaining momentum.

ISTs have proven capable of predicting treatment responses and determining optimal dosages tailored to specific disease characteristics. A notable example is the VICTRE study, which used a synthetic cohort of 2,986 virtual patients to compare digital mammography with digital breast tomosynthesis for detecting breast lesions. The findings closely aligned with those of a traditional trial involving 400 participants, demonstrating the efficacy of digital tools. The FDA supports these methods, emphasizing their value in regulatory evaluations and providing guidance for medical device development.

Synthetic control arms are further advancing ISTs by accurately predicting clinical outcomes. For example, a synthetic control group of 68 patients facilitated the extension of alectinib for non-small cell lung cancer across Europe. Similarly, synthetic controls played a role in expanding Palbociclib's use for HR-positive, HER2-negative advanced breast cancer in men and fast-tracking the approval of blinatumomab for a rare subtype of acute lymphoblastic leukaemia. These examples underscore ISTs as cost-effective tools that enhance clinical trial efficiency and regulatory processes.

Synthetic controls have been used across diverse conditions, including cancer, hepatitis C, and rheumatoid arthritis. By integrating multimodal data – such as clinical, genomic, radiomic, and socio-economic inputs – ISTs leverage AI-driven analyses to improve predictive reliability. Continuously updated with new patient data, AI-enhanced ISTs build on prior simulations, offering ever-improving accuracy.

The design of ISTs depends on faithfully replicating human physiology and pathology, accounting for genetic and environmental factors. Recent applications include in silico models that predict patient responses to immunotherapies, refining clinical trial frameworks. DTs designed for clinical trials feature advanced analytics and AI-driven decision-making tools, enabling real-time adaptive support and simulating human responses with remarkable accuracy.

9.5.1.4 Transforming Medical Device Development

The field of medical device design and testing is embracing digitally customizable DT models, particularly for organ-specific applications. A key initiative, the "SIMULIA Living Heart" project, was launched in 2014 by Dassault Systèmes in collaboration with the FDA. This project aims to create a validated virtual heart model to simulate drug interactions and enhance cardiac device development. Partnerships with companies like Philips and Siemens Healthineers have expanded the project's scope, accelerating advancements in device design. Similarly, FEops integrates AI with heart DT models to improve the management of structural heart diseases. Their platform, FEops HEARTguide, supports procedures such as transcatheter aortic valve implantation (TAVI) and left atrial appendage occlusion (LAAO) and is globally available, including in the European Union and the United States.

DT technology has also been applied to optimize lung ventilation strategies. Projects like Project BreathEasy developed virtual lung models during the COVID-19 pandemic to improve treatment. The concept extends to initiatives like the Living Brain, which aims to monitor neurodegenerative diseases, laying the groundwork for comprehensive human body DT models. These developments promise personalized care and improved resource management in healthcare.

9.5.1.5 Revolutionizing Drug Discovery and Development

Drug discovery traditionally requires significant time and resources, with high failure rates. Digital technologies, such as computer-aided drug discovery (CADD) and DT modelling, are reshaping this process. By integrating ML, methods like structure-based and ligand-based drug discovery streamline high-throughput screening of ADME-Tox properties (absorption, distribution, metabolism, excretion, and toxicity). DTs simulate biochemical reactions, reducing the time and costs associated with drug development.

In silico methods have contributed to breakthroughs, including the development of HIV inhibitors and anti-cancer agents. Leading pharmaceutical companies, such as GSK and Takeda, utilize DTs to optimize manufacturing, improve laboratory workflows, and accelerate the drug development pipeline, ultimately benefiting public health.

9.5.1.6 Enhancing Coordinated Care and Treatment Planning

DTs in healthcare are revolutionizing care delivery by simulating patient physiology, hospital workflows, and resource allocation. These applications fall into three main categories:

Process Optimization: DTs improve ICU monitoring and trauma management by enabling early detection of anomalies and coordinating timely responses.

Product Design: Embedded DTs in wearable devices facilitate real-time health management, as seen in frameworks like CloudDTH for elderly care and DTCoach for dementia support.

System Performance: Hospital DT systems, such as GE's Care Command, optimize resource distribution and reduce patient risk during crises, such as the COVID-19 pandemic.

By combining real-time data with predictive analytics, DTs are transforming patient care and hospital operations, minimizing medical errors, and boosting efficiency.

In summary, ISTs and DTs are reshaping healthcare by improving clinical trial design, medical device development, drug discovery, and coordinated care. These technologies promise significant advancements in personalized medicine, operational efficiency, and regulatory processes.

9.5.2 Surgical Planning and Simulation

DTs minimize surgical risks by simulating anatomical structures and procedures:

Cardiac Surgery: Heart models assist in complex procedures like valve replacements by simulating prosthesis options and implant depths.

Orthopaedic Surgery: DTs guide implant selection and postoperative strategies tailored to individual patients. Advanced AI techniques simulate bone surgeries, predicting outcomes more accurately.

Cancer Treatment: For metastatic spine cancer, DTs estimate fracture risks post-radiotherapy, aiding in treatment planning.

By offering preoperative insights, DTs enhance precision and safety in surgeries.

9.5.2.1 Biomanufacturing Process

Biomanufacturing uses biological systems to produce medicines, vaccines, and tissues. Advanced technologies enhance efficiency:

Predictive Modelling: In silico techniques, such as those employed by In Silico Biotechnology AG, optimize biopharmaceutical production through simulations.
AI and IoT: Real-time monitoring and predictive adjustments improve quality control and resource management.
Industry 4.0: This revolution integrates robotics and AI to automate complex tasks, enabling flexible, scalable production of custom medical products.

These innovations ensure efficient, high-quality biopharmaceutical manufacturing.

9.6 CLINICAL APPLICATIONS OF DIGITAL TWINS

DTs bridge traditional medicine with precision healthcare, offering organ-specific and disease-specific insights. Applications include

Device Testing: Simulating medical devices like pacemakers for improved performance.
Drug Modelling: Enhancing pharmacokinetics and pharmacodynamics studies.
Comprehensive Models: Multi-scale frameworks analyse disease progression and anatomical interactions across time and context.

These advancements support personalized care and refine medical research methodologies.

9.6.1 Challenges in Clinical Translation and Development of Digital Twins (DTs) for Health

Data Acquisition and Integration: One of the main challenges in creating a DT for health is acquiring accurate, real-time data from multiple sources and ensuring seamless integration. Health data comes from various sources like EHRs, medical imaging, wearable devices, and genetic databases. These data are often stored in different formats and systems, making it difficult to integrate them into a unified DT model. Overcoming interoperability issues is a key challenge, as it requires standardizing different data formats and medical coding systems. Additionally, real-time data streaming and synchronization are essential for accurate DT models, but this constant data flow can be difficult to manage.

Data Privacy and Security: Since DTs rely on extensive health data, including sensitive personal health information, privacy and security are paramount. Safeguarding patient data against unauthorized access, breaches, or misuse is a major concern. Compliance with data protection regulations like Health Insurance Portability and Accountability Act (HIPAA) and General Data Protection Regulation (GDPR) adds complexity to this process. Measures such as data encryption, secure storage, and ensuring compliance with these regulations are crucial to protect patient confidentiality while enabling access to necessary health data.

Data Quality and Accuracy: The reliability of DTs in healthcare heavily depends on the quality and precision of the data used. Inaccurate or incomplete data can lead to flawed representations and unreliable insights. However, health data is frequently scattered across various healthcare-providers, making it challenging to compile comprehensive datasets. Additionally, inaccuracies from sensors, individual patient variability, or data entry errors can introduce noise into the data. DTs also require longitudinal data that reflects changes over time, but such data is often limited or incomplete. Inconsistencies and missing information can significantly impede the development of an accurate DT, and ensuring data quality from diverse sources over time remains a major challenge.

Data Bias and Fairness: Data bias is another significant challenge when developing DTs for health. The accuracy and effectiveness of a DT depend on the quality, completeness, and representativeness of the data used to create it. If the data used to build DTs is biased – such as being skewed towards specific demographics or conditions – it can result in inaccurate or unfair models. For instance, health data might reflect biases related to race, gender, or other factors, which can lead to suboptimal recommendations. This could exacerbate existing health disparities, making it crucial to ensure that DTs are built on balanced datasets to avoid discrimination and ensure fairness in healthcare recommendations.

Creating and maintaining effective DTs for healthcare involves overcoming challenges related to data acquisition, privacy, quality, and bias. Ensuring accurate, secure, and representative data is essential for building reliable DT models that can help improve healthcare outcomes.

9.7 ETHICAL CONSIDERATIONS IN DIGITAL TWINS FOR HEALTH

Developing DTs for health brings about several important ethical considerations. These include

Informed Consent: Obtaining clear and informed consent from individuals is crucial before collecting their data for DT development. Individuals must fully understand how their data will be used, stored, and shared.

Data Ownership and Control: There needs to be clarity around who owns the health data used to create DTs and who controls access to it. This includes considering the rights of individuals to control their personal health data.

Patient Autonomy: Individuals should retain control over decisions regarding their health data and DT models. Ensuring patient autonomy is important for maintaining trust and ethical standards.

Healthcare Equity: It is vital that DTs do not exacerbate existing health disparities. Developers must ensure that DT technology is accessible to all, regardless of socio-economic or demographic factors.

Ethical Data Sharing: When sharing data for DT development, ethical guidelines should be established to ensure secure, responsible data handling, including data anonymization to protect patient privacy.

Data Accuracy and Bias Prevention: Ensuring that DTs are built on accurate data is essential to prevent bias and avoid harmful effects. Inaccurate or biased models can lead to unequal healthcare recommendations, further deepening health inequities.

9.8 MODELLING CHALLENGES IN DIGITAL TWINS FOR HEALTH

Modelling the human body and behaviours presents significant challenges. Human bodies are complex systems influenced by numerous dynamic factors, and creating models that accurately reflect this complexity is difficult. Furthermore, human behaviours in real-world environments involve intricate causal relationships that must be captured in DTs.

Socio-ethical Considerations: Before deploying DTs for healthcare, socio-ethical concerns must be addressed. While DTs can bring societal benefits and equalize access to healthcare, they can also perpetuate inequality. For instance, patterns identified in population data may lead to segmentation, where certain groups may not benefit equally from DT technology. The high cost of developing DT systems and privacy concerns are critical factors that could widen the socio-economic gap, leading to further digital divides.

Future Modelling Focus: Future efforts will likely focus on developing multi-scale models that capture data from various levels of observation, such as disease progression or anatomical systems, across different time frames and contexts.

9.9 COMPUTING INFRASTRUCTURE FOR DIGITAL TWINS

The successful integration of DTs in healthcare relies on advancements in

Computing Power: High-performance systems handle complex simulations and real-time data.

5G Networks: Enhanced data transfer speeds improve DT responsiveness.

AR/VR: Immersive tools facilitate better interaction with DTs for medical research.

Blockchain: Ensures secure, decentralized data storage.

Investments in these areas will enhance DT capabilities and accuracy.

9.10 BUSINESS MODELS FOR DIGITAL TWIN HEALTH PLATFORMS

Sustainable business models are crucial for widespread DT adoption in healthcare. Early implementation by government agencies and military organizations for monitoring specialized populations could pave the way. Long-term, collaborative efforts across sectors will drive the development of equitable, scalable DT systems that revolutionize healthcare delivery.

9.11 ADDRESSING CHALLENGES IN DIGITAL TWIN IMPLEMENTATION FOR HEALTHCARE

The development and successful integration of DTs in healthcare require a collaborative, multidisciplinary effort. Effective teamwork among healthcare practitioners, data scientists, technologists, policymakers, patients, and advocates is vital to establish robust frameworks, standards, and best practices for designing, deploying, and utilizing DTs in the healthcare domain.

9.11.1 Governance and Regulatory Oversight

Robust governance mechanisms are necessary to safeguard the rights of individuals represented by DTs, ensure data privacy and security, and maintain fairness and transparency in data use. Adhering to applicable regulatory standards and ethical principles, while maintaining accountability, is key to ensuring the responsible and equitable application of DT technology. These efforts are crucial for fostering trust among stakeholders and supporting the ethical use of DTs in healthcare.

9.11.2 Achieving Data Interoperability

Healthcare data is often generated from diverse sources and stored in varying formats, posing challenges for integration into unified DT models. Addressing this requires establishing interoperability protocols and standards to facilitate smooth data exchange and integration across healthcare systems, devices, and platforms. Standardization ensures that disparate data sources can be effectively combined, enabling the creation of accurate and comprehensive DTs.

9.11.3 Building Infrastructure for DT Implementation

The adoption of DT technology in healthcare demands significant computational capabilities and robust infrastructure to manage the increasing volume, speed, and complexity of health data. Healthcare organizations need to invest in scalable, reliable computing systems capable of processing and analysing extensive datasets in real time. Such infrastructure is essential to ensure DTs remain effective as the complexity of healthcare data continues to grow.

Successfully addressing these challenges involves establishing strong governance frameworks, setting data interoperability standards, and investing in advanced infrastructure. These foundational steps are critical to integrating DTs into healthcare systems, paving the way for enhanced patient care and improved operational efficiency.

9.12 FUTURE OUTLOOK OF DIGITAL TWINS IN HEALTHCARE

With each patient being unique, collecting deep digital phenotypes is essential for developing accurate, patient-specific DTs. This review explores the growing utility of DTs across various healthcare sectors and highlights emerging use cases. Many of these applications are being realized through consortium-driven approaches that leverage collective resources and advanced technologies to maximize the potential of DTs in medicine and beyond.

DTs have the potential to revolutionize various facets of patient care, healthcare operations, and medical research. The integration of DT technology in healthcare is anticipated to drive substantial transformation in community health systems, fundamentally reshaping patient care delivery. Over the next six years, DT adoption in healthcare facilities is expected to accelerate significantly. Furthermore, in the coming years, DT technology is projected to see extensive application in precision medicine and precision nutrition, paving the way for highly individualized therapeutic and dietary interventions.

9.13 SUMMARY

DTs are virtual replicas of physical objects or systems that allow real-time tracking, analysis, and continuous improvement. In healthcare, DTs create a dynamic representation of medical entities – be it patients, devices, or entire hospital systems – by capturing and updating data through embedded sensors and IoT devices. Advanced tools like ML and big data analytics then analyse this data, providing valuable insights.

In patient care, DTs enable personalized treatments by incorporating an individual's unique factors, including genetics, physiology, family history, and lifestyle habits. This technology empowers patients to access reliable medical advice, book appointments, transfer prescriptions, and review doctors' notes and appointment recordings online, making care more accessible and tailored to individual needs.

For healthcare facilities, DTs offer a comprehensive view of operations. Hospitals can simulate changes to departments or resource allocation, visualize the impact on workflows, and optimize layouts and asset distribution in response to these simulations. With powerful data inputs like real-time locating systems, DTs help forecast equipment availability and resource needs, ensuring more efficient healthcare delivery.

As the healthcare sector rapidly digitizes, patients are evolving from passive recipients of care to informed participants seeking value. The widespread use of smartphones and digital access to medical information empowers patients and their families to make more informed healthcare choices. Overall, DT technology is reshaping healthcare, driving both operational improvements and enhancing patient engagement and personalization.

PRACTICE QUESTIONS

1. What is a digital twin, and how does it function within the healthcare sector?
2. Describe how digital twin technology can enhance patient care. Provide examples related to personalized treatments and patient engagement.
3. Explain the role of sensors, IoT devices, and data analytics in creating a digital twin. How do these technologies contribute to making the digital twin "live" and "dynamic"?
4. How do digital twins allow healthcare providers to simulate changes in hospital operations and assess potential impacts on workflows and resource management?
5. Discuss the significance of data interoperability and computational infrastructure in the successful deployment of digital twins in healthcare. Why are these elements critical?
6. What are some ethical and privacy considerations associated with using digital twins in healthcare? How can these be addressed?
7. Explain the importance of data quality and real-time synchronization in the effectiveness of digital twins for healthcare. What challenges might arise in maintaining these standards?
8. How can digital twins contribute to personalized medicine? Give examples of patient characteristics that can be integrated into a digital twin model to tailor treatment plans.
9. In what ways has the increasing availability of smartphones and digital platforms changed the role of patients in the healthcare process with respect to digital twins?
10. Describe how digital twins can improve operational efficiency in healthcare institutions. Include examples such as resource forecasting and asset management.
11. What role does collaboration between healthcare professionals, data scientists, and technologists play in the development and implementation of digital twins?
12. How might future advancements in high-performance computing, 5G, and augmented/virtual reality further enhance the capabilities of digital twins in healthcare?
13. Discuss the concept of data interoperability in the context of digital twins. Why is it challenging, and what standards are necessary to improve data integration?
14. What are some potential business models that could support the adoption and growth of digital twins in healthcare? Who are likely early adopters of this technology?
15. How might digital twins help address healthcare inequities, and what measures can be taken to ensure they do not exacerbate them?

BIBLIOGRAPHY

Ahmadian, H. et al. A digital twin for simulating the vertebroplasty procedure and its impact on mechanical stability of vertebra in cancer patients. *Int J. Numer. Method. Biomed. Eng.* 38, e3600 (2022).

Ahmadian, H. et al. Toward an artificial intelligence-assisted framework for reconstructing the digital twin of vertebra and predicting its fracture response. *Int J. Numer Method Biomed. Eng.* 38, e3601 (2022).

Ai, M. B. Mind Bank AI www.mindbank.ai/ (2022).

Angulo, C. Gonzalez-Abril L.; Raya, C.; Ortega, J. A. A proposal to evolving towards digital twins in healthcare. In *Proceedings of the International Work-Conference on Bioinformatics and Biomedical Engineering*, Granada, Spain, 418–427 (2020).

Antonelli, L., Guarracino, M. R., Maddalena, L. & Sangiovanni, M. Indhupriya Subramanian, Srikant Verma, Shiva Kumar, Abhay Jere, Krishanpal Anamika, Integrating imaging and omics data: A review. *Biomed. Signal Process. Control* 52, 264– 280 (2019).

Aubert, K. et al. Development of digital twins to optimize trauma surgery and postoperative management. A case study focusing on tibial plateau fracture. *Front. Bioeng. Biotechnol.* 9, 722275 (2021).

Babylon Health www.emed.com/uk (2023).

Badano, A. et al. Evaluation of digital breast tomosynthesis as replacement of full-field digital mammography using an in silico imaging trial. *JAMA Netw. Open* 1, e185474 (2018).

Baillargeon, B., Rebelo, N., Fox, D. D., Taylor, R. L. & Kuhl, E. The Living Heart Project: A robust and integrative simulator for human heart function. *Eur. J. Mech. A Solids* 48, 38– 47 (2014).

Beiwe, B. http:// beiwe.wpeng ine.com (2023).

Bruynseels, K., Santoni de Sio, F. & van den Hoven, J. Digital twins in health care: ethical implications of an emerging engineering paradigm. *Front. Genet. 9,* 31 (2018).

Dang, J. et al. Predictive modeling in neurocritical care using causal artificial intelligence. *World J. Crit. Care Med* 10, 112–119 (2021).

Digipredict. *Digipredict Consortium* www.digi predict.eu/outcomes/ (2022).

DigiTwin. Medical Augmented Intelligence and DigiTwin www.mai.ai/digitwin/ (2022).

Elkefi, S. & Asan, O. Digital twins for managing health care systems: *rapid literature review. J. Med. Internet Res.* 24, e37641 (2022).

Erol, T., Mendi, A. F. & Doğan, D. Digital Transformation Revolution with Digital Twin Technology 2020 4th International Symposium on Multidisciplinary Studies and Innovative Technologies (ISMSIT), 1–7, Istanbul, Turkey, (2020). https://doi.org/10.1109/ISMS IT50672.2020.9254 288.

Ferdousi R., Laamarti F., Hossain M. A., Yang C. & El Saddik A. *Digital twins for well-being: an overview.* https://doi.org/10.12688/digitaltwin.17475.2 (2022).

Fisher, C. K. et al. Machine learning for comprehensive forecasting of Alzheimer's disease progression. *Sci. Rep.* 9, 13622 (2019).

Gargalo, C. L. et al. *In Digital Twins: Tools and Concepts for Smart Biomanufacturing* (eds Christoph H., Ralf P., & Johannes M) 1–34, Springer International Publishing (2021).

Hay, M., Thomas, D. W., Craighead, J. L., Economides, C. & Rosenthal, J. Clinical development success rates for investigational drugs. *Nat. Biotechnol.* 32, 40–51 (2014).

He, X. et al. Towards a shape-performance integrated digital twin for lumbar spine analysis. *Digital Twin* 1, 8 (2021).

Healthineers, S. The value of digital twin technology, www.siem ens-healthineers.com/en-us/services/value-partnerships/asset-center/white-papers-articles/value-of-digitaltwin-technology.

Hernigou, P., Safar, A., Hernigou, J. & Ferre, B. Subtalar axis determined by combining digital twins and artificial intelligence: influence of the orientation of this axis for hindfoot compensation of varus and valgus knees. *Int Orthop.* 46, 999– 1007, (2022). https://doi.org/10.1007/s00264-022-05311-6.

Human Digital Twin, O. R. C. *Human Digital Twin, OnePlanet Research Center* (accessed 12/27/2022).

Insel, T. R. Digital phenotyping: technology for a new science of behavior. *JAMA* 318, 1215–1216 (2017).

Kamel Boulos, M. N. & Zhang, P. Digital twins: from personalised medicine to precision public health. *J. Pers. Med.* 11 (2021) https://doi.org/10.3390/jpm11080745.

Kolla, L., Gruber, F. K., Khalid, O., Hill, C. & Parikh, R. B. The case for AI-driven cancer clinical trials – The efficacy arm in silico. *Biochim Biophys. Acta Rev. Cancer* 1876, 188572 (2021).

Le, T., Su, S. & Shahriyari, L. Investigating optimal chemotherapy options for osteosarcoma patients through a mathematical model. *Cells* 10, (2021) https://doi.org/10.3390/cells10082009.

Lexma. Project Breath Easy Lexma Digital Twin Lung, http:// projec tbre athe asy.org/(2023).

Li, X. et al. A dynamic single cell-based framework for digital twins to prioritize disease genes and drug targets. *Genome Med.* 14, 48 (2022).

Liu Y., Lin. Z. & Yuan Y. A novel cloud-based framework for the elderly healthcare services using digital twin. *IEEE Access,* 49088–49101 (2019).

Lopez-Minguez, J. R., Martinez-Caceres, G., Gonzalez-Fernandez, R., Nogales-Asensio, J. M. & Millan-Nunez, V. Applications designed to successfully implant in challenging left atrial appendage occlusion cases: a new tool for the interventional cardiologist. *Int. J. Cardiovasc. Imag.* 37, 2749–2751 (2021).

Marti-Bonmati, L. et al. PRIMAGE project: predictive in silico multiscale analytics to support childhood cancer personalised evaluation empowered by imaging biomarkers. *Eur. Radio. Exp.* 4, 22 (2020).

Merkel, M. J. et al. Statewide real-time tracking of beds and ventilators during coronavirus disease 2019 and beyond. *Crit. Care Explor* 2, e0142 (2020).

Obaid, D. R., Smith, D., Gilbert, M., Ashraf, S. & Chase, A. Computer simulated "Virtual TAVR" to guide TAVR in the presence of a previous Starr-Edwards mitral prosthesis. *J. Cardiovasc. Comput. Tomogr.* 13, 38–40 (2019).

Onscale, L. Digital twins of lungs to improve COVID-19 patient outcomes, https://onscale.com/blog/project-bre athe asy-digital-twins-of-lungs-to-improve-covid-19-patients-outcomes/ (2022).

Park, S., Park, C. H., Choi, D. H., Hong, J. K. & Lee D. Y. Bioprocess digital twins of mammalian cell culture for advanced biomanufacturing. *Curr. Opin. Chem. Eng.* 33, (2021). https://doi.org/10.1016/j.coche.2021.100702.

Parrott, A., Umbenhauer, B. & Warshaw, L. Digital Twins Bridging the Physical and Digital. Deloiyye Insights. www2.deloitte.com/us/en/insig hts/focus/tech-trends/2020/digital-twin-applications-bridging-the-physical-and digital.html (2020).

Popa, E. O., van Hilten, M., Oosterkamp, E. & Bogaardt, M. J. The use of digital twins in healthcare: socio-ethical benefits and socio-ethical risks. *Life Sci. Soc. Policy* 17, 6 (2021).

Rifaioglu, A. S. et al. Recent applications of deep learning and machine intelligence on in silico drug discovery: methods, tools and databases. *Brief. Bioinform* 20, 1878–1912 (2019).

Robinson, P. N. Deep phenotyping for precision medicine. *Hum. Mutat.* 33, 777– 780 (2012).

Shaker, B., Ahmad, S., Lee, J., Jung, C. & Na, D. In silico methods and tools for drug discovery. *Comput Biol. Med* 137, 104851 (2021).

Stahlberg, E. A. et al. Exploring approaches for predictive cancer patient digital twins: opportunities for collaboration and innovation. *Front. Digit. Health* 4, 1007784 (2022).

Sun, T. et al. The digital twin: a potential solution for the personalized diagnosis and treatment of musculoskeletal system diseases. *Bioeng. (Basel)* 10, 627, (2023). https://doi.org/ 10.3390/ bio engi neer ing1 0060 627.

Sun, T., He, X., Song, X., Shu, L. & Li, Z. The digital twin in medicine: a key to the future of healthcare? *Front Med (Lausanne)* 9, 907066 (2022).

Systems, D. *The Living Brain Project*. (2022).

The Cleveland Clinic. Randomized Controlled Trial of Digital Twin Precision Treatment: A Novel Whole Body Digital Twin Enabled Precision Treatment for Type 2 Diabetes (TPT), https:// cli nica ltri als.gov/ ct2/ show/ NCT0 5181 449 (2023).

Trobinger M. et al. A dual docto-patient twin paradigm for transparent remote examination, diagnosis, and rehabilitation. IEEE/ RSJ International Conference on Intelligent Robots and Systems (IROS), 2933– 2940 (2021).

Twin, I. D. IBM Digital Twin www.ibm.com/products/maximo/predictive-maintenance?cm_sp=ibmdev--developer-articles--product (2023).

Watson, E. R., Taherian Fard, A. & Mar, J.C. Computational methods for single-cell imaging and omics data integration. *Front. Mol. Biosci.* 8, https://doi.org/ 10.3389/fmolb.2021.768 106 (2021).

Wickramasinghe, N. et al. Digital twins to enable better precision and personalized dementia care. *JAMIA Open* 5, ooac072 (2022).

Wu, C. et al. MRI-based digital models forecast patient-specific treatment responses to neoadjuvant chemotherapy in triple-negative breast cancer. *Cancer Res* 82, 3394– 3404 (2022).

Zhou, C. et al. Virtual patients for mechanical ventilation in the intensive care unit. *Comput. Methods Prog. Biomed.* 199, 105912 (2021).

Section IV

Trustworthy AI in Smart Defence CPS Design

10 Security-Aware Computing for UAVs in Defence

10.1 INTRODUCTION

Unmanned aerial vehicles (UAVs) have a long history in military applications, evolving significantly since their inception. In this chapter, we will look into security aspect of modern UAVs in defence. The concept of UAVs dates back to the early 20th century, when they were initially conceived for reconnaissance purposes. During the First World War, both Britain and the United States developed pilotless vehicles such as Britain's Aerial Target and the American Kettering Bug. Although these early models showed promise in flight tests, they were not used operationally during the war. Figure 10.1 shows Churchill watching preparations being made for the launch of a pilotless target drone – Queen Bee. The Cold War era marked a period of substantial innovation and strategic development for unmanned vehicles. Military powers sought advantages in intelligence, reconnaissance, and combat scenarios, leading to the deployment of more advanced UAVs [Total Military Insight 2024]. The evolution of UAVs continued through the 20th century, transitioning from simple radio-controlled vehicles to sophisticated platforms capable of carrying sensors, electronic receivers, transmitters, and offensive ordnance. By the 21st century, the UAV technology became very much sophisticated and advanced. In the next section, we will briefly look at various modern applications of UAVs.

10.1.1 Unmanned Aerial Vehicles (UAV) in Defence Applications

UAVs have become indispensable tools in modern defence operations, providing versatile capabilities across a range of mission-critical applications. The key areas where UAVs contribute significantly to defence are

- **Surveillance and Reconnaissance:** UAVs play a crucial role in gathering intelligence, monitoring hostile territories, and providing real-time situational awareness. Their ability to operate in high-risk environments without endangering human lives makes them ideal for reconnaissance missions. Examples include the use of MQ-9 Reaper drones for intelligence gathering in military operations [BBC 2022].

DOI: 10.1201/9781003489986-14

FIGURE 10.1 Prime Minister Winston Churchill and Captain David Margesson, watching preparations being made for the launch of a De Havilland Queen Bee seaplane L5984 from its ramp.

- **Combat Operations:** Armed UAVs, such as the Bayraktar TB2 or Predator drones, are deployed for precision strikes, close air support, and suppression of enemy defences. Their low radar visibility and high manoeuvrability enable effective combat engagement with minimal risk to personnel. Figure 10.2 shows a MQ-9 Reaper drone equipped with multiple cameras and sensor. It can travel at the maximum speed of 463 km/hr and carry global positioning system (GPS) or laser-guided missiles. Jordan [2021] envisaged an exploratory scenario with UAVs commanding the airspace.
- **Search and Rescue (SAR):** In disaster and conflict zones, UAVs are deployed for search and rescue missions. They provide aerial imagery, identify survivors, and deliver essential supplies to inaccessible areas. Lightweight UAVs equipped with thermal imaging cameras have been particularly effective in locating survivors under debris [Lyu et al. 2023].
- **Ad-hoc Communication Networks:** UAVs can establish temporary communication networks in regions with compromised infrastructure. This capability is particularly valuable for coordinating defence operations in remote or war-torn areas. Swarm-based UAVs have demonstrated success in providing resilient communication frameworks during tactical missions [Sherman et al. 2024].

FIGURE 10.2 MQ-9 Reaper drone equipped with multiple cameras and sensors has a maximum speed of 463 km/h and is capable of carrying GPS or laser-guided missiles and bombs [BBC 2022].

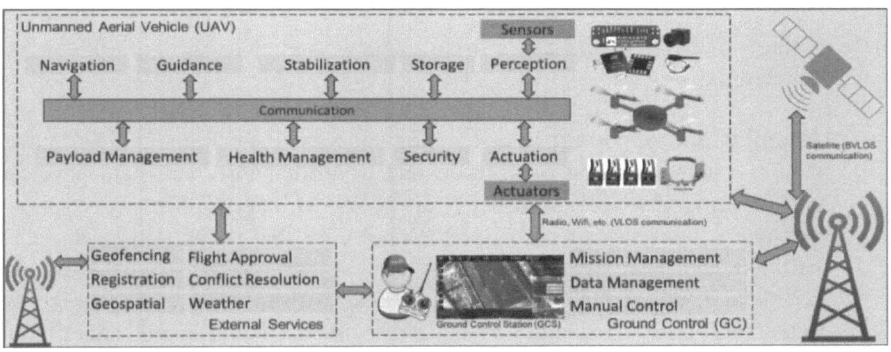

FIGURE 10.3 Reference architecture of a typical UAV ecosystem [Hussein and Nouacer 2022].

10.2 THREAT LANDSCAPE FOR UAV SYSTEMS

The increasing adoption of UAVs in both civilian and military domains has brought about a host of security concerns. This section provides an in-depth exploration of the vulnerabilities, emerging threats, and adversarial scenarios relevant to UAV systems.

10.2.1 Vulnerabilities in UAV Systems

The UAV systems are vulnerable due to their reliance on interconnected hardware, software, limited resources, and communication networks. Reference architecture (RA) is commonly used to understand and explain the vulnerabilities of systems. It depicts the attack surface and dependencies of each components. Figure 10.3 shows an RA of a typical drone ecosystem. Understanding the attack surface gives a clear idea on how an attacker may compromise the mission, drone, or both. Also, if a new component were to be added, how does it affect the overall security and attack surface. Now, let's look at the primary vulnerabilities faced by the UAVs.

- **Communication Links:** UAVs depend on communication channels such as GPS, radio-frequency (RF), and telemetry for navigation and control. These links are susceptible to jamming, spoofing, and signal interception, which can lead to loss of control or redirection of the UAV [Warner et al. 2012].
- **Onboard Systems:** Sensors, actuators, and control units are prone to cyber and physical attacks. Malware injection into onboard systems can compromise mission-critical operations, as demonstrated in research on UAV malware [Wang et al. 2023, Niu et al. 2021].
- **Ground Control Stations (GCS):** The GCS serves as the central hub for UAV operation, making it a prime target for cyber attacks such as distributed denial of service (DDoS) or unauthorized access [Mitchell and Chen 2014].

10.2.2 Emerging Threats to UAVs

The evolving threat landscape has introduced new challenges for securing UAV systems, including

- **Jamming and Spoofing:** GPS jamming and spoofing can disrupt UAV navigation systems, causing them to deviate from their intended flight paths or crash [Humphreys 2012, Warner et al. 2012]. Military GPS is encrypted so it may not be easily spoofed like the civil GPS, but it can still be jammed [Humphreys 2012].
- **Artificial Intelligence (AI)-Driven Attacks:** Advanced threats leverage AI to launch sophisticated attacks, such as swarm-based coordinated attacks that exploit UAV network vulnerabilities [Sherman et al. 2024].
- **Physical Attacks:** UAVs, especially those operating in hostile environments, are vulnerable to physical interception, sabotage, or destruction using counter-UAV technologies [Sherman et al. 2024].
- **Adversarial Attacks:** The machine learning (ML) software in the UAVs can be subjected to adversarial attacks that can alter the decision made by UAVs [Khadka et al. 2024].

10.3 SECURITY-AWARE COMPUTING FRAMEWORK

To develop a security aware computing framework for UAVs, we need to look at the information security triad. The information security triad consists of three components, namely – *confidentiality*, *integrity*, and *availability* (CIA) as shown in Figure 10.4. It is imperative to maintain all three aspects of the CIA triad for all purposes. Any breach at any of the attack surface at any point in the mission can lead to the threats defined in the previous section: leaking of the secure information, delaying or jeopardizing the mission, or failing the mission.

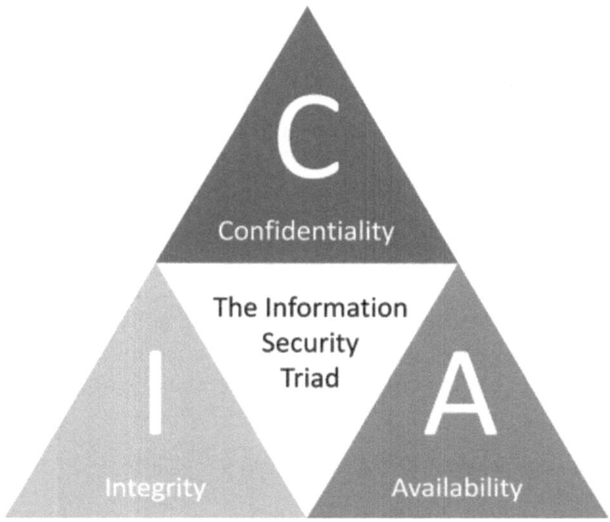

FIGURE 10.4 The information security triad comprising confidentiality, integrity, and availability [Nikander et al. 2020].

10.3.1 Hardware Security

Strategies including hardware-level security (e.g., tamper-proof modules), software-level approaches (e.g., secure boot, encrypted firmware), and network-level protections should be prioritized for ensuring the UAVs are safe to operate and cannot be tampered with. Similarly concepts such as "security by design" should be used while designing and developing the software. This would ensure the system is secure from the design rather than adding cybersecurity as an add-on.

10.3.2 Resilient Machine Learning

While using ML on board of the UAVs, there are many adversarial attacks that can thwart the security of the UAVs and compromise the mission. Resilient ML-based approaches should be used to make sure adversarial attacks do not pose a threat [Khadka et al. 2024].

10.3.3 Secure Communication Protocols

Implementation of end-to-end encryption algorithms can secure the data during communication. The transfer of data could be a control signal between the ground station to UAVs or raw data from the UAVs to the ground station. In addition if there are multiple UAVs, the communication could be between them. UAVs for their purpose of flying need to be lightweight, making them resource constrained. This means the security algorithm they choose may not be suitable for the purpose. The encryption algorithm can be chosen dynamically based on the environment similar to our previous work in space system [Sthapit et al. 2022].

10.3.4 Intrusion Detection and Prevention Systems (IDPS)

While we ensure best measure is taken during design and implementation, UAVs may still be under attack due to new vulnerabilities or new attack vectors. So it is essential to have intrusion detection and prevention system (IDPS) that can monitor the network of any suspicious activities. IDPS can be of signature-based or anomaly-based IDPS methods.

10.4 ASSURANCE AND VALIDATION

This section discusses the importance of ensuring that UAV systems meet stringent security and operational requirements. Assurance and validation are critical to building trust in UAV systems for defence applications by adhering to standards, employing advanced simulation environments, and establishing quantifiable metrics.

10.4.1 Security Standards and Frameworks

Compliance with established security standards and frameworks is essential for ensuring the security of UAV systems. Key considerations include

- **NATO UAV Security Guidelines:** NATO's guidelines for securing UAVs outline critical defence-specific requirements for operational security. These include robust communication encryption, secure data handling, and incident response protocols to mitigate cyber and physical threats.
- **ISO/IEC 27001 and 15408 Standards:**
  - ISO/IEC 27001 focuses on establishing information security management systems (ISMS) applicable to UAVs' operational data and communications.
  - ISO/IEC 15408 (Common Criteria) offers a framework for evaluating the security of UAV systems and their components, ensuring they meet specific security targets.
- **NIST Cybersecurity Framework (CSF):** The NIST CSF provides a flexible approach to securing UAV operations through its core functions – identify, protect, detect, respond, and recover. This is particularly relevant for integrating UAVs into broader military networks.

These frameworks ensure that UAVs meet baseline security requirements while providing a structured approach to evaluate compliance and resilience.

10.4.2 Simulation and Testing Environments

Simulation and testing environments play a critical role in validating the security and operational readiness of UAV systems. Key approaches include

- **Digital Twins for UAV Security Validation:** Digital twins provide a virtual replica of UAV systems, enabling real-time monitoring, simulation of cyber attacks, and testing of mitigation strategies. This approach helps identify vulnerabilities without risking physical assets.

- **Penetration Testing:** Penetration testing simulates cyber attacks to uncover potential vulnerabilities in UAV systems. This includes testing:
  - Communication links for susceptibility to jamming or spoofing.
  - Onboard systems for unauthorized access to sensors and actuators.
  - Ground control stations for vulnerabilities in command and control protocols.
- **Vulnerability Assessment:** Comprehensive vulnerability assessments evaluate UAV software, hardware, and network components. These assessments focus on:
  - Identifying known vulnerabilities using databases such as Common Vulnerabilities and Exposures (CVE).
  - Assessing the potential impact of exploitation on mission-critical operations.
- **Red Team/Blue Team Exercises:** Defence organizations can conduct adversarial simulations (red team) against defensive measures (blue team) to test the robustness of UAV security strategies in realistic scenarios.
- **Hardware-in-the-Loop (HIL) Testing:** HIL testing integrates physical UAV hardware into simulation environments, allowing for a combination of virtual and real-world testing to assess system performance under various threat scenarios. MATLAB[1] and other softwares allow HIL testing.

These environments enable iterative testing and refinement, ensuring that UAV systems are prepared for both anticipated and emergent threats.

10.4.3 Metrics for Security Evaluation

Developing metrics for security evaluation is essential for quantifying the robustness and resilience of UAV systems. Key metrics include

- **Risk-Based Assessment Models:** Risk assessments evaluate potential threats, vulnerabilities, and impacts on UAV operations. Techniques include
  - **Likelihood-Impact Matrix:** Prioritizes threats based on their likelihood and potential impact.
  - **Attack Surface Quantification:** Measures the extent of exposed vulnerabilities across UAV components.
- **Resilience Metrics:** Resilience is assessed based on:
  - **Mean Time to Detect (MTTD) and Mean Time to Respond (MTTR):** Evaluates the efficiency of detection and response mechanisms.
  - **Recovery Time Objective (RTO):** Measures how quickly UAV systems can recover from a cyber attack or failure.
  - **Fault Tolerance Threshold:** Tests the system's ability to operate under partial failure or attack.

- **Security Scoring Systems:** Scoring systems provide a quantifiable measure of overall security readiness. Examples include
  - **Common Vulnerability Scoring System (CVSS):** Rates the severity of vulnerabilities.
  - **Custom Defence-Specific Metrics:** Tailored scoring based on mission-critical parameters and adversarial threat models.

These metrics ensure a systematic evaluation of UAV security, providing actionable insights to improve system robustness and inform risk mitigation strategies.

10.5 CASE STUDY: SECURING A TACTICAL UAV FOR MILITARY OPERATIONS

10.5.1 Introduction

This case study explores the security challenges and solutions involved in deploying a tactical UAV for military reconnaissance in a conflict zone. The UAV, designated as *Phoenix-X*, is tasked with providing real-time surveillance and intelligence while operating in a hostile environment with advanced adversarial threats, including electronic warfare and cyber attacks.

10.5.2 Scenario Overview

The mission involves deploying *Phoenix-X* to monitor and gather intelligence on enemy troop movements and critical infrastructure over a period of 48 hours. The UAV operates autonomously but maintains intermittent communication with the ground control station (GCS) to transmit reconnaissance data. The operational environment is characterized by:

- **Adversarial Threats:**
  - GPS jamming and spoofing attempts.
  - RF signal interception and malware injection attempts targeting the UAV's onboard systems.
  - Physical counter-UAV measures, such as directed-energy weapons.
- **Environmental Challenges:**
  - Harsh weather conditions that may affect sensor reliability.
  - Limited bandwidth for secure communication.

10.5.3 Identified Vulnerabilities

An initial risk assessment revealed the following vulnerabilities:

- Susceptibility to GPS spoofing during navigation, leading to potential mission failure.
- Weak encryption in telemetry data, increasing the risk of interception and manipulation.
- Lack of real-time anomaly detection, allowing malware to persist undetected.

10.5.4 Security Measures Implemented

To address these vulnerabilities, a multi-layered security strategy was deployed:

1. **Navigation Security:**
   - Implemented a multi-sensor navigation system combining GPS, inertial navigation, and visual odometry to mitigate spoofing attempts.
   - Integrated a GPS authentication protocol that detects and rejects spoofed signals [Humphreys 2012].
2. **Data Encryption:**
   - Upgraded telemetry and video feed encryption to use AES-256, ensuring data confidentiality during transmission.
   - Established dynamic key exchange protocols using elliptic curve cryptography to protect against key compromise.
3. **Intrusion Detection System (IDS):**
   - Deployed a lightweight IDS capable of detecting abnormal patterns in network traffic and onboard system behaviour.
   - Implemented ML-based anomaly detection to identify potential malware or unauthorized access.
4. **Physical Security:**
   - Incorporated lightweight anti-drone countermeasures, including evasive manoeuvres and infrared shielding, to reduce detectability.
   - Hardened critical components with tamper-proof hardware to prevent physical sabotage.

10.5.5 Mission Outcomes

The *Phoenix-X* UAV successfully completed the mission with the following results:

- Detected and mitigated three GPS spoofing attempts using multi-sensor validation.
- Prevented interception of telemetry data through secure encryption protocols, maintaining the integrity of transmitted intelligence.
- Identified and neutralized a malware attempt targeting the UAV's onboard camera system using the IDS.
- Avoided detection and interception by adversarial forces through physical countermeasures and infrared shielding.

10.5.6 Lessons Learned

The mission highlighted the importance of adopting a layered security approach for UAVs. Key takeaways include

- The need for robust, redundant navigation systems to ensure mission continuity in the face of GPS-related threats.
- The critical role of encryption and secure communication protocols in protecting sensitive operational data.

- The effectiveness of ML-based anomaly detection in identifying emerging cyber threats.
- The necessity of physical hardening to protect UAVs from adversarial countermeasures.

10.5.7 Conclusion

This case study demonstrates that a comprehensive, multi-layered security framework is essential for ensuring the success of UAV missions in defence. By addressing both cyber and physical threats, defence organizations can enhance the resilience and operational reliability of UAV systems.

10.6 CONCLUSION

The increasing reliance on UAVs in defence applications has brought unparalleled advantages in surveillance, combat, and operational efficiency. However, this technological advancement is accompanied by significant security challenges, necessitating the development of comprehensive, security-aware frameworks. This chapter has explored the multifaceted threat landscape, vulnerabilities, and emerging risks facing UAV systems, along with strategies to enhance their resilience.

Securing UAVs in defence requires a holistic approach that combines technological innovation, rigorous testing, and adherence to established standards. As UAV technology continues to evolve, ongoing research and collaboration among academia, industry, and defence organizations will be pivotal in addressing emerging threats and ensuring the safe and effective deployment of UAVs in critical defence operations. This chapter serves as a foundation for understanding the complex interplay between UAV capabilities and security requirements, paving the way for future advancements in security-aware UAV computing.

PRACTICE QUESTIONS

Applying Security Principles to UAV Systems

This exercise is designed to test your understanding of the security principles and challenges associated with UAVs in defence applications. Use the concepts and frameworks discussed in this chapter to answer the following questions.

I. Threat Identification

   Identify potential vulnerabilities for a UAV operating in the following scenarios:

   1. A UAV conducting a reconnaissance mission in a GPS-degraded environment.
   2. A UAV transmitting real-time video feeds over an unsecured network.
   3. A UAV deployed in a hostile area with known adversarial AI capabilities.

Questions:

- What are the key vulnerabilities in each scenario?
- Propose one security measure to mitigate each identified vulnerability.

II. Security Framework Design

You are tasked with designing a security-aware computing framework for a fleet of UAVs deployed in a swarm for coordinated military operations. Address the following:

1. How would you ensure secure communication among the UAVs in the swarm?
2. What measures would you implement to detect and respond to a compromised UAV in the fleet?
3. How can you protect the ground control station from cyberattacks?

Deliverable: Provide a brief outline of your framework, highlighting key components such as encryption methods, intrusion detection systems, and resilient navigation techniques.

III. Case Study Analysis

Review the case study on *Phoenix-X* from this chapter and answer the following:

1. What were the primary threats faced by *Phoenix-X* during its mission?
2. Which implemented security measures were most effective, and why?
3. Suggest one additional measure that could further enhance the UAV's security.

Submission:
Write a 1–2 page(s) response addressing the exercise questions. Include diagrams or references to support your answers where applicable.

NOTE

1 MathWorks, Inc. www.mathworks.com

BIBLIOGRAPHY

Anita Khadka, Saurav Sthapit, Gregory Epiphaniou, and Carsten Maple. Resilient machine learning: Advancement, barriers, and opportunities in the nuclear industry. *ACM Computing Surveys*, 56(9):1–29, October 2024. ISSN 0360-0300, 1557-7341. doi: 10.1145/3648608.

BBC. Combat Drones: We are in a new era of warfare – Here's why. *BBC News*, February 2022.

J. Sterling Warner, Roger G. Johnston, and Cpp Los Alamos. A simple demonstration that the global positioning system (GPS) is vulnerable to spoofing. 2012. URL: https://api.semanticscholar.org/CorpusID:198960039.

Javier Jordan. The future of unmanned combat aerial vehicles: An analysis using the Three Horizons framework. *Futures*, 134:102848, December 2021. ISSN 0016-3287. doi: 10.1016/j.futures.2021.102848.

Jussi Nikander, Onni Manninen, and Mikko Laajalahti. Requirements for cybersecurity in agricultural communication networks. *Computers and Electronics in Agriculture*, 179:105776, 2020. doi: 10.1016/j.compag.2020.105776.

IWM. A Brief History of Drones. www.iwm.org.uk/history/a-brief-historyof-drones.

Mahmoud Hussein and Réda Nouacer. Reference architecture specification for drone systems. *Microprocessors and Microsystems*, 95:104705, November 2022. ISSN 0141–9331. doi: 10.1016/j.micpro.2022.104705. URL: https://doi.org/10.1016/j.micpro.2022.104705.

Michelle Sherman, Sihua Shao, Xiang Sun, and Jun Zheng. Counter UAV swarms: Challenges, considerations, and future directions in UAV warfare. *IEEE Wireless Communications,* 32:1–7, 2024. ISSN 1558-0687. doi: 10.1109/MWC.003.2400047.

Mingyang Lyu, Yibo Zhao, Chao Huang, and Hailong Huang. Unmanned aerial vehicles for search and rescue: A survey. *Remote Sensing*, 15(13):3266, January 2023. ISSN 2072-4292. doi: 10.3390/rs15133266.

Robert Mitchell and Ing-Ray Chen. A survey of intrusion detection techniques for cyber-physical systems. *ACM Computing Surveys*, 46(4):1–29, March 2014. ISSN 0360-0300. doi: 10.1145/2542049. URL: https://doi.org/10.1145/2542049..

Saurav Sthapit, Subhash Lakshminarayana, Ligang He, Gregory Epiphaniou, and Carsten Maple. Reinforcement learning for security-aware computation offloading in satellite networks. *IEEE Internet of Things Journal*, 9(14):12351–12363, July 2022. ISSN 2327-4662. doi: 10.1109/JIOT.2021.3135632.

Todd Humphreys. Statement on the vulnerability of civil unmanned aerial vehicles and other systems to civil GPS spoofing. Technical report, 2012.

Total Military Insight. The evolution and significance of the history of military drones, July 2024.

Weina Niu, Jian'An Xiao, Xiyue Zhang, Xiaosong Zhang, Xiaojiang Du, Xiaoming Huang, and Mohsen Guizani. Malware on internet of uavs detection combining string matching and Fourier transformation. *IEEE Internet of Things Journal,* 8(12):9905–9919, 2021. doi: 10.1109/JIOT.2020.3029970.

Zhaoxuan Wang, Yang Li, Shihao Wu, Yuan Zhou, Libin Yang, Yuan Xu, Tianwei Zhang, and Quan Pan. A survey on cybersecurity attacks and defenses for unmanned aerial systems. *Journal of Systems Architecture*, 138:102870, 2023. ISSN 1383-7621. doi: https://doi.org/10.1016/j.sysarc.2023.102870. URL: www.sciencedirect.com/science/article/pii/S1383762123000498.

11 Developing Trustworthy AI for Security-Sensitive Defence Applications

11.1 INTRODUCTION

Artificial intelligence (AI) has rapidly become a foundation in modern defence strategies. It offers capabilities that range from data analysis to autonomous operations. It is revolutionizing the defence sector by enhancing both strategic and tactical capabilities utilizing various advanced technologies including machine learning algorithms, data analytics, and autonomous systems, and enables the rapid processing and interpretation of vast amounts of data generated from diverse sources such as satellites and sensors. This heightened data-processing capability enhances situational awareness, allowing for more informed and timely decision-making in complex and dynamic environments [1]. For example, AI-driven technologies facilitate predictive maintenance of critical equipment, optimizing logistics and reducing downtime, which is essential for maintaining operational readiness [2]. Furthermore, autonomous systems, including drones and unmanned vehicles, extend the reach and efficiency of military operations while minimizing human risk in hazardous missions [3].

AI plays a pivotal role in identifying and mitigating cyber threats in real time, thereby safeguarding sensitive defence infrastructure and national security assets. Additionally, AI enhances training real-time programmes by providing realistic and adaptive training environments that improve the effectiveness of military personnel. However, the integration of AI in defence also brings forth significant ethical and strategic considerations, such as ensuring accountability in autonomous weapon systems and preventing the escalation of conflicts driven by AI-enabled technologies. It is equally important to focus on the responsible and ethical deployment of AI to harness its benefits and safeguard human lives while mitigating potential risks [4]. Overall, AI multi-faceted applications are not only augmenting the capabilities of defence forces but also reshaping the strategic landscape, making it imperative to develop robust frameworks that ensure AI systems are trustworthy, reliable, and aligned with ethical standards. Figure 11.1 presents some of the key defence system applications that have adopted AI.

With the advancement of the AI, it has found application in almost sector, as it has been mentioned earlier, including the defence sector. However, considering the high stake and criticality of the sector, the reliability and integrity of AI systems should be at the highest level. Therefore, a trustworthy AI system should provide

DOI: 10.1201/9781003489986-15

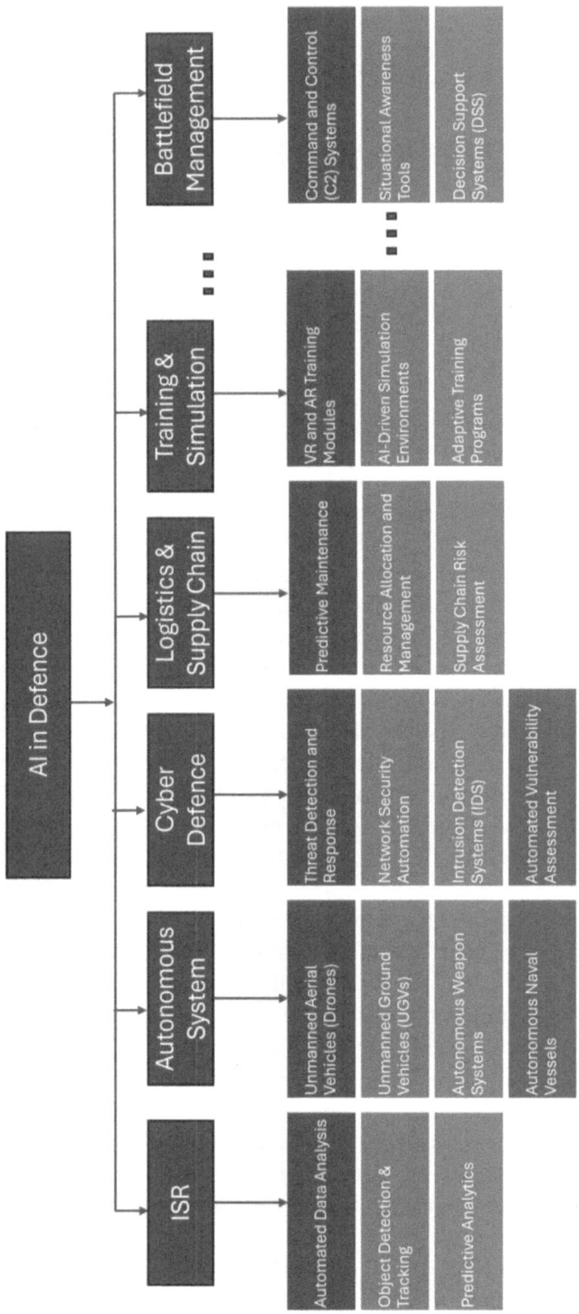

FIGURE 11.1 List of applications of AI in defence.

TABLE 11.1
Critical Attributes of Trustworthy AI in Defence Applications

Attribute	Description
Explainability	Ability to understand and interpret AI decision-making processes
Robustness	Resilience against adversarial attacks and unexpected inputs
Security	Protection of AI systems from cyber threats and unauthorized access
Reliability	Consistent performance and accuracy in varying operational environments
Ethical compliance	Adherence to ethical standards and legal frameworks governing AI usage

- **Operational Reliability:** Consistent performance under diverse and adverse conditions.
- **Decision Integrity:** Accurate and unbiased decision-making processes.
- **Security Resilience:** Protection against adversarial attacks and unauthorized access.
- **Ethical Compliance:** Adherence to legal and moral standards governing military operations [5].

The implications of untrustworthy AI in defence are severe, potentially leading to mission failures, unintended escalations, and ethical breaches [5]. Therefore, Table 11.1 presents key critical attributes of trustworthy AI in defence applications.

11.2 DEVELOPING TRUSTWORTHY AI FOR DEFENCE

Defence environments are inherently high-stakes, where AI systems operate under constant threat of adversarial actions, including *electronic warfare*, where disruption of communication and sensor system can be catastrophic; *cyber attacks*, where critical AI algorithm and sensitive data can be hacked; and *physical threats*, where damage to hardware and infrastructure supporting AI system can be possible. For example, the Stuxnet attack was a cyber weapon that demonstrated the potential for malware to target and disrupt critical infrastructure, highlighting the need for resilient AI systems in defence [6].

11.2.1 Adversarial Attacks and Robustness

Adversarial attacks represent a significant challenge to the integrity and reliability of AI systems, especially within the high-stakes and security-sensitive environments such as the defence sector [7, 8]. These attacks involve the intentional manipulation of input data through subtle perturbations that are often imperceptible to human eyes but can deceive AI models into predicting wrong outcomes [9, 10]. Therefore, adversarial attacks can have severe consequences, especially in defence applications, such as the misidentification of targets, the compromise of surveillance systems, or the disruption of autonomous operations [2]. The inherent vulnerability of AI models, particularly deep neural networks, to these attacks highlights the critical need for enhancing system robustness and resilience [11]. While robustness refers to the ability of AI systems to maintain their performance and reliability in the face of adversarial inputs and unforeseen perturbations [12], resilience refers to the ability to recover from an unanticipated accident or change [13]. Reference [9] highlighted the challenging task of defending against perturbation in deep learning systems, as the internal representation of models and decision-making processes can be exploited by attackers easily. They have proposed a taxonomy of threat models in deep learning systems (see Figure 11.2), depicting how adversarial attacks can be crafted in a way that forces the model to make mistakes depending upon the adversarial capabilities, which could have serious implication in security-sensitive applications, including autonomous driving and fraud detection.

To address these vulnerabilities, various strategies have been explored, including adversarial training, where models are exposed to adversarial examples during the training phase to improve their resilience [2]. For example, defensive distillation and the implementation of certified defences are additional methods aimed at reducing the susceptibility of AI models to adversarial manipulations [14, 15]. Hence, ensuring robustness and resilience not only protects AI systems from malicious exploitation but also enhances their trustworthiness and reliability in critical defence operations [13, 16]. As adversarial attack methodologies continue to advance, the defence sector must prioritize the development and integration of robust AI frameworks to safeguard national security and maintain operational effectiveness [10].

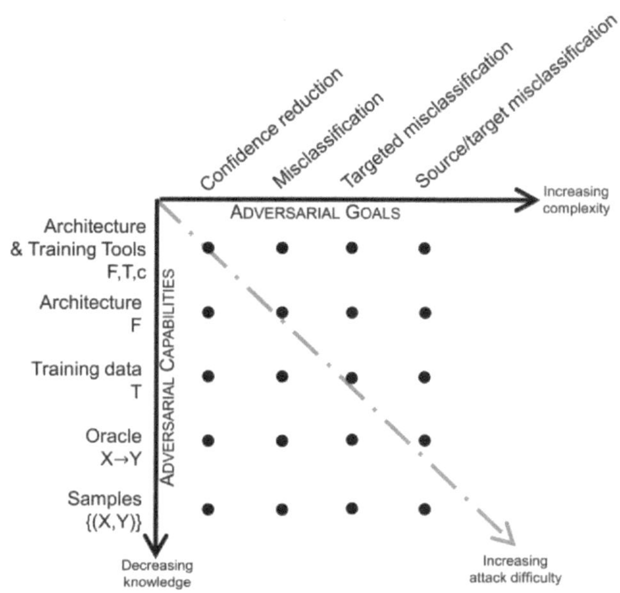

FIGURE 11.2 Adversarial threat modelling taxonomy [9].

11.2.2 Principles of Trustworthy AI

The development of trustworthy AI systems is crucial, particularly within the defence sector where the stakes are exceptionally high and human lives can be at stake. Therefore, trustworthy AI is outlined by a set of fundamental principles that ensure these systems operate reliably, ethically, and transparently while safeguarding national security interests [17]. There are six principles for trustworthy AI according to the Deloitte Trustworthy AI framework [18], and Figure 11.3 exhibits all the principles of the framework.

- **Explainability and Transparency** are the cornerstone of trustworthy AI, ensuring that AI-driven decisions and actions can be understood and interpreted by human operators, thereby facilitating accountability and oversight [19]. This transparency is crucial in defence applications where decisions may have significant ethical and operational consequences.

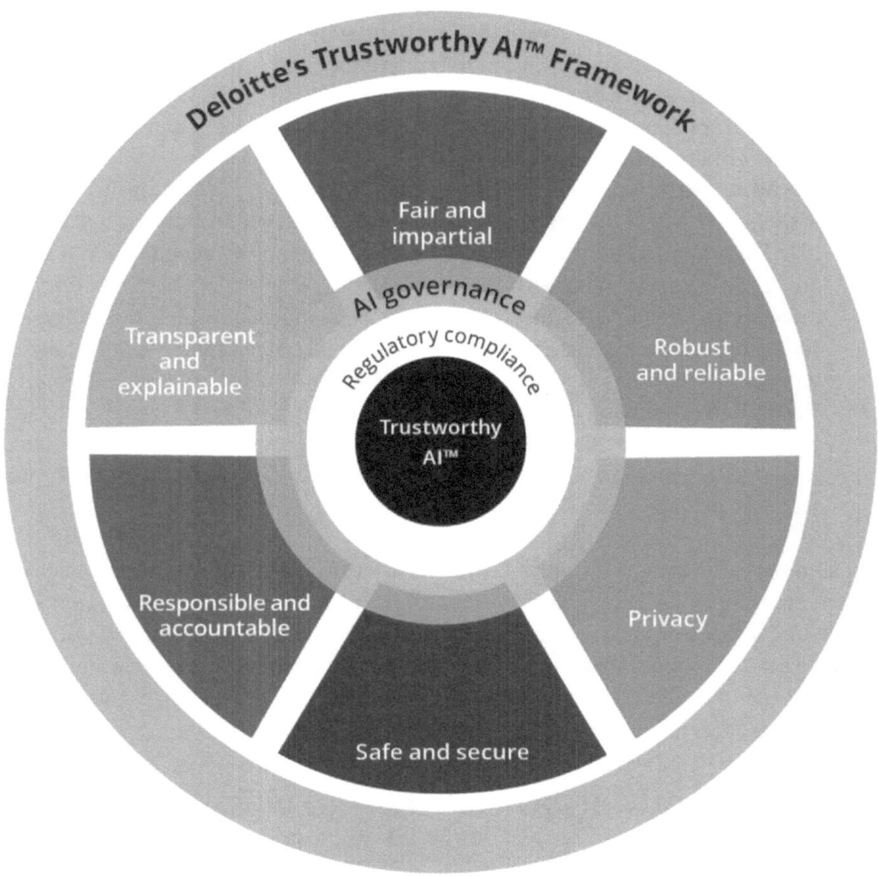

FIGURE 11.3 Deloitte's Trustworthy AI framework [18].

- **Robustness and Reliability** are equally critical, as AI systems must consistently perform accurately under diverse and potentially adversarial conditions [1]. In defence scenarios, where environments are dynamic and unpredictable, AI systems must resist manipulation and maintain functionality despite attempts at disruption or adversarial attacks [2]. Robustness enhancement involves implementing rigorous testing and validation protocols, as well as developing adaptive learning mechanisms that allow AI systems to respond effectively to unforeseen challenges [16].
- **Safety and Security** are another essential principle, comprising the protection of AI systems from cyber threats, unauthorized access, and exploitation [11]. In the defence sector, securing AI systems ensures the integrity of sensitive operations and prevents adversaries from undermining critical defence infrastructure through cyber attacks. This involves employing advanced encryption techniques, access controls, and continuous monitoring to detect and mitigate potential vulnerabilities [10].
- **Fairness and Impartiality** are achievable through ethical compliance which mandates that AI systems adhere to legal standards, particularly in autonomous weapon systems and surveillance technologies [20]. It is crucial to ensure principles such as fairness, non-discrimination, and respect for human rights[1] in the design and deployment of AI technologies [5]. This ethical framework is vital to prevent biases in AI decision-making processes that could lead to unjust or harmful outcomes [21].
- **Accountability and Governance** frameworks are necessary to outline responsibility and oversight for AI-driven actions within defence operations [4]. Therefore, establishing clear lines of accountability maintains that there are mechanisms in place to audit and evaluate AI systems [22]. Additionally, effective governance structures also facilitate collaboration among technologists, policymakers, and military leaders to align AI development with strategic defence objectives [23].
- **Privacy Preservation** plays an integral role in trustworthy AI, especially when handling sensitive or classified data within defence contexts [12]. Implementing robust data governance policies and privacy-enhancing technologies can protect personal and operational data against breaches and misuse.[2].

In summary, the principles of trustworthy AI – explainability, reliability, security, fairness, accountability, and privacy preservation – collectively ensure that AI systems deployed in defence are not only technologically advanced but also aligned with ethical standards and operational requirements. Thus, adhering to these principles is essential to harness the full potential of AI in enhancing defence capabilities while mitigating risks associated with its deployment [16, 20].

11.3 METHODOLOGIES FOR DEVELOPING TRUSTWORTHY AI

Developing trustworthy AI systems for defence applications requires a multi-faceted approach that integrates robust security measures, rigorous testing protocols, and ethical governance frameworks. A foundational methodology is to build secure AI development lifecycle (SDL), which embeds security considerations at every stage of AI system development, from initial planning and requirement analysis to deployment and maintenance [24]. This lifecycle approach ensures that potential vulnerabilities are identified and mitigated early, reducing the risk of adversarial exploitation and enhancing the overall resilience of AI systems. Threat modelling is a crucial component of this lifecycle, a proactive process that anticipates potential adversarial actions and designs countermeasures accordingly [8]. Therefore, AI practitioners and developers can focus on building secure AI systems against both cyber and physical attacks by systematically identifying and addressing security threats.

Another pivotal methodology is the implementation of resilient and robustness measures, for example, adversarial training involves exposing AI models to adversarial examples during the training phase, thereby enhancing their ability to withstand and correctly interpret manipulated inputs [8]. This technique not only improves the robustness of AI systems against known adversarial strategies but also fosters the development of more resilient models capable of adapting to novel attack vectors [2, 8, 14]. Additionally, methods such as defensive distillation and ensemble adversarial training have been proposed to further mitigate the susceptibility of AI models to adversarial manipulations [11, 15]. These strategies collectively contribute to building AI systems that maintain high performance and reliability even under adversarial conditions, which is indispensable for defence applications where operational integrity is important.

Several other factors can contribute to building a trustworthy AI system. One of the concepts is *explainable AI (XAI)* which ensures the trustworthiness of AI systems. XAI focuses on making AI decision-making processes transparent and interpretable to human operators [19]. In defence contexts, where decisions can have significant ethical and operational consequences, the ability to understand and validate AI-driven outcomes are crucial for accountability and oversight [25]. Techniques such as feature attribution, saliency maps, and rule extraction enable developers and operators to discern the factors influencing AI decisions, which increase trust and facilitate informed decision-making [25].

Likewise, *data governance and privacy preservation* also play an integral part in the development of trustworthy AI systems especially for defence as the stake is high in terms of human lives and the economy. Therefore, implementing robust data governance policies that ensure the quality, integrity, and security of data used in AI models [26] is imperative. In defence applications, where data often contains classified or personal information, maintaining stringent data protection standards is essential to prevent espionage, data leaks, and misuse [12]. Therefore, robust techniques like differential privacy, data anonymization, secure multi-party computation, and other advanced security and privacy-preserving techniques are employed for the protection of sensitive information from unauthorized access and breaches [12]. AI-based systems are dynamic and iterative. The traditional approach of security and privacy preservation of the systems which are normally added when the systems are developed will not be appropriate as adversaries can do damage at any components of AI systems. Therefore, security-by-design [27] and privacy-by-design [28] concepts should be followed at the start of the designing stage of the systems.

Moreover, the establishment of accountability and governance frameworks is crucial for the ethical and responsible deployment of AI in defence. These frameworks describe clear lines of responsibility, establish oversight mechanisms, and enforce compliance with ethical and legal standards [4]. Organizations can ensure that AI systems operate within defined ethical boundaries, uphold human rights, and adhere to international laws governing military operations by implementing comprehensive governance structures [20]. Therefore, regular audit trails and regulatory compliance checks further reinforce accountability, enabling continuous monitoring and evaluation of AI systems to identify and amend deviations from established norms [4].

Finally, cross-disciplinary collaboration among technologists, ethicists, legal experts, and defence personnel plays an imperative role for developing trustworthy AI systems [21]. This collaborative approach ensures that AI development is informed by diverse perspectives, technical robustness, ethical considerations, and legal compliance [21]. With the integration of expert knowledge from various fields, organizations can address the complex challenges associated with deploying AI in defence. Thereby enhancing the overall trustworthiness and effectiveness of AI systems [23].

In summary, the methodologies for developing trustworthy AI in defence applications comprise a comprehensive collection of secure development practices, adversarial resilience strategies, explainable AI techniques, data governance, robust accountability frameworks, and interdisciplinary collaboration. These methodologies collectively ensure that AI systems are not only technologically advanced but also reliable, secure, and ethically aligned and trustworthy, thereby safeguarding national security and operational integrity in defence operations [2, 4, 19].

11.4 EMERGING TECHNOLOGIES AND TRENDS

The advancement of AI is continually evolving, driven by rapid progression in various technological domains. One of the most significant emerging technologies is *quantum computing*, which promises to enhance AI system's processing capabilities [29]. This advancement is particularly pertinent to defence applications, where real-time data processing and decision-making are critical for operational effectiveness and national security. *Neuromorphic computing* is another pivotal trend which mimics the human brain's architecture to facilitate more efficient and adaptive AI systems [30]. For instance, AI can improve the performance of autonomous systems, such as drones and unmanned vehicles [31]. Edge AI is also gaining popularity where AI processing needs to occur close to data sources to reduce latency and enhance operational efficiency [32]. With the edge technology, military assets can perform on-site data analysis, leading to faster decision-making and reduced reliance on centralized computing resources, which is crucial in remote or hostile environments [33]. Additionally, the convergence of AI with the internet of military things (IoMT) is creating a more interconnected and responsive defence ecosystem [33]. This integration facilitates seamless communication and data sharing among various military assets, enhancing situational awareness and coordination during operations.

Another emerging trend is the advancement of human–AI teaming, which focuses on optimizing the collaboration between human operators and AI systems [34]. In defence applications, effective human–AI teaming can leverage the strengths of both humans and machines, resulting in more accurate threat assessments, improved mission planning, and enhanced overall operational effectiveness [35].

The rise of federated learning and differential privacy techniques is also noteworthy, as these methodologies enable the development of AI models that preserve data privacy and security while still benefiting from collaborative learning across multiple platforms [12]. In defence contexts, where data sensitivity and confidentiality are paramount, these techniques ensure that AI systems can learn from vast and diverse datasets without compromising the integrity or privacy of the information [26].

In summary, the integration of emerging technologies such as quantum computing, neuromorphic computing, and edge AI, coupled with evolving trends like human–AI teaming and advanced explainability models, is significantly enhancing the capabilities and trustworthiness of AI systems in the defence sector. These advancements not only improve operational efficiency and decision-making but also address critical challenges related to data privacy, security, and ethical compliance [22].

11.5 CONCLUSION

Developing trustworthy AI for security-sensitive defence applications is imperative to ensure operational effectiveness, ethical compliance, and national security. To develop AI systems to meet the security and rigorous demands of the defence sector, the foundational principles of trustworthy AI of explainability, reliability, security, and ethical governance need to be implemented. However, there are several challenges in defence applications due to its sensitive environments, therefore it necessitates a multidisciplinary approach, combining technological innovation, ethical deliberation, and robust policy frameworks through collaboration. The continuous cooperation between technologists, policymakers, ethicists, and defence personnel is essential to advance trustworthy AI.

NOTES

1 International Committee of the Red Cross. The impact of gender and rase bias in AI. https://blogs.icrc.org/law-and-policy/2018/08/28/impact-gender-race-bias-ai/

2 National Institute of Standards and Technology. Privacy Framework. www.nist.gov/

REFERENCES

1. D. Amodei, C. Olah, J. Steinhardt, P.F. Christiano, J. Schulman, and D. Mané, *Concrete problems in ai safety*, *ArXiv abs/1606.06565,* 2016, Available at https://api.semanticscholar.org/CorpusID:10242377.

2. A. Madry, A. Makelov, L. Schmidt, D. Tsipras, and A. Vladu, *Towards deep learning models resistant to adversarial attacks, ArXiv abs/1706.06083,* 2017, Available at https://api.semanticscholar.org/CorpusID:3488815.
3. P. Scharre, *Army of None: Autonomous Weapons and the Future of War*, Tantor Audio, Old Saybrook, CT, 2018.
4. K. Shahriari and M. Shahriari, *IEEE standard review—Ethically aligned design: A vision for prioritizing human wellbeing with artificial intelligence and autonomous systems*, in *2017 IEEE Canada International Humanitarian Technology Conference (IHTC)*. 2017, pp. 197–201.
5. H.L.E.G. on AI (AI HLEG), *Ethics guidelines for trustworthy AI*, Available at https://ec.europa.eu/digital-single-market/en/news/ethics-guidelines-trustworthy-ai. Accessed: 2024-9-23.
6. K. Zetter, *Countdown to Zero Day: Stuxnet and the Launch of the World's First Digital Weapon*, Crown Publishing Group, USA, 2014.
7. C. Szegedy, W. Zaremba, I. Sutskever, J. Bruna, D. Erhan, I.J. Good-fellow, and R. Fergus, *Intriguing properties of neural networks, CoRR abs/1312.6199,* 2013, Available at https://api.semanticscholar.org/CorpusID:604334.
8. I.J. Goodfellow, J. Shlens, and C. Szegedy, *Explaining and harnessing adversarial examples, CoRR abs/1412.6572,* 2014, Available at https://api.semanticscholar.org/CorpusID:6706414.
9. N. Papernot, P. McDaniel, S. Jha, M. Fredrikson, Z.B. Celik, and A. Swami, *The limitations of deep learning in adversarial settings*, in *2016 IEEE European Symposium on Security and Privacy (EuroS&P)*. 2016, pp. 372–387.
10. N. Carlini and D.A. Wagner, *Towards evaluating the robustness of neural networks, CoRR abs/1608.04644,* 2016, Available at http://arxiv.org/abs/1608.04644.
11. N. Papernot, P. McDaniel, X. Wu, S. Jha, and A. Swami, *Distillation as a defense to adversarial perturbations against deep neural networks*, in *2016 IEEE Symposium on Security and Privacy (SP)*. 2016, pp. 582–597.
12. C. Dwork and A. Roth, *The algorithmic foundations of differential privacy*, Foundations and Trends Theoritical Computer Science 9 (2014), pp. 211–407, Available at https://doi.org/10.1561/0400000042.
13. A. Khadka, S. Sthapit, G. Epiphaniou, and C. Maple, *Resilient machine learning: Advancement, barriers, and opportunities in the nuclear industry, ACM Computing Survey*, 56 (2024), Available at https://doi.org/10.1145/3648608.
14. N. Papernot, P. Mcdaniel, A. Sinha, and M.P. Wellman, *Towards the science of security and privacy in machine learning, ArXiv abs/1611.*03814, 2016, Available at https://api.semanticscholar.org/CorpusID:6726938.
15. F. Tramèr, A. Kurakin, N. Papernot, D. Boneh, and P. Mcdaniel, *Ensemble adversarial training: Attacks and defenses, ArXiv abs/1705.07204,* 2017, Available at https://api.semanticscholar.org/CorpusID:21946795.
16. M. Brundage, S. Avin, J. Clark, H. Toner, P. Eckersley, B. Garfinkel, A. Dafoe, P. Scharre, T. Zeitzoff, B. Filar, H. Anderson, H. Roff, G.C. Allen, J. Steinhardt, C. Flynn, S.O. hÉigeartaigh, S. Beard, H. Belfield, S. Farquhar, C. Lyle, R. Crootof, O. Evans, M. Page, J. Bryson, R. Yampolskiy, and D. Amodei, *The Malicious Use of Artificial Intelligence: Forecasting, Prevention, and Mitigation*, 2018, Available at www.repository.cam.ac.uk/handle/1810/275332.
17. H. Liu, Y. Wang, W. Fan, X. Liu, Y. Li, S. Jain, Y. Liu, A. Jain, and J. Tang, *Trustworthy AI: A computational perspective, ACM Transactions on Intelligent System and Technology*, 14 (2022), Available at https://doi.org/10.1145/3546872.

18. Deloitte, *Deloitte's trustworthy AI™ framework and the white house blueprint for an ai bill of rights* (2022). Available at www2.deloitte.com/us/en/pages/consulting/articles/ai-bill-of-rights.html, accessed: 2024-09-24.
19. D. Gunning, E. Vorm, J.Y. Wang, and M. Turek, *Darpa's explainable ai (XAI) program:* A retrospective, Applied AI Letters, 2 (2021), p. e61, Available at https://onlinelibrary.wiley.com/doi/abs/10.1002/ail2.61.
20. J.P. Affairs, *AI ethical principles – Highlighting the progress and future of responsible AI in the DoD* (2021). Available at www.defense.gov/News/Releases/release/article/2091996/dod-adopts-ethical-principles-for-artificial-intelligence/#:~:text=Responsible.,unintended%20bias%20in%20AI%20capabilities .
21. M. Taddeo and L. Floridi, *Regulate artificial intelligence to avert cyber arms race*, *Nature* 556 (2018), pp. 296–298.
22. NATO, *Emerging and disruptive technologies* (2024). Available at www.nato.int/cps/en/natohq/topics_184303.htm.
23. C. Hofmann and U. Schneckener, *Engaging non-state armed actors in state- and peace-building: Options and strategies*, *International Review of the Red Cross* 93 (2011), pp. 603–621.
24. Microsoft, *Secure Development Lifecycle (SDL) Practices*, 2021.
25. F. Doshi-Velez and B. Kim, *Towards a rigorous science of interpretable machine learning, arXiv: Machine Learning,* 2017, Available at https://api.semanticscholar.org/CorpusID:11319376.
26. N.I. of Standards and T. (NIST), *Privacy framework*, 2020, Available at www.nist.gov/privacy-framework.
27. C. Kern, *Secure by design at google*, Technical Report, Google Security Engineering, 2024.
28. A. Cavoukian, *Privacy by Design: The 7 Foundational Principles*, Information and Privacy Commissioner of Ontario, 2009.
29. J. Preskill, *Quantum computing in the NISQ era and beyond, Quantum* 2 (2018), p. 79, Available at http://dx.doi.org/10.22331/q-2018-08-06-79.
30. P.A. Merolla, J.V. Arthur, R. Alvarez-Icaza, A.S. Cassidy, J. Sawada, F. Akopyan, B.L. Jackson, N. Imam, C. Guo, Y. Nakamura, B. Brezzo, I. Vo, S.K. Esser, R. Appuswamy, B. Taba, A. Amir, M.D. Flickner, W.P. Risk, R. Manohar, and D.S. Modha, *Artificial brains. A million spiking-neuron integrated circuit with a scalable communication network and interface*, *Science* 345 (2014), pp. 668–673.
31. C.D. Schuman, S.R. Kulkarni, M. Parsa, J.P. Mitchell, P. Date, and B. Kay, *Opportunities for neuromorphic computing algorithms and applications*, *Nature Computational Science* 2 (2022), pp. 10–19.
32. A. King, *Digital targeting: artificial intelligence, data, and military intelligence*, *Journal of Global Security Studies* 9 (2024), p. ogae009, Available at https://doi.org/10.1093/jogss/ogae009.
33. R. Singh and S.S. Gill, Edge AI: A survey, Internet of Things and Cyber-Physical Systems 3 (2023), pp. 71–92, Available at www.sciencedirect.com/science/article/pii/S2667345223000196.
34. J.Y.C. Chen and M.J. Barnes, *Human–agent teaming for multirobot control: A review of human factors issues*, *IEEE Transactions on Human–Machine Systems* 44 (2014), pp. 13–29.
35. A.R. Wagner, *The role of trust and relationships in human-robot social interaction*, 2009, Available at https://api.semanticscholar.org/CorpusID:147269411.

12 Smart Camera Network in Smart Defence Applications

12.1 INTRODUCTION

The government and defence organizations have used surveillance technology for a long time; in some cases pioneering and driving new technologies that have then found usage in other sectors. Traditional surveillance systems, characterized by static cameras and manual monitoring, often suffer from limitations such as delayed response times, low-resolution imagery, and an inability to process large volumes of data efficiently (Hu et al. 2004). The evolution towards smart camera networks represents a significant leap forward, integrating sophisticated sensors, artificial intelligence (AI), and advanced communication protocols to create systems capable of autonomous decision-making and real-time data analysis (Valera and Velastin 2005).

Smart camera networks are instrumental in defence applications due to their enhanced capabilities in monitoring, threat detection, and situational awareness (Rowe and Bares 2012). These networks provide defence agencies with the tools necessary to protect borders, secure military installations, and gather critical intelligence. By leveraging AI and machine learning algorithms, smart cameras can identify potential threats more accurately and respond promptly, thereby enhancing the overall effectiveness of defence operations (Girshick 2015) (see, e.g., Figure 12.1). This chapter aims to provide a comprehensive analysis of smart camera networks within the context of defence applications. We will delve into different generations of surveillance, the technological components, explore defence use cases, discuss key enabling technologies, address the challenges faced in deployment and operation, and consider future trends that may shape the evolution of these systems. We look at *person reidentification* in detail as a case study for the smart camera network use case. We then close the chapter with the future trends and developments of this interesting technology.

12.2 GENERATIONS OF VISUAL SURVEILLANCE

The visual surveillance technology can be classified into four distinct generations based on the technology it used, as shown in Figure 12.2.

The first generation of visual surveillance systems was characterized by analogue and wired technologies. Analogue closed-circuit television (CCTV) cameras

 DOI: 10.1201/9781003489986-16

FIGURE 12.1 Usage of camera in future soldier's uniform. (Ministry of Defence 2015.)

The UK Ministry of Defence's prediction of future soldier's uniform that equips camera on it, connected back to the military base to analyse the footage on the ground.

captured images that were transmitted to a control room and recorded on analogue media like videotapes. Operators in the control room had to continuously monitor these video feeds to detect anomalies or security breaches. Since the system was entirely analogue, the signals lacked encryption, resulting in reduced security. The analogue nature also made wireless transmission impractical, as signals could be easily intercepted or tampered with, preventing first-generation surveillance systems (1GSS) from being wireless. Moreover, computer vision (CV) algorithms were not utilized at all.

In the second generation of surveillance systems (2GSS), analogue feeds were digitized and stored on devices such as hard disk drives. This digitization allowed for

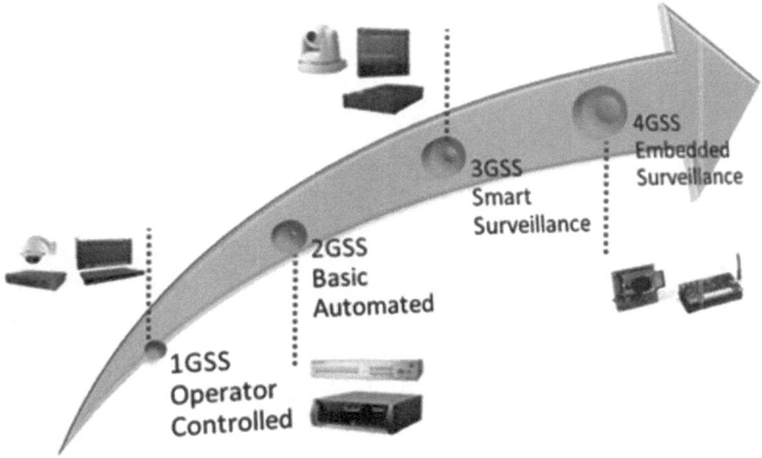

FIGURE 12.2 Evolution of surveillance technology.

the application of basic CV algorithms for target detection and tracking, enhancing the system's overall efficiency. However, 2GSS faced two significant problems: first, the detection and tracking algorithms were not robust enough; second, scaling the system to accommodate a large number of sensors was difficult due to centralized processing.

Advancements in electronics and CV addressed some of these issues in the third generation surveillance system (3GSS), enabling smart and wide-area surveillance systems. In addition to red–green–blue (RGB) sensor data, other modalities like audio inputs were incorporated, resulting in more robust detection and tracking. Processing tasks were also delegated to individual cameras or dedicated devices, alleviating the central processor and allowing for broader area surveillance systems.

In the fourth generation, the focus shifted towards distributed embedded devices with wireless communication capabilities that are easily deployable, scalable, and intelligent enough to execute robust CV algorithms (Al Najjar et al. 2014). Such systems need to be robust and highly fault-tolerant. The growth in hardware technologies has facilitated this development; for example, the interest in smartphones over the last two decades has driven significant advancements in mobile processors, making them powerful, efficient, and affordable. Similarly, compact embedded devices like Raspberry Pi[1] and NVIDIA Jetson Tx2[2] have entered the market, featuring multi-core central processing units (CPUs) and multi-core graphics processing units (GPUs) while maintaining a small form factor and low cost.

12.3 OVERVIEW OF SMART CAMERA NETWORKS

Smart camera networks consist of interconnected cameras equipped with advanced processing capabilities, sensors, and communication interfaces (Chen et al. 2016). These networks transcend traditional surveillance by enabling on-site data processing, reducing the need for centralized analysis, and allowing for immediate action based on real-time information (Nixon and Aguado 2019). The primary components include high-resolution cameras, various sensors (e.g., thermal, infrared, motion detectors), communication modules (wired or wireless), and embedded processing units that often utilize edge computing technologies (Shi et al. 2016). Smart camera networks offer significant improvements over conventional systems:

Enhanced Analytics and Automated Responses: AI algorithms analyse patterns, recognize objects, and detect anomalies, reducing the reliance on human monitoring and increasing the accuracy of threat detection (Ren et al. 2015).

Scalability and Flexibility: Modular designs allow for easy expansion and customization to meet specific defence requirements, ensuring adaptability to various operational environments (Hampapur et al. 2005).

12.3.1 Sensors and Imaging Technologies

Camera are the essential component of the smart camera network. There are different cameras that can be used in the smart camera network. High-resolution optical sensors capture detailed images, while thermal imaging allows for detection in low-light or obscured environments (Pavlidis and Levine 2017). Thermal cameras extend operational effectiveness during nighttime or in dark conditions (Ridao et al. 2015). Figure 12.4 is an example of thermal image showing cars, humans, and a person on skateboard. The bounding boxes show the detection of humans and vehicles as well as the confidence of detection. However, thermal cameras are associated with low-resolution images. Multi-spectral imaging can detect a range of electromagnetic frequencies, providing additional layers of information (Zhang and Li 2018). Smart camera is associated with the onboard processing capability that enables immediate analysis of captured data, facilitating prompt detection and response to threats without the latency associated with transmitting data to centralized servers (Satyanarayanan 2017). For example, when searching for a person of interest, images without humans can be quickly discarded with onboard human detection.

12.3.2 Communication Protocols

Multiple smart cameras may be connected to one another directly or via the base (see Figure 12.3). Efficient data transmission is facilitated through protocols that may be wired (Ethernet, fibre optics) or wireless (Wi-Fi, cellular networks, military-grade communication systems) (Akyildiz et al. 2002). The integration of the internet of things (IoT) principles enables seamless device connectivity and data exchange across the network (Atzori et al. 2010). 5G allows low-latency communication allowing the cameras to quickly transmit and receive data (Tehrani et al. 2014).

Impact of Next-Generation Networks: The deployment of 5G networks offers higher data transmission speeds, lower latency, and increased device connectivity, all of which enhance the capabilities of smart camera networks (Chen et al. 2017). Based on the camera network, a person of interest can be tracked from one camera to another.

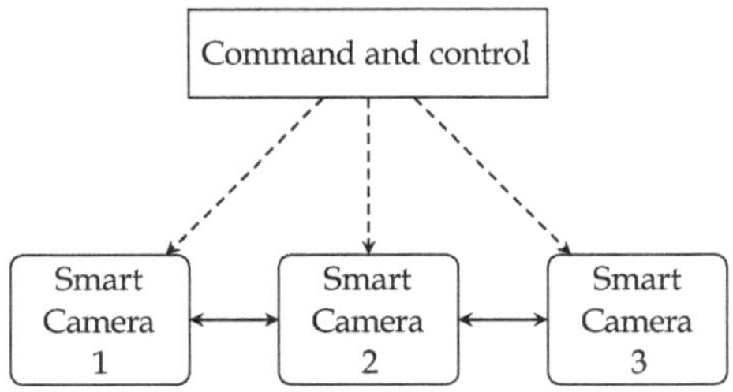

FIGURE 12.3 Network of smart cameras. A "smart camera" is a term used to describe camera as well as its processing capabilities.

12.3.3 Command and Control (CnC)

The command and control is the high-level entity that may control the operation of the cameras. Dependent on the cameras, they may be point tilt and zoom (PTZ) or mobile cameras. The CnC may control them based on the information. They may also further process the camera data and instruct others based on the content. In some cases, it is in the interest of saving energy to offload computation to the backup server (Sthapit et al. 2019).

12.4 ALGORITHMS USED IN SMART CAMERA NETWORK

In general, the algorithms in the camera network would depend on the application. However, there are certain CV algorithms that are useful for these systems. OpenCV is one of the most used libraries for CV. For more details on OpenCV, the readers are directed to (Bradski and Kaehler 2011).

- **Image Preprocessing:** The raw images captured by the camera may have to be preprocessed before being suitable for further usage. The preprocessing algorithms generally include algorithms such as *colour correction*, *image cropping*, *image rotation*, *histogram equalization*, *greyscaling* etc.
- **Object Detection:** Algorithms in this category can detect presence of certain objects. For example, in Figure 12.4, multiple vehicles and humans are detected. The detection algorithms can be various such as Viola–Jones detector for detecting faces or deep neural network (DNN) based detectors such as Alexnet (Krizhevsky et al. 2017) that can detect many categories of objects. Listing 12.1 shows a complete Python code for detecting faces on images using Viola–Jones algorithm in OpenCV. OpenCV needs to be installed to try it.

```
#!/ usr/ bin / env python

import numpy as np
import cv2

# local modules
```

```
from video import create_capture
from common import clock, draw_str

help_message = '''
USAGE: facedetect. py [-- cascade <cascade_fn>]
     [-- nested - cascade <cascade_fn >] [<video_
     source>]
'''

def detect(img, cascade):
    rects = cascade. detectMultiScale (img, scale
        Factor =1.3, min Neighbors =4, min Size =(30,
        30), flags = cv2. CASCADE_SCALE_IMAGE)
    if len (rects) == 0:
      return []
  rects [:, 2:] += rects [:,:2]
  return rects

def draw_rects(img, rects, color):
  for x1, y1, x2, y2 in rects:
     cv2. rectangle (img, (x1, y1), (x2, y2),
     color, 2)

if  name  == 'main':
  import sys, getopt
  print help_message

  args, video_src = getopt. getopt( sys. argv [1:],
  '', [' cascade =', 'nested - cascade ='])
  try :
    video_src = video_src [0]
  except:
    video_src = 0
args = dict( args)
cascade_fn = args. get('-- cascade', "../../ data /
haarcascades/ haarcascade_frontalface_alt. xml")
nested_fn = args. get('-- nested - cascade', "../
../ data / haarcascades/ haarcascade_eye. xml")

cascade = cv2. Cascade Classifier (cascade_fn)
nested = cv2. Cascade Classifier (nested_fn)

cam = create_capture ( video_src, fallback ='
    synth: bg =../ cpp / lena. jpg: noise =0.05')

while True :
```

```
    ret, img = cam. read ()
    gray = cv2. cvtColor(img, cv2. COLOR_BGR 2 GRAY)
    gray = cv2. equalize Hist(gray)

    t = clock ()
    rects = detect(gray , cascade)
    vis = img. copy ()
    draw_rects(vis, rects, (0, 255, 0))
    for x1, y1, x2, y2 in rects:
        roi = gray [y1: y2, x1: x2]
        vis_roi = vis[y1: y2, x1: x2]
        subrects = detect(roi. copy (), nested)
        draw_rects(vis_roi, subrects, (255, 0, 0))
    dt = clock () - t

    draw_str(vis, (20, 20), 'time:%.1 f ms' % (dt *1000))
    cv2. imshow ('facedetect', vis)

    if 0 xFF & cv2. waitKey (5) == 27:
        break
cv2. destroy AllWindows ()
```

LISTING 12.1 Python code for detecting face using Viola–Jones algorithm on OpenCV. (From OpenCV samples).

- **Object Tracking:** Objects that are detected can be tracked with time if they are in motion. The tracking algorithm can locate the object at any given time. If the object is moved out of the view of the camera, it may be tracked on other camera views. However, for tracking objects in multiple cameras, it is essential to establish if it is the same or a different object. If the object is a person, we can use person reidentification algorithms to confirm they are same or different persons. This is described in detail in Section 12.6.

12.5 DEFENCE APPLICATIONS OF SMART CAMERA NETWORKS

- **Border Surveillance:** Border surveillance is an essential part of national security, tasked with preventing illegal crossings, smuggling, and infiltration. Traditional surveillance methods often fall short in covering extensive and remote border areas, leading to vulnerabilities that adversaries can exploit. Smart camera networks can address these challenges by integrating imaging, unmanned aerial vehicles (UAVs), and communication system to create a comprehensive monitoring system. Cameras deployed along border lines capture

FIGURE 12.4 Thermal image containing cars, humans, and person on skateboard.

detailed images and videos, enabling accurate detection of movements even in harsh environmental conditions. Thermal camera can be used to monitor the activities during night, whereas the inclusion of UAVs provides aerial surveillance capabilities, allowing for flexible and extensive coverage of areas that are otherwise difficult to monitor from the ground. Additionally, satellite imagery can complement these efforts by offering a macro-perspective of border regions, facilitating the tracking of movements over large geographical areas. The network can be used to track people and assets on the border.

- **Urban Warfare and Counterterrorism:** Smart camera networks offer unparalleled capabilities in maintaining public safety and combating terrorism within densely populated and infrastructure-rich environments. Urban settings present unique challenges, including the need for continuous monitoring of public spaces, critical infrastructure, and large public events. Smart camera networks deployed in these areas utilize high-density camera installations, facial recognition systems, and behavioural analysis algorithms to detect and respond to suspicious activities effectively.

High-density deployment of cameras ensures comprehensive coverage of high-traffic areas such as city centres, transportation hubs, and government buildings, minimizing blind spots and enhancing situational awareness. Facial recognition technologies

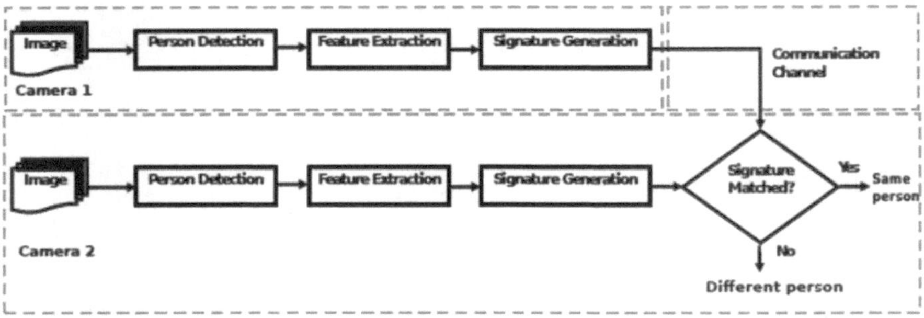

FIGURE 12.5 Person reidentification workflow. (Sthapit, 2018.)

enable the identification of individuals of interest by matching captured images against databases of known suspects, aiding in the swift identification and apprehension of potential threats. Behavioural analysis algorithms further enhance threat detection by monitoring patterns of movement and identifying anomalies that may indicate malicious intent, such as unexpected loitering or unusual crowd formations.

12.6 CASE STUDY: PERSON REIDENTIFICATION IN A SMART CAMERA NETWORK

Person reidentification (PRID) is associating people across camera views at different locations and times (Gong et al. 2014). For defence application it holds a significant application as it can help the authorities to locate and track person of interest. As the field of view (FOV) of the cameras can be non-overlapping, the background and pose can change, and the target can be occluded which makes PRID challenging. A particular individual can look dissimilar in different views, while different individuals can look similar from different angles. Figure 12.5 shows a block diagram of PRID for two cameras. As the cameras are smart, they have the ability to process the images they capture and proceed as necessary. In PRID, as Figure 12.5 shows, the smart camera needs to reidentify the same individual across the camera network. First they need to detect a person in the image. If the image does not contain any person, it can be safely discarded. However, when the image contains a person, features are extracted from the person's image to generate a unique signature. This unique signature is transmitted over the network to compare with other signatures to determine if the two images are of the same person.

Reference	Algorithm	Remarks
Ma et al. (2012)	BiCov	Biologically inspired filters (Gabor filters)
Bazzan et al. (2014)	i SDALF	Feature design with recurrent high-structured patches (RHSP) and MSCR
Wang and Zhao (2014)	Spatio-temporal	Spatial and temporal features made up of colour histograms and edge information
Zhao et al. (2014)	U. Saliency	Features using colour and SIFT, patch making weighted by saliency of the patch

Reference	Algorithm	Remarks
Bai et al. (2020)	Deep-Person	Long short-term memory (LSTM) based Re-ID method, using contextual information to enhance feature discriminability, improving performance in various reidentification tasks
Davis et al. (2007)	ITML	Mahalanobis metric learning
Li et al. (2013)	LAFT	Mahalanobis metric learning
Kosting et al. (2012)	eKr ISSME	Generic feature using colour and LBP, Mahalanobis metric learning
Herman et al. (2017)	sTriplet Loss	Triplet-based metric learning that optimises anchor, positive, and negative samples to create discriminative feature embeddings
Sun et al. (2018)	PCB	Part-based convolutional baseline model splits the image into parts, capturing detailed features for better performance in complex Re-ID scenarios
Chen et al. (2019)	ABD-Net	Attention-based deep network integrates attentive and discriminative feature learning for robust Re-ID
Ye et al. (2021)	TransReID	Transformer-based architecture that captures long-range dependencies and global context for better person reidentification

12.7 FUTURE TRENDS AND DEVELOPMENTS

The landscape of smart camera networks is continually evolving, driven by technological advancements and emerging defence needs (Xu et al. 2017). Real-world implementations of smart camera networks in urban settings demonstrate their effectiveness in enhancing security and counterterrorism efforts. For instance, London's Metropolitan Police utilizes an extensive network of smart cameras equipped with facial recognition and behavioural analysis capabilities to monitor and secure public spaces (Live Facial Recognition). These systems exemplify how smart camera networks can be leveraged to maintain public safety, manage large crowds during events, and ensure the security of critical infrastructure.

Looking ahead, advancements in AI and deep learning promise to further enhance the capabilities of smart camera networks in both border surveillance and urban security. Predictive analytics will enable defence agencies to anticipate threats before they materialize, while autonomous systems will facilitate more efficient and effective responses to detected threats. The ongoing development of 5G and 6G will also play a crucial role, providing higher data transmission speeds and lower latency, which are essential for real-time monitoring and rapid data processing.

Federated learning is a notable development in machine learning in recent times that preserves user data in the local devices. This may be useful in future to maintain the user privacy while also tracking people for defence purpose.

12.8 CONCLUSION

Smart camera networks have become an indispensable component of modern defence strategies. Their ability to provide real-time surveillance, enhanced threat detection, and improved situational awareness significantly bolsters national security efforts. By integrating advanced technologies such as AI, edge computing, and secure communication protocols, these networks address the complex challenges faced in defence operations.

While the benefits are substantial, challenges related to technical limitations, environmental conditions, and ethical considerations must be carefully managed. Ongoing research and development, coupled with robust policy frameworks, are essential to optimize the effectiveness of smart camera networks and mitigate associated risks.

Looking forward, continued innovation will expand the capabilities of these systems. Defence agencies must remain agile and forward-thinking, embracing new technologies and adapting to evolving threats to maintain a strategic advantage.

NOTES

1 Raspberry Pi. www.raspberrypi.com
2 Nvidia .www.nvidia.com/en-us/autonomous-machines/embedded-systems/jetson-tx2/

BIBLIOGRAPHY

Ian F. Akyildiz, Weilian Su, Yogesh Sankarasubramaniam, and Erdal Cayirci. A survey on wireless sensor networks. *IEEE Communications Magazine*, 40(8):102–114, 2002.

Mayssaa Al Najjar, Milad Ghantous, and Magdy Bayoumi. Video surveillance for sensor platforms, vol. 114, *Lecture Notes in Electrical Engineering*. Springer, New York, NY, 2014. ISBN 978-1-4614-1856-6 978-1-4614-1857-3. doi: 10.1007/978-1-4614-1857-3.

Luigi Atzori, Antonio Iera, and Giacomo Morabito. The internet of things: A survey. *Computer Networks*, 54(15):2787–2805, 2010.

Xiang Bai, Mingkun Yang, Tengteng Huang, Zhiyong Dou, Rui Yu, and Yongchao Xu. Deep-person: Learning discriminative deep features for person re-identification. *Pattern Recognition*, 98:107036, 2020. ISSN 0031-3203. doi: https://doi.org/10.1016/j.patcog.2019.107036. URL: www.sciencedirect.com/science/article/pii/S0031320319303395.

Loris Bazzan, Marco Cristani, and Vittorio Murino. SDALF: Modeling human appearance with symmetry-driven accumulation of local features. In Shaogang Gong, Marco Cristani, Shuicheng Yan, and Chen Change Loy, editors, *Person Re-Identification*, pages 43–69. Springer, London, 2014. ISBN 978-1-4471-6295-7 978-1-4471-6296-4. doi: 10.1007/978-1-4471-6296-4_3.

Gary R. Bradski and Adrian Kaehler. *Learning OpenCV: Computer vision with the OpenCV library*. Software That Sees. O'Reilly, Beijing, 1. edn., [nachdr.] edition, 2011. ISBN 978-0-596-51613-0.

Mao Chen, Bo Hu, Weixin Jiang, Min Chen, and Yuanqing Li. A survey of the applications of machine learning in networking. *Annals of Telecommunications*, 71(7):435–444, 2016.

Ming Chen, Yuanqing Li, Xiao Zhang, and Shiwen Liu. Caching in 5G networks: A survey. *IEEE Access*, 5:5870–5889, 2017.

Tianlong Chen, Sen Ding, Weijian Xie, Lu Yuan, and Jiashi Wang. Abdnet: Attentive but diverse person re-identification. In *Proceedings of the IEEE/CVF international conference on computer vision*, pages 8351–8361, 2019.

Jason V. Davis, Brian Kulis, Prateek Jain, Suvrit Sra, and Inderjit S. Dhillon. Information-theoretic metric learning. In *Proceedings of the 24th International Conference on Machine Learning – ICML'07*, pages 209–216, Corvalis, Oregon, 2007. ACM Press. ISBN 978-1-59593-793-3. doi: 10.1145/1273496.1273523.

Ross Girshick. Fast R-CNN. In *Proceedings of the IEEE International Conference on Computer Vision*, pages 1440–1448, 2015.

Shaogang Gong, Marco Cristani, Shuicheng Yan, and Chen Change Loy, editors. *Person Re-Identification*. Springer, London, 2014. ISBN 978-1-4471-6295-7 978-1-4471-6296-4. doi: 10.1007/978-1-4471-6296-4.

Arun Hampapur, Lowell Brown, Jonathan Connell, Ahmet Ekin, Norbert Haas, Ming Lu, Heinz Merkl, and Sharath Pankanti. Smart video surveillance: exploring the concept of multiscale spatiotemporal tracking. *IEEE Signal Processing Magazine*, 22(2):38–51, 2005.

Alexander Hermans, Lucas Beyer, and Bastian Leibe. In defense of the triplet loss for person re-identification. In *arXiv preprint arXiv:1703.07737*, 2017.

Weiming Hu, Tieniu Tan, Liang Wang, and Stephen Maybank. A survey on visual surveillance of object motion and behaviors. *IEEE Transactions on Systems, Man, and Cybernetics, Part C (Applications and Reviews)*, 34(3):334–352, 2004.

M. Kostinger, M. Hirzer, P. Wohlhart, P. M. Roth, and H. Bischof. Large scale metric learning from equivalence constraints. In *2012 IEEE Conference on Computer Vision and Pattern Recognition*, pages 2288–2295, Providence, RI, June 2012. IEEE. ISBN 978-1-4673-1228-8 978-1-4673-1226-4 978-1-4673-1227-1. doi: 10.1109/CVPR.2012.6247939.

Alex Krizhevsky, Ilya Sutskever, and Geoffrey E. Hinton. ImageNet classification with deep convolutional neural networks. *Communications of the ACM*, 60(6):84–90, May 2017. ISSN 00010782. doi: 10.1145/3065386.

Zhen Li, Shiyu Chang, Feng Liang, Thomas S. Huang, Liangliang Cao, and John R. Smith. Learning locally-adaptive decision functions for person verification. In *2013 IEEE Conference on Computer Vision and Pattern Recognition*, pages 3610–3617, Portland, OR, June 2013. IEEE. ISBN 978-0-7695-4989-7. doi: 10.1109/CVPR.2013.463.

Live Facial Recognition. www.met.police.uk/advice/advice-and-information/fr/facial-recognition-technology/.

Bingpeng Ma, Yu Su, and Frederic Jurie. BiCov: A novel image representation for person re-identification and face verification. In *Proceedings of the British Machine Vision Conference 2012*, pages 57.1–57.11, Surrey, 2012. British Machine Vision Association. ISBN 978-1-901725-46-9. doi: 10.5244/C.26.57.

Ministry of Defence. MOD unveils futuristic uniform design, September, 2015. www.gov.uk/government/news/mod-unveils-futuristic-uniform-design.

Mark Nixon and Alberto S Aguado. *Feature Extraction and Image Processing for Computer Vision*. Academic Press, Cambridge, CA, 2019.

Ioannis T. Pavlidis and Jacob A. Levine. Thermal imaging for surveillance applications. *IEEE Signal Processing Magazine*, 14(6):17–19, 2017.

Shaoqing Ren, Kaiming He, Ross Girshick, and Jian Sun. Faster R-CNN: Towards real-time object detection with region proposal networks. In *Advances in Neural Information Processing Systems*, pages 91–99, 2015.

Pablo Ridao, Kurt Barnard, and Carlos Pecharromán. Nighttime vehicle re-identification based on deep learning. In *International Conference on Image Processing (ICIP)*, pages 2349–2353. IEEE, 2015.

Neil C. Rowe and John E. Bares. Smart camera networks in virtual battlefield simulation. *IEEE Aerospace and Electronic Systems Magazine*, 27(6):4–13, 2012.

Mahadev Satyanarayanan. The emergence of edge computing. *Computer*, 50(1):30–39, 2017.

Weisong Shi, Jie Cao, Quan Zhang, Youhu Li, and Lanyu Xu. Edge computing: Vision and challenges. *IEEE Internet of Things Journal*, 3(5):637–646, 2016.

Saurav Sthapit. *Computation Offloading for Algorithms in Absence of the Cloud*. PhD thesis, The University of Edinburgh, November 2018.

Saurav Sthapit, John Thompson, Neil M. Robertson, and James R. Hopgood. Computational load balancing on the edge in absence of cloud and fog. *IEEE Transactions on Mobile Computing*, 18(7):1499–1512, July 2019. ISSN 1558-0660. doi: 10.1109/TMC.2018.2863301.

Yifan Sun, Liang Zheng, Yi Yang, Qi Tian, and Shengjin Wang. Beyond part models: Person retrieval with refined part pooling (and a strong convolutional baseline). In *Proceedings of the European conference on computer vision (ECCV)*, pages 480–496, 2018.

Mohsen Nader Tehrani, Murat Uysal, and Halim Yanikomeroglu. Device-to-device communication in 5G cellular networks: Challenges, solutions, and future directions. *IEEE Communications Magazine*, 52(5):86–92, May 2014. ISSN 0163-6804, 1558-1896. doi: 10.1109/MCOM.2014.6815897.

María Valera and Sergio A Velastin. Intelligent distributed surveillance systems: A review. *IEE Proceedings-Vision, Image and Signal Processing,* 152(2):192–204, 2005.

Paul Viola and Michael J Jones. Robust real-time face detection, page 18, 2004.

Xiaogang Wang and Rui Zhao. Person re-identification: System design and evaluation overview. In Shaogang Gong, Marco Cristani, Shuicheng Yan, and Chen Change Loy, editors, *Person Re-Identification*, pages 351–370. Springer, London, 2014. ISBN 978-1-4471-6295-7 978-1-4471-6296-4. doi: 10.1007/978-1-4471-6296-4_17.

Xiaowei Xu, Qingshan Yu, Zhigang Ma, Zhiqiang Li, and Josef Kittler. Deep learning for free-viewpoint video synthesis and person re-identification. *IEEE Transactions on Circuits and Systems for Video Technology*, 27(3):613–627, 2017.

Mang Ye, Jianbing Shen, Gao Lin, Tao Xiang, Ling Shao, and Steven C. H. Hoi. Deep learning for person re-identification: A survey and outlook. *IEEE Transactions on Pattern Analysis and Machine Intelligence*, 44(6):2872–2893, 2021.

Qi Zhang and Hui Li. Advances in multi-spectral imaging for defense applications. *IEEE Sensors Journal*, 18(12):4856–4864, 2018.

Rui Zhao, Wanli Ouyang, and Xiaogang Wang. Person re-identification by saliency learning. *arXiv:1412.1908 [cs]*, December 2014.

Section V

AI in Infrastructure and Transportation Management

13 Smart Heating and Cooling Controller Design for Smart Infrastructure Management

13.1 INTRODUCTION

Heating, ventilation, and air conditioning (HVAC) of the homes require one of the major usages of energy, and approximately 30% of the energy is spent on it globally (Agency 2023). As the world becomes increasingly urbanized and digitalized, the demand for smart infrastructure continues to grow. Manual and simple rule-based controllers can result in energy wastage. This can result in reduced thermal comfort and increased energy usage. Therefore, optimizing the HVAC controller can offer a huge financial benefit to the consumer as well as raise their comfort levels. As more and more households adopt the smart HVAC, it will minimize the global energy usage or at least balance the equations for the increasing number of new houses. In addition, networking smart homes has even better energy savings than isolated systems (King and Perry, 2017). Thus, it is also a high priority amongst governments as they pledge to realize the net-zero commitments. In this chapter, we look at the brief history, different components, publically available datasets of smart HVAC. We conclude the chapter with a use case of smart HVAC.

13.2 HISTORY

Since the ancient times, humans have used fireplaces and open hearths as a means to warm their dwellings. The Romans developed an advanced way of heating their homes with a hypocaust system. Hypocaust systems heated their homes using hot air circulated under the floors and spaces in the wall providing efficient and wide-spread warmth. During the middle ages, fireplaces were supplemented with chimneys improving ventilation and reducing indoor smoke. In the 17th and 18th century, wood and coal stoves were introduced that were more efficient and controlled means of heating homes. Then in the 19th century, radiators and central heating systems were introduced. The cast iron radiators became a common component of central heating system carrying hot water and steam supplied by the boilers and distributed via pipes and ducts to various rooms in the house.

DOI: 10.1201/9781003489986-18

13.3 THERMAL COMFORT AND ITS EVALUATION

Thermal comfort is defined as the state of mind that expresses satisfaction with the surrounding thermal environment. It is a critical consideration in the design and operation of HVAC systems, particularly in smart homes. Ensuring thermal comfort involves maintaining an indoor environment that balances air temperature, humidity, airflow, and other factors to suit the preferences of occupants. Thermal comfort is influenced by several interrelated environmental and personal factors:

- **Air Temperature:** The most intuitive factor, representing the warmth or coolness of the air.
- **Relative Humidity:** Impacts the ability of sweat to evaporate, affecting how hot or cold a person feels.
- **Air Velocity:** Movement of air influences heat transfer between the body and the environment.
- **Radiant Temperature:** Heat radiated from surrounding surfaces, such as walls, floors, and furniture.
- **Clothing Insulation:** The thermal resistance provided by clothing, which affects heat retention.
- **Metabolic Rate:** The rate of heat production in the human body due to activity level.

13.3.1 THERMAL COMFORT STANDARDS

Standards such as American Society of Heating, Refrigerating, and Air-Conditioning Engineers (ASHRAE) (Jayaraman et al. 2009–2010) provide guidelines for achieving and evaluating thermal comfort. These standards consider the physical and psychological responses of occupants to the thermal environment. Thermal comfort can be quantitatively and qualitatively evaluated using predicted mean vote and predicted percentage dissatisfied metrics. Figure 13.1 shows the scale and corresponding metric scores.

Predicted Mean Vote (PMV) and Predicted Percentage Dissatisfied (PPD):

- **PMV Index:** Predicts the average thermal sensation of a large group of people on a scale ranging from –3 (cold) to +3 (hot), with 0 representing neutrality.

$$\mathrm{PMV} = f(M, W, I_{\mathrm{cl}}, T_{\mathrm{a}}, T_{\mathrm{r}}, v, p_{\mathrm{a}}) \tag{13.1}$$

where
– M: Metabolic rate
– W: External work
– I_{cl}: Clothing insulation
– T_{a}: Air temperature
– T_{r}: Mean radiant temperature

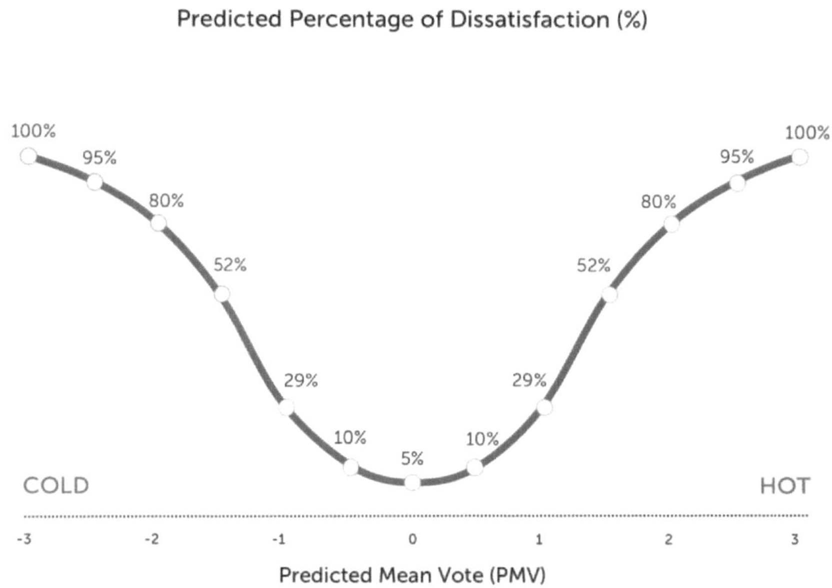

FIGURE 13.1 Evaluating thermal comfort using predicted mean vote and predicted percentage of dissatisfaction. (https://a.storyblok.com/f/52232/1968x1256/8d44cc2709/screen-shot-2019-01-23-at-9-53-12-am.png).

– v: Air velocity
– p_a: Partial pressure of water vapour

- **PPD Index:** Represents the percentage of people likely to feel dissatisfied with the thermal environment, calculated based on PMV. Lower PPD values indicate better thermal comfort.

Although PMV and PPD are the most common metrics, there are several many metrics to measure thermal comfort. See Python library *PyThermalComfort* for a collection.[1]

13.4 AI-BASED SMART SOLUTIONS

Over the recent years, there has been significant interest in creating smart homes and their automation. The term smart home is generic and an umbrella term that may include many intelligent solutions. However, the salient benefits of all smart home solutions include one or more of the following.

13.4.1 Energy Efficiency

One of the primary benefits of smart heating and cooling systems is their ability to optimize energy consumption. By using real-time data from IoT sensors and predictive algorithms, these systems adjust heating and cooling based on actual needs rather than fixed schedules or preset temperatures, which are common in traditional systems.

For example, a smart thermostat equipped with motion sensors can detect room occupancy and only activate climate control in rooms that are occupied, significantly reducing unnecessary energy consumption. Research shows that such systems can reduce HVAC energy usage from 10% to 30%.

Smart systems also adjust based on outdoor weather conditions. For instance, if outdoor temperatures are mild, the system may use natural ventilation or adjust setpoints to reduce the demand on HVAC units. Studies in offices and commercial spaces indicate that this approach can further reduce energy consumption by 5–15% (Wang et al. 2022).

13.4.2 Cost Savings

The reduction in energy usage directly translates to financial savings for both residential and commercial users. In residential applications, smart thermostats like Nest or Ecobee can reduce energy bills by up to £120 annually, according to estimates from the U.S. Environmental Protection Agency (EPA) (U.S. Department of Energy 2023). For commercial buildings, the cost savings are even more significant, especially in large facilities with complex HVAC needs.

In addition to direct energy cost savings, peak demand reduction also results in financial benefits. Some utility companies offer dynamic pricing or demand-response programmes, where users can earn credits or rebates by reducing HVAC load during peak hours. Smart heating and cooling systems automatically adjust during these times to minimize energy costs, potentially saving users up to 10% on energy bills annually (Smith et al. 2023).

Case studies from large office buildings and manufacturing facilities using smart HVAC systems report an average of 20–25% reduction in energy expenses, demonstrating substantial ROI over time. Companies with multiple facilities have also seen substantial savings, as aggregated data enables them to optimize energy use across locations by analysing usage patterns and making data-driven adjustments (Jones and Rodriguez 2022).

13.4.3 Enhanced Comfort and Convenience

Smart heating and cooling systems are designed to offer users a high degree of personalization, making it possible to tailor indoor climates to individual comfort preferences. Unlike traditional systems that rely on a single thermostat, smart systems use zonal heating and cooling, allowing users to set different temperatures for different rooms or zones based on usage and preferences.

Geo-fencing technology enhances this convenience further by automatically adjusting settings when users enter or leave their homes. For instance, a system can start pre-heating a home in winter as the user approaches, ensuring comfort upon arrival without needing to run the heating continuously. Machine learning algorithms within smart systems learn user behaviour over time, providing temperature and humidity levels that align with past preferences. For example, the system may learn that certain family members prefer cooler temperatures at night and will adjust accordingly. These features ensure comfort without requiring users to make manual adjustments, making the climate control experience seamless and stress-free. Voice assistant integration (e.g., Amazon Alexa, Google Assistant) further enhances convenience by allowing users to change settings or get updates on system performance through simple voice commands. In a study of homes using smart heating systems with voice assistance, 75% of users reported increased satisfaction and ease of use (Huang and Smith 2022).

13.4.4 Environmental Sustainability

Smart heating and cooling systems play a crucial role in reducing greenhouse gas (GHG) emissions. Buildings account for approximately 40% of global energy consumption, much of which is attributed to HVAC systems (International Energy Agency 2021). By optimizing energy use, smart systems help reduce carbon emissions and support sustainability goals.

For instance, studies indicate that a typical home equipped with a smart thermostat and occupancy sensors can reduce CO_2 emissions by around 300–400 kg annually. In larger commercial buildings, emissions reductions can reach thousands of kilograms per year, contributing significantly to organizational sustainability targets (Zhou et al. 2022).

Beyond energy efficiency, many smart systems are also compatible with renewable energy sources like solar or wind. These systems can optimize energy use based on renewable energy availability, such as prioritizing solar power during peak sunlight hours. This integration not only reduces reliance on fossil fuels but also improves the building's overall energy efficiency and contributes to grid stability.

13.4.5 Health and Safety

Maintaining optimal indoor climate conditions can positively impact health and well-being. Smart heating and cooling systems contribute to indoor air quality management by controlling humidity levels, filtering air, and providing adequate ventilation. Proper humidity control, in particular, can prevent the growth of mould and reduce dust mites, which are common allergens.

Research from the ASHRAE has shown that maintaining humidity between 40–60% can reduce respiratory symptoms and decrease the spread of airborne pathogens. Smart systems maintain these levels by adjusting humidity settings based on real-time data from humidity sensors, ensuring a healthier indoor environment (ASHRAE 2023).

In commercial settings, such as healthcare facilities, data centres, or laboratories, maintaining precise temperature and humidity levels is critical. Smart systems automatically adjust to preserve these optimal conditions, protecting both equipment and individuals. For example, in healthcare, where precise climate control can influence patient recovery, smart HVAC systems are invaluable for maintaining consistent conditions.

13.4.6 Data-Driven Insights for Continuous Improvement

Smart heating and cooling systems also provide valuable data insights that can inform better decision-making and continuous improvement. By collecting data on energy usage patterns, peak demand times, and environmental conditions, these systems offer users (or facility managers) insights into how to further improve efficiency and reduce costs.

Data-driven insights can reveal usage trends that might otherwise go unnoticed, such as identifying times of day when heating or cooling demand is highest. This information enables users to make informed choices about further optimizing their energy consumption. For example, facility managers may adjust operational hours or adjust maintenance schedules to prevent high demand during peak times.

In industrial settings, data insights also help identify system inefficiencies or wear, prompting preventive maintenance before issues become costly or disruptive. This aspect of smart systems is crucial in facilities like manufacturing plants or warehouses, where climate control is integral to protecting equipment and products.

13.5 SENSORS AND TECHNOLOGIES

Internet of things (IoT) devices play a crucial role in smart heating and cooling systems by providing real-time data and control capabilities. Figure 13.2 shows a general schematics of smart home. It contains two rooms fitted with different sensors that measure the temperature, humidity, occupancy, and light levels. The data is fed to the HVAC controller which can be centralized to the home. It can be extended to a collection of homes where the controller can control more than one home as well. Based on the sensor data, the controller can vary the thermal environment of the room. For example, using the occupancy sensor, only the rooms with people need to be maintained. Temperature and fan speed can be lowered or raised based on thermal comfort.

13.5.1 Temperature and Humidity Sensors

Temperature and humidity sensors are the backbone of smart HVAC systems. They continuously monitor environmental conditions and provide real-time feedback to the control system. These sensors are placed both indoors and outdoors to optimize system performance.

- **Indoor Sensors:** Monitor room temperature and humidity to ensure optimal comfort levels. Help identify temperature variations across rooms for zonal control. Detect anomalies such as excessive heat gain or loss due to insulation issues.
- **Outdoor Sensors:** Provide data about external weather conditions like temperature, humidity, and dew point. Enable predictive heating or cooling, especially when integrated with weather forecasts.

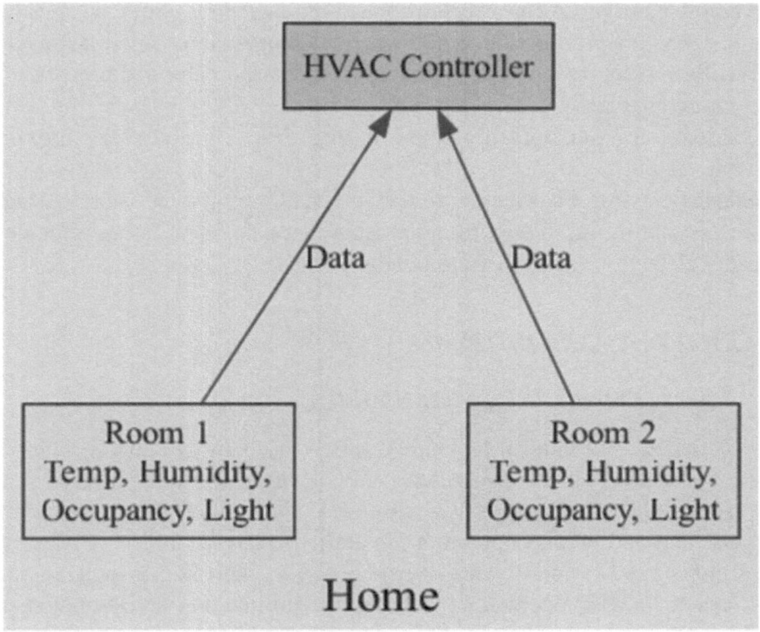

FIGURE 13.2 Schematic of a smart home with two rooms and a central controller.

13.5.2 Occupancy Sensors

Occupancy sensors are crucial for tailoring HVAC operations to actual room usage, ensuring energy is not wasted in unoccupied spaces. There are three main types of sensors that can be used, listed further. The camera obviously has concerns for privacy especially when used within the commercial building and infrastructures.

- **Passive Infrared (PIR) Sensors:** Detect motion based on infrared heat signatures. Commonly used in offices and residential settings.
- **Ultrasonic Sensors:** Detect movement using sound waves, suitable for spaces with minimal line-of-sight.
- **Camera-Based Sensors:** Provide advanced detection, distinguishing between humans and objects. Used in high-end smart systems with privacy-preserving features.

13.5.3 Smart Thermostats

Smart thermostats are advanced control devices that adapt to user preferences and schedules. They integrate with sensors and external data sources to create a highly responsive HVAC system.

- **Learning Algorithms:** Analyse user schedules and behaviours to automate temperature adjustments. Example: Nest Learning Thermostat adjusts settings after observing patterns for a few weeks.

- **Remote Control:** Allow users to control settings via smartphone apps or voice assistants. Integration with platforms like Google Home or Amazon Alexa.
- **Weather Integration:** Use weather forecasts to predict and prepare for temperature changes.
- **Geofencing:** Automatically adjust settings based on the user's proximity to home.
- **Energy-Saving Features:** Provide detailed reports on energy usage and suggest optimization opportunities. Some models, like Ecobee, include occupancy detection for further efficiency.

13.6 CONTROL ALGORITHMS

13.6.1 Model Predictive Controller (MPC) for Smart Homes

MPC is a sophisticated control technique widely used in dynamic systems where optimization and adaptability are critical. MPC leverages a mathematical model of the system to predict future states and optimize control inputs over a given horizon, ensuring efficient and precise operation. Its ability to handle multi-variable systems with constraints makes it particularly well-suited for smart home applications, such as HVAC systems. MPC operates by solving an optimization problem at each time step to determine the optimal control inputs. The key mathematical components are as follows:

System Model: MPC relies on a mathematical model of the system, typically expressed as a discrete-time state-space model:

$$\vec{x}(k+1) = \boldsymbol{A}\vec{x}(k) + \boldsymbol{B}\vec{x}(k) \tag{13.2}$$

where

- $\vec{x}(k)$ is the state vector at time step k (e.g., room temperature, humidity)
- $\vec{u}(k)$ is the control input vector (e.g., heating/cooling power)
- $\boldsymbol{A}$ and $\boldsymbol{B}$ are system matrices defining the dynamics

Objective Function: The optimization problem minimizes a cost function J over a prediction horizon N:

$$J = {}^{I::\,iN-1}_{\;\;i=0}\vec{x}(k+i \mid k) - \vec{x}_{\text{ref}}(k+i)^2_{\boldsymbol{Q}} + u(k+i \mid k)^2_{\boldsymbol{R}} \tag{13.3}$$

where

- $\vec{x}_{\text{ref}}(k+i)$ is the desired reference trajectory (e.g., target room temperature)
- $\boldsymbol{Q}$ and $\boldsymbol{R}$ are weighting matrices balancing state tracking and input effort.

Constraints: MPC incorporates system constraints, such as

$$\vec{x}_{\min} \leq \vec{x}(k+i|k) \leq \vec{x}_{\max}, \vec{u}_{\min} \leq \vec{u}(k+i \mid k) \leq \vec{u}_{\max} \tag{13.4}$$

These constraints ensure physical feasibility, such as maintaining temperatures within comfort bounds or limiting HVAC power usage.

Controller: At each time step k, MPC solves the optimization problem:

$$\min_{\vec{u}(k),\ldots,\vec{u}(k+N-1)} J \tag{13.5}$$

subject to system dynamics and constraints. Only the first control input $\vec{u}(k)$ is applied, and the process repeats at the next step with updated state measurements.

13.6.2 Reinforcement Learning (RL)–Based Solution

Reinforcement learning (RL) is a powerful AI technique that has shown immense potential in optimizing smart HVAC systems (Ding et al. 2024). Unlike traditional rule-based controllers, which rely on predefined rules and parameters, RL-based systems adapt dynamically to their environment, learning optimal strategies through trial and error. Approaches using RL for HVAC control include these works (Ding et al. 2024, Bayer and Pruckner 2022, Lissa et al. 2020).

RL operates on the principle of an agent interacting with an environment to achieve a specific goal. The key components of an RL framework in the context of smart HVAC systems are

- **Agent:** The HVAC control system, which decides actions such as heating or cooling power levels.
- **Environment:** The building's thermal conditions, influenced by factors such as weather, occupancy, and internal heat sources.
- **State:** The current conditions of the environment, represented by variables such as indoor temperature, humidity, occupancy status, and outdoor temperature.
- **Action:** Possible adjustments the HVAC system can make, such as changing the thermostat setting, airflow distribution, or switching heating/cooling modes.
- **Reward:** A feedback signal that quantifies the system's performance. For smart HVAC systems, this could be a combination of:
  - Maintaining thermal comfort for occupants. This could be predicted with the PMV and PPD metric described in Section 13.3
  - Minimizing energy consumption
  - Reducing operational costs

The agent's objective is to maximize the cumulative reward over time, balancing comfort and energy efficiency.

13.6.2.1 RL Algorithm Types for HVAC

Different RL algorithms can be employed based on the complexity of the HVAC system:

- **Q-Learning:** A value-based method that works well for discrete action spaces. For example, the system can decide between predefined temperature settings.
- **Deep RL:** Combines RL with deep learning to handle large, continuous state and action spaces. This is particularly useful in smart homes with many sensors and controls.
- **Policy Gradient Methods:** Focus on learning the optimal policy directly and are well-suited for continuous action spaces, such as adjusting HVAC power levels in fine increments.

13.6.2.2 Challenges and Considerations

While RL has significant potential, there are challenges to consider:

- **Exploration versus Exploitation:** The agent must balance trying new strategies (exploration) with leveraging known successful actions (exploitation).
- **Training Time:** RL systems often require substantial training data to converge to optimal policies. This may include developing a simulator that mimics real world environment with sensors and actuators. The RL agent can be safely trained in the simulator and transferred to real-world application.
- **System Constraints:** Ensuring actions that remain physically feasible (e.g., avoiding sudden temperature changes that strain HVAC equipment).
- **Safety and Robustness:** Incorporating mechanisms to avoid catastrophic failures during the learning phase.

13.7 REFIT DATASET

As discussed previously, approaches such as RL requires data and simulator. There are several real-world datasets publicly available. For example, the REFIT dataset (https://repository.lboro.ac.uk/articles/dataset/REFIT_Smart_Home_dataset/2070091) and IDEAL Household Energy Dataset (https://datashare.ed.ac.uk/handle/10283/3647) (Pullinger et al. 2021). We briefly describe the REFIT dataset:

The REFIT dataset is for 20 homes located near to the town of Loughborough in the East Midlands region of the United Kingdom. A building survey was carried out at each home, collecting data on building geometry, construction materials, occupancy, and energy services. Each home has a selection of the following sensors and devices installed:

- **CurrentCost mains clamps**, to measure household mains electrical power load (data available from Strathclyde University, see further).
- **Replacement gas meters**, to measure household mains gas consumption.
- **Hobo pendant or Hobo U12 sensors** to measure room air temperature, relative humidity, and light level.

TABLE 13.1
Dataset Statistics

Feature	Numbers
Homes	20
Rooms	389
Radiators	252
Showers	34
Appliances	618
Light bulbs	672
Fixed heaters	19
Surfaces	2,237
Openings	970
Sensors	1,567
Variables recorded by sensors and devices	2,457
Time series readings	25,312,397

- **iButton temperature sensors** to measure radiator surface temperature.
- **CurrentCost individual appliance monitors**, to measure plug electrical power loads (data available from Strathclyde University, see further).
- **RWE Smart Home devices** including programmable thermostatic radiator valves, interior and exterior motion detectors, door and window opening sensors, and smoke alarms.
- **British Gas Hive programmable thermostats**.

In addition, climate data was collected at the Loughborough University campus weather station (Table 13.1).

13.8 CONCLUSION

The exploration of smart heating and cooling systems within this chapter highlights the transformative potential of intelligent technologies in modernizing HVAC systems. By leveraging advanced sensors, control algorithms, like MPC and RL, and integration with smart home frameworks, these systems achieve significant improvements in energy efficiency, cost savings, and occupant comfort. Furthermore, the integration of data-driven insights, environmental sustainability, and health-focused optimizations underscores the holistic impact of smart HVAC systems on both residential and commercial applications.

To further enhance the capabilities of smart HVAC systems, integrating digital twin technology is a promising direction. A digital twin is a virtual representation of a physical system that mirrors its operations in real time, enabling advanced simulations, monitoring, and predictive control. Beyond individual homes, digital twins can scale to represent entire buildings or city-wide HVAC networks. Digital twins can analyse historical data and operational trends to predict when HVAC components are likely to fail, enabling proactive maintenance. This reduces downtime and extends the lifespan of HVAC equipment. It also enables coordinated energy management across multiple systems, contributing to urban sustainability goals.

NOTE

1 Python Package Index. https://pypi.org/project/pythermalcomfort/

BIBLIOGRAPHY

American Society of Heating, Refrigerating, and Air-Conditioning Engineers (ASHRAE). Guidelines on indoor humidity for healthier environments. *Healthcare Facilities Journal*, 18:203–214, 2023. doi: 10.1080/12345678. 2023.1140234.

Daniel Bayer and Marco Pruckner. Enhancing the performance of multi-agent reinforcement learning for controlling hvac systems. In *2022 IEEE Conference on Technologies for Sustainability (SusTech)*, pages 187–194. IEEE, 2022.

Xianzhong Ding, Zhiyu An, Arya Rathee, and Wan Du. A safe and data-efficient model-based reinforcement learning system for hvac control. *arXiv preprint arXiv:2407.12195*, 2024.

X. Huang and R. Smith. User satisfaction and smart heating systems: A survey analysis. *Energy Research Journal,* 72:290–304, 2022. doi: 10.1016/ j.enres.2022.004567.

International Energy Agency. Energy Efficiency. Technical report, 2023.

International Energy Agency. The role of smart systems in reducing global carbon emissions. *IEA Annual Report*, 2021. URL www.iea.org/energy-system/decarbonisation-enablers/digitalisation. Accessed: 2023-10-05.

Nadar R. Jayaraman, Byron W. Jones, Jay A. Kohler, Carol E. Marriott, Merle F. McBride, Frank Myers, Janice C. Peterson, Douglas T. Reindl, Lawrence J. Schoen, Boggarm S. Setty, Bodh R. Subherwal, James R. Tauby, James K. Vallort, William F. Walter, Michael W. Woodford, Craig P. Wray, Wayne R. Reedy, and Thomas E. Watson. ASHRAE Standards Committee Report – Energy-Efficient Design of Low-Rise Residential Buildings, 2009–2010.

T. Jones and M. Rodriguez. Reducing commercial energy consumption with smart HVAC systems. *Energy Efficiency Journal*, 48:89–104, 2022. doi: 10.1007/s12053-022-9874-0.

Jen King and Christopher Perry. Smart buildings: Using smart technology to save energy in existing buildings. *Smart Buildings*, 2017.

Paulo Lissa, Michael Schukat, and Enda Barrett. Transfer learning applied to reinforcement learning-based hvac control. *SN Computer Science*, 1(3):127, 2020.

Martin Pullinger, Jonathan Kilgour, Nigel Goddard, Niklas Berliner, Lynda Webb, Myroslava Dzikovska, Heather Lovell, Janek Mann, Charles Sutton, Janette Webb, et al. The ideal household energy dataset, electricity, gas, contextual sensor data and survey data for 255 UK homes. *Scientific Data*, 8(1):146, 2021.

J. Smith, R. Patel, and K. Johnson. Cost benefits of smart thermostats in demand-response programs. *Journal of Energy Management*, 54:112–130, 2023. doi: 10.5555/jem.2023.321654.

U.S. Department of Energy. Smart thermostats and energy savings: An overview. 2023. URL www.energy.gov. Accessed: 2023-10-05.

L. Wang, D. Taylor, and J. Kim. Impact of weather-responsive HVAC systems in commercial spaces. *Energy and Buildings Journal*, 145:210–225, 2022. doi: 10.1016/j.enbuild.2022.145001.

Y. Zhou, L. Chen, and H. Wang. Energy efficiency analysis of smart heating and cooling systems. *Energy and Buildings*, 123:45–58, 2022. doi: 10. 1016/j.enbuild.2022.123456.

14 Resilient CPS Design for Connected and Autonomous Vehicles in Smart Transportation

14.1 INTRODUCTION

The rapid evolution of connected and autonomous vehicles (CAV) has ushered in a new era of smart transportation, driven by advancements in cyber-physical systems (CPSs). CPS enables seamless integration of computational algorithms, physical systems, and networked communications to achieve complex functionalities. Despite these advancements, resilience in the design of CPSs for CAVs remains a critical concern, given the complexity of their operation in dynamic environments and their exposure to cyber threats. For example, security researchers Charlie Miller and Chris Valasek demonstrated a remote hacking attack on a Jeep Cherokee exploiting the vulnerabilities in its Uconnect infotainment system [Miller 2019]. The researchers were able to take full control of the vehicle remotely over the internet and demonstrated their ability to

- kill the engine while the car was driving,
- disable the brakes,
- take control of the steering (only at low speeds), and
- alter the vehicle's speedometer and entertainment system.

It was possible by exploiting the infotainment system which was connected to the vehicle's controller area network (CAN) bus and sending malicious commands on the CAN bus. Figure 14.1 shows the CAN bus of the vehicle. This chapter explores the principles, challenges, and methodologies for designing resilient CPS in the context of CAVs. It highlights the importance of robust design paradigms to ensure safety, security, and reliability in smart transportation systems.

14.2 BACKGROUND

CAVs rely on a diverse array of sensors and algorithms to perceive their environment, make decisions, and execute actions. These components work together to ensure safe, reliable, and efficient operation. This section provides an overview of the primary sensors and typical algorithms utilized in CAVs. However, before delving into the sensors and respective algorithms, we need to understand different levels of autonomy in autonomous driving and what is expected in each of these levels.

DOI: 10.1201/9781003489986-19

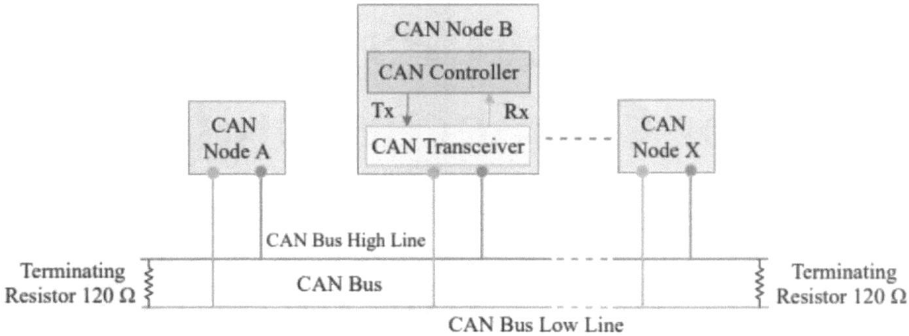

FIGURE 14.1 CAN bus of a vehicle; it is the information highway of the vehicles. (From Aldhyani and Alkahtani [2022].)

14.2.1 Levels of Autonomy

CAVs are classified into different levels of autonomy based on the degree of driver intervention and system automation. Society of Automotive Engineers (SAE) defined them from Level 0 (no automation) to Level 5 (full automation):

- **Level 0: No Automation** – The driver is fully responsible for all aspects of vehicle control, with the system providing only warnings or momentary assistance.
- **Level 1: Driver Assistance** – Basic driver assistance systems, such as adaptive cruise control or lane-keeping assistance, are available, but the driver must remain engaged and supervise the vehicle's actions.
- **Level 2: Partial Automation** – The vehicle can perform some automated functions, such as steering and acceleration, simultaneously. However, the driver must monitor the environment and be prepared to take control.
- **Level 3: Conditional Automation** – The vehicle can make decisions and perform driving tasks under specific conditions. Human intervention is required only when the system requests it.
- **Level 4: High Automation** – The vehicle can handle all driving tasks within defined conditions (e.g., geofenced areas or specific weather scenarios) without human intervention.
- **Level 5: Full Automation** – The vehicle is entirely autonomous and capable of operating under all conditions without any driver input.

14.2.2 Sensors in CAVs

Sensors are the cornerstone of a CAV's ability to interact with its surroundings. Figure 14.2 depicts an autonomous vehicle with different sensors. The most commonly used sensors in CAVs are also listed as follows:

- **Light Detection and Ranging (LiDAR):** LiDAR sensors emit laser beams to measure distances and create high-resolution 3D maps of the vehicle's environment. They are particularly effective for object detection and localization.

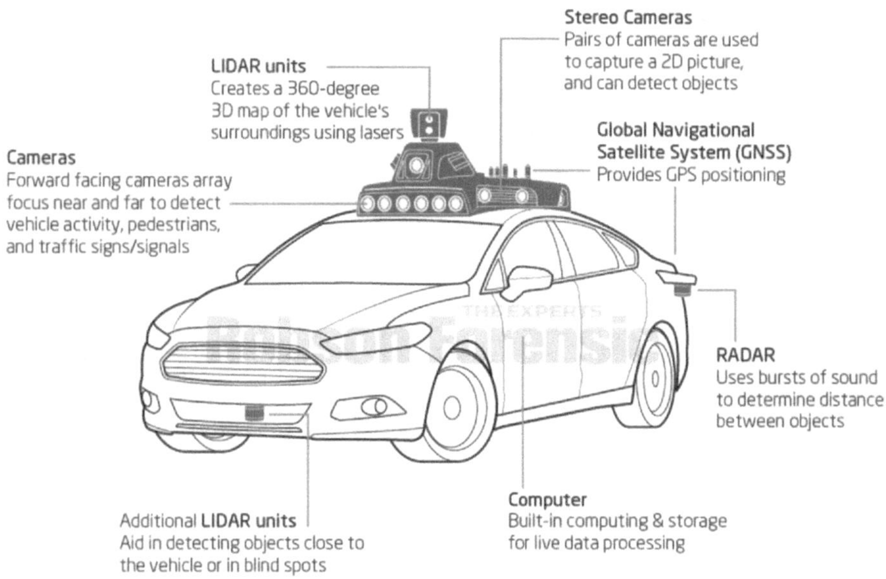

FIGURE 14.2 Autonomous vehicle with sensors. (From www.robsonforensic.com/articles/autonomous-vehicles-sensors-expert.)

- **Radio Detection and Ranging (Radar):** Radar sensors use radio waves to detect objects and measure their speed, distance, and direction. They are commonly used for adaptive cruise control, collision avoidance, and blind spot detection.
- **Cameras:** Cameras capture visual data and are used for lane detection, object recognition, traffic sign identification, and pedestrian tracking. Both monocular and stereo cameras are employed depending on the application. In addition to the red, green, and blue (RGB) cameras, thermal cameras can be used as well. Thermal cameras can be used in low-light scenarios for object detection.
- **Ultrasonic Sensors:** Ultrasonic sensors measure distances to nearby objects using sound waves. They are typically used for parking assistance and low-speed manoeuvring.
- **Inertial Measurement Units (IMUs):** IMUs consist of accelerometers and gyroscopes to measure the vehicle's acceleration and orientation. They are critical for stability control and navigation.
- **Global Navigation Satellite System (GNSS):** GNSS provides location data and is essential for navigation and route planning. When combined with real-time kinematics (RTK), it offers high-precision positioning.
- **Vehicle-to-Everything (V2X) Communication:** V2X technologies enable data exchange between vehicles, infrastructure, and other road users. They provide situational awareness and facilitate cooperative driving strategies. Figure 14.3 shows multiple vehicles and different types of connections.

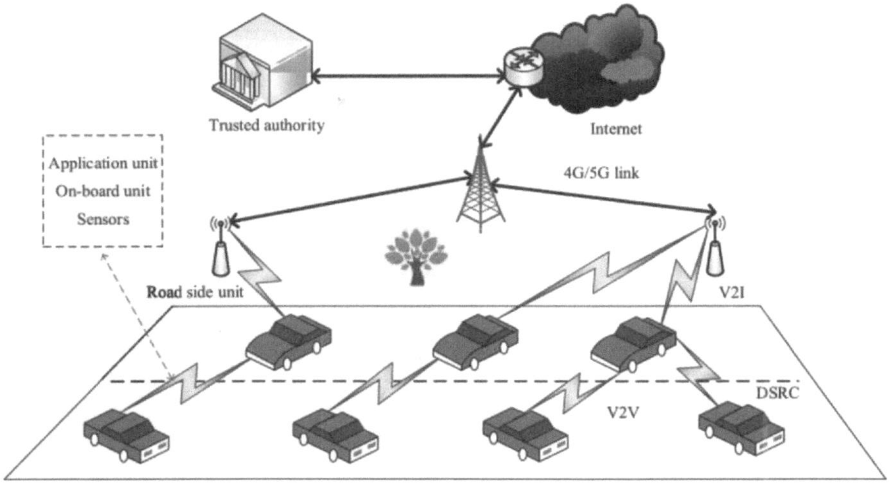

FIGURE 14.3 CAV ecosystem with roadside units, terrestrial network. (From Sun et al. [2022].)

14.2.3 Algorithms in CAVs

The sensors in CAVs generate vast amounts of data, which are processed using advanced algorithms to enable autonomous decision-making. Some of the most commonly used algorithms include

- **Perception Algorithms:** These algorithms process sensor data to detect and classify objects, lanes, and traffic signals. Techniques include computer vision, deep learning (e.g., convolutional neural networks), and sensor fusion.
- **Localization Algorithms:** Localization involves determining the precise position of the vehicle within its environment. Algorithms such as simultaneous localization and mapping (SLAM) and particle filters are commonly used.
- **Path Planning Algorithms:** These algorithms compute optimal routes and manoeuvres while avoiding obstacles. Examples include A* (A-star), rapidly-exploring random trees (RRT), and dynamic programming-based approaches.
- **Control Algorithms:** Control algorithms manage the execution of planned routes and manoeuvres by adjusting throttle, brake, and steering. Proportional–integral–derivative (PID) controllers and model predictive control (MPC) are widely applied.
- **Sensor Fusion Algorithms:** Sensor fusion combines data from multiple sensors to provide a comprehensive understanding of the environment. Kalman filters, extended Kalman filters (EKF), and deep-learning-based fusion methods are commonly employed.
- **Decision-Making Algorithms:** These algorithms enable CAVs to make high-level decisions, such as when to merge lanes or yield to pedestrians.

Techniques include Markov decision processes (MDPs), reinforcement learning, and heuristic-based approaches.

- **Predictive Algorithms:** Predictive models forecast the behaviour of other road users to ensure safe interactions. Recurrent neural networks (RNNs) and trajectory prediction models are frequently used.

14.2.4 Integration of Sensors and Algorithms

The integration of sensors and algorithms is crucial for the effective operation of CAVs. Sensor fusion ensures that data from multiple sources is combined to reduce uncertainty and improve situational awareness. Moreover, real-time data processing and edge computing are essential to meet the latency requirements of autonomous systems. The development of robust algorithms and the optimization of sensor configurations are ongoing areas of research, with the ultimate goal of achieving reliable, safe, and efficient operation across all levels of autonomy.

14.2.5 Datasets

Next we briefly describe popular public datasets available for CAV research.

- **KITTI Vision Benchmark Suite:** One of the most comprehensive datasets for CAV research, KITTI includes data from LiDAR, cameras, and GPS collected in urban and rural areas. It offers benchmarks for tasks such as object detection, stereo vision, and optical flow [Geiger et al. 2012].
  Website: www.cvlibs.net/datasets/kitti/
- **nuScenes Dataset:** nuScenes provides a large-scale dataset with sensor data from LiDAR, radar, and cameras, along with annotated 3D bounding boxes and semantic maps. It is designed for urban driving scenarios.
  Website: www.nuscenes.org/
- **Waymo Open Dataset:** A high-quality dataset collected by Waymo's autonomous vehicles, featuring multi-camera and LiDAR data with object detection, tracking, and segmentation annotations.
  Website: https://waymo.com/open/
- **Oxford RobotCar Dataset:** This dataset includes over 1000 km of driving data from Oxford, UK, captured under varying weather and traffic conditions [Maddern et al. 2017]. It provides LiDAR, GPS, IMU, and camera data, making it ideal for localization and mapping research.
  Website: https://robotcar-dataset.robots.ox.ac.uk/
- **ApolloScape Dataset:** Developed by Baidu, ApolloScape offers large-scale semantic segmentation and 3D localization data for high-definition mapping.
  Website: http://apolloscape.auto/
- **Argoverse Dataset:** Argoverse provides annotated trajectories, map information, and 3D tracking for autonomous driving scenarios [Chang et al. 2019]. It is particularly useful for path planning and trajectory prediction.
  Website: www.argoverse.org/

- **V2V-Sim:** A simulation-based dataset for vehicle-to-vehicle (V2V) communication research. It includes scenarios for cooperative perception and decision-making in connected environments.
  Website: https://github.com/sakshir0/V2V-Simulator-App
- **Car Hacking Dataset:** A dataset for research on adversarial attacks and defences in connected vehicles. It includes simulated attack scenarios such as denial of service (DoS) and spoofing attacks.
  Website: https://ocslab.hksecurity.net/Datasets/car-hacking-dataset

14.2.6 Synthetic Data and Simulation Datasets

Apart from the datasets that collected real data, there are simulators and synthetic dataset that can be used in CAV research. They are particularly useful when the algorithms require training as more data can be generated at will. Also, some of the scenarios are hard to replicate in real life, but easily done in simulators.

- **CARLA Simulator:** CARLA is an open-source simulator for autonomous driving research. It generates synthetic data for testing algorithms in various weather, lighting, and traffic conditions [Dosovitskiy et al. 2017].
  Website: https://carla.org/
- **Comma2k19 Dataset:** A dataset from Comma.ai providing driving data collected from a vehicle equipped with a camera and GPS. It is useful for lane-keeping and behavioural cloning tasks.
  Website: https://github.com/commaai/comma2k19

14.3 RESILIENCE IN CPS DESIGN: CONCEPTS AND FRAMEWORKS

The integration of CPS into CAVs has transformed the landscape of smart transportation by enhancing safety, increasing efficiency, and providing personalized user experience. However, this integration also introduces complexities and vulnerabilities into the systems therefore the need for the development of resilient system designs. The ability of a system to resist, adapt to, and recover from disruption while maintaining critical functionalities is referred to as resilience. This section explores various threats prevalent in the CAV ecosystem and the fundamental concepts of resilience in CPS design. We examine various frameworks that guide resilience implementation and discuss metrics for evaluating CAVs resilience to build smart transportation.

14.3.1 Cybersecurity Threats

The interconnected nature of CAVs makes them susceptible to cyber attacks, such as spoofing, denial of service (DoS) attacks, and ransomware. Figure 14.4 shows the potential attack surfaces in a vehicle. Resilient CPS must address these threats to safeguard the system's integrity. Sun et al. [2022] classifies the attacks on the autonomous vehicles into three categories:

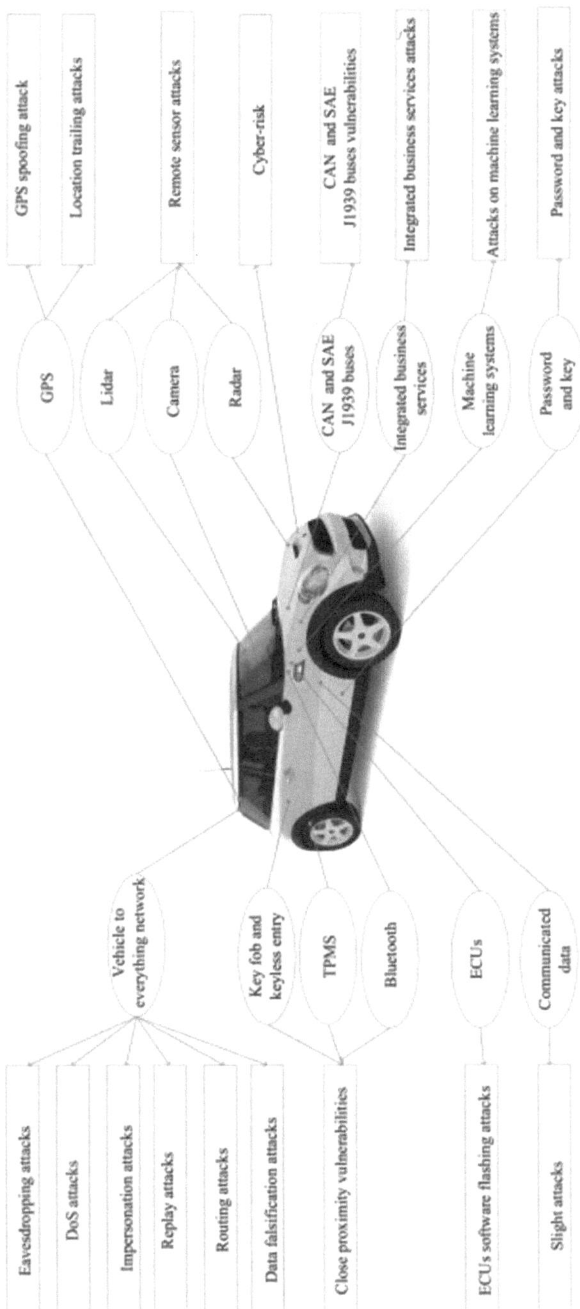

FIGURE 14.4 Attack surface of a typical autonomous vehicle. (From Sun et al. [2022].)

14.3.1.1 In-vehicle Network Attacks

In-vehicle network attacks target the internal communication systems of a vehicle, exploiting vulnerabilities in protocols and hardware components. These include:

- **Remote Sensor Attacks:** Attackers inject malicious data into sensors (e.g., LiDAR, radar) to manipulate vehicle responses, such as activating emergency braking unnecessarily.
- **GPS Spoofing:** Manipulating GPS signals to mislead navigation systems, potentially rerouting vehicles into dangerous or unauthorized areas.
- **CAN and SAE J1939 Vulnerabilities:** Exploiting weaknesses in vehicle communication protocols to gain unauthorized access, eavesdrop, or inject malicious commands.
- **Electronic Control Units (ECUs) Software Flashing Attacks:** Unauthorized reprogramming of ECUs, leading to compromised safety-critical systems like braking or acceleration.

14.3.1.2 Vehicle-to-Everything (V2X) Network Attacks

V2X communication connects vehicles with other vehicles (V2V), infrastructure (V2I), and external networks. These networks introduce new vulnerabilities, such as:

- **DoS Attacks:** Overloading communication channels, causing delays in critical messages, potentially leading to collisions.
- **Impersonation and Replay Attacks:** Spoofing identities or replaying valid data packets to manipulate V2X communications, such as creating false traffic alerts.
- **Routing and Data Falsification Attacks:** Compromising routing protocols or injecting false information (e.g., fake accident alerts) to disrupt traffic flow.
- **Eavesdropping:** Intercepting sensitive data, such as authentication keys or private messages, exposing vehicles to further attacks.

14.3.1.3 Other Attacks

Other attacks encompass broader cybersecurity threats that may target infrastructure, AI models, or the integrity of data:

- **Infrastructure Attacks:** Compromising backend cloud servers or roadside infrastructure, disrupting communication or essential services.
- **Attacks on Machine Learning Models:** Data poisoning or adversarial attacks on AI-based systems that vehicles rely on for decision-making, leading to misclassifications or errors.
- **Slight and Signal Manipulation Attacks:** Introducing subtle deviations in sensor or communication data to mislead systems without triggering alarms.
- **Cyber-Risk in Vehicles:** A dynamic and complex phenomenon involving interconnected threats, emphasizing the need for holistic risk management strategies.

14.3.2 Defining Resilience in CPS

Resilience is a multifaceted concept that extends beyond reliability and robustness. In the context of CPS, resilience provides the capacity of a system to anticipate, absorb, adapt to, and recover from adverse events. This allows continuous operation under both normal and abnormal conditions. While in the context of CAV, building system resilient is important due to the critical nature of transportation systems and the potential consequences of system failures. In this section, we look into various key features of resilience such as robustness, adaptability, fault tolerance, recoverability, and anticipation, discussing their significance in CAVs.

- Robustness refers to a system's ability to operate correctly under a wide range of conditions, including unexpected inputs or stressful environmental factors, without significant degradation in performance [Wang et al. 2015]. In CAVs, it is critical because vehicles must function reliably under varying road conditions, weather scenarios, and traffic situations. Robust systems can handle anomalies such as sensor noise, signal interference, and fluctuating network connectivity. For example,
  (i) A CAV operating in adverse weather conditions, such as heavy rain or fog, may experience reduced sensor accuracy. Robust perception systems employ sensor fusion techniques, combining data from radar, LiDAR, cameras, and ultrasonic sensors to compensate for the limitations of any single sensor modality [Chen et al. 2015b]. This ensures the vehicle can still detect obstacles and navigate safely despite degraded sensor inputs.
  (ii) The control algorithms in CAVs must handle unexpected vehicle dynamics, such as sudden tyre pressure loss or slippery road surfaces. Robust control strategies, like model predictive control (MPC), can adjust the vehicle's trajectory in real time to maintain stability and safety [Falcone et al. 2007].
- Adaptability is the capacity of a system to modify its operations in response to changes in the environment or system conditions, ensuring continued functionality and optimized performance [Saleh et al. 2010]. CAVs operate in dynamic environments where traffic patterns, road conditions, and user preferences can change rapidly. An adaptable system can adjust its behaviour to accommodate these changes without manual intervention. For example, if a CAV encounters unexpected traffic congestion due to an accident, an adaptable navigation system can recalculate the route in real time to find an alternative path, minimizing delays [Katrakazas et al. 2015].
- Fault tolerance is the ability of a system to continue operating properly in the event of the failure of some of its components [Pradhan 1996]. Faults in CAVs can arise from hardware failures, software bugs, or communication breakdowns. A fault-tolerant system can detect such faults and reconfigure itself to maintain essential functions. For example: (a) CAVs often incorporate redundant hardware components, such as multiple processors or parallel computing units. If one processor fails, another can take over the tasks without

interrupting the vehicle's operation [Yoon et al. 2016]. (b) Communication systems in CAVs use error-detecting codes and protocols to identify and correct data transmission errors. For instance, cyclic redundancy checks (CRC) can detect corrupted messages, prompting retransmission [Khan and Salah 2018].

- Recoverability is the efficiency and speed with which a system can restore normal operations after experiencing a disruption or failure [Laprie 2008]. Quick recovery is vital in transportation systems to minimize the impact of failures on safety and traffic flow. Recoverability involves not only fixing the issue but also resuming operations at an acceptable performance level. For example, (a) automatic system restarts: If a software component in the vehicle's operating system crashes, the system can automatically restart the affected modules without requiring a full system reboot, reducing downtime [Sundaram et al. 2013].
- Anticipation involves the proactive identification of potential threats or failures before they occur, allowing the system to take preventive measures [Hollnagel 2011]. Anticipatory capabilities are crucial for preempting situations that could lead to accidents or system failures. This involves predictive analytics and continuous monitoring to foresee and mitigate risks. For example, (a) the system can predict when a part is likely to fail and schedule maintenance before a breakdown occurs, preventing unexpected disruptions by analysing data from vehicle components [Grall et al. 2002]. (b) Machine learning algorithms can monitor sensor data and operational patterns to detect anomalies indicative of cyber attacks or system malfunctions. For instance, unusual communication patterns might signal a cyber intrusion attempt, prompting security protocols to activate [Kwon et al. 2013].

  (iii) Advanced driver-assistance systems (ADAS) use predictive models to anticipate pedestrian movements or sudden stops by other vehicles, allowing the CAV to adjust speed or change lanes proactively [Hu et al. 2018]. The attack types, impacts, attack methods, and features of cyber attacks in autonomous vehicles are shown in Table 14.1.

Let us discuss a case study of the integration of key features in CAVs where we plan to enhance the overall reliability and safety of CAVs. Consider a scenario where a CAV is driving on a highway when heavy snowfall begins unexpectedly. The vehicle's robustness allows it to continue operating despite reduced visibility and sensor interference from snowflakes. Its adaptability enables the adjustment of speed and following distance to suit slippery road conditions. If a camera sensor becomes obstructed by snow, fault tolerance mechanisms switch reliance to other sensors like radar. In case the obstruction leads to partial system failure, recoverability features allow the system to clear the sensor (e.g., activating a defroster or wiper) and restore full functionality. Throughout the journey, the anticipation capability uses weather forecasts and sensor data to predict worsening conditions, potentially re-routing to safer roads or advising the driver to take manual control. The summary of key features importance in CAVs is given in Table 14.2.

TABLE 14.1
Types, Impacts, Attack Methods, and Features of Cyber Attacks in Autonomous Vehicles

Types	Impacts	Attack Methods	Features
	In-vehicle network attacks		
Remote sensor attacks	Regard injected signals as legal signals	Various attack vectors	Cannot distinguish the legal input and deceived input
GPS spoofing attacks	Manipulate the received GPS signals arbitrarily	Broadcast fake GPS signals with high strength	Ignore true GPS signals
Location trailing attacks	Discover behaviours and activities of the vehicle, the profile of the driver	Relate pseudonymous positions to specific vehicles	A single pseudonym is not sufficient to protect vehicles' privacy
Close proximity vulnerabilities	The system or communication mechanism may be attacked	Bluetooth or keyless entry and ignition systems	The attack may be primitive and happen occasionally
CAN and SAE J1939 buses vulnerabilities	Unauthorized nodes can listen to messages in the bus	Exploit arbitration mechanism	The adversary can insert unauthorized frames into the network
ECUs software flashing attacks	Disclose or corrupt information, degrade hardware performance, trick a remote device	Reverse engineering, code modification, fuzzing attacks, phishing attacks	Use unpatched vulnerabilities to gain access
Integrated business services attacks	Gain client-level or system level access to other vehicle systems	Phish attacks or malicious codes	The adversary uses client-level access to implement another attack
	Vehicle-to-everything network attacks		
DoS attacks	Delay and interfere with the receiver's response	Insert useless messages, create issues on network nodes	Dangerous and fatal
Impersonation attacks	Send fake messages, spoof modified received messages, drop critical messages	Use another identity or fake identity	Different identities are spoofed simultaneously
Replay attacks	Confuse authorities, mislead entire traffic, damage transportation safety	Record and retransmit valid packets later	Happen in authorization and key agreement protocols

(continued)

TABLE 14.1 (Continued)
Types, Impacts, Attack Methods, and Features of Cyber Attacks in Autonomous Vehicles

Types	Impacts	Attack Methods	Features
Routing attacks	Disturb normal routing processes, drop passing packets	A single compromised node or some cooperative nodes	Exploit the drawbacks and vulnerabilities of routing protocols
Data falsification attacks	Mislead receivers' reactions and result in fatalities	Send or broadcast false information and safety alerts	A false alert can clear the forward road; a fake accident message can cause traffic jams
Eavesdropping attacks	Leak actual identity and private key	Overhear the wireless medium	The victim is not conscious of the attack
Password and key attacks	Crack password or key	Mechanisms tested with different values until compromised	Attacks can be classified into three categories
		Other attacks	
Infrastructure attacks	Cut off communication in infrastructure	Exploit vulnerabilities in message transfer	Related to cloud services
Slight attacks	Cause hazardous conditions and serious safety problems	N/A	Communicated data deviates slightly but does not exceed thresholds
Attacks on machine learning systems	Modify training algorithms and data	Data poisoning attacks, escape attacks, model steal attacks, model inference attacks	Happen in the phases of data collection, training, and prediction
Cyber-risk	Complex phenomenon	N/A	Dynamic

Source: Sun et al. (2022).

14.3.3 Framework for Resilience: Resilience Analysis Grid (RAG)

Over the years, several frameworks have been developed to guide the design and assessment of resilience in CPS. These frameworks provide structured approaches to address the challenges that may arise in creating resilient CAV systems. We describe the resilience analysis grid (RAG) introduced by Hollnagel [2011].

14.3.3.1 Introduction

The RAG is a conceptual framework developed by Hollnagel to assess and enhance the resilience of complex socio-technical systems, including CPS in CAVs [Hollnagel 2015]. RAG provides a structured approach to evaluate how systems can anticipate, monitor, respond to, and learn from disruptions.

TABLE 14.2
Summary of Key Features Importance in CAVs

Feature	Definition	Importance in CAVs
Robustness	Ability to maintain performance under varying conditions	Determine reliable operation despite environmental or system variations
Adaptability	Capacity to adjust operations in response to changes	Allows handling of dynamic environments and user preferences
Fault tolerance	Ability to continue functioning despite component failures	Maintaining essential functions even when parts of the system fail
Recoverability	Efficiency in restoring normal operations after disruptions	Minimizes downtime and impact of failures on system performance
Anticipation	Proactive identification and mitigation of potential threats	Prevents incidents by forecasting and addressing issues before they occur

14.3.3.2 Fundamental Concepts of RAG

RAG is based on four essential system capabilities, namely anticipate, monitor, respond, and learn. These capabilities form a continuous loop, enabling systems to adapt and maintain functionality under varying conditions.

1. **Anticipate:** It is the ability to foresee potential future events, opportunities, threats, and conditions, and prepare accordingly [Hollnagel 2015] utilizing predictive analytics to forecast traffic conditions, weather changes, and potential hazards. For example, machine learning algorithms predict traffic congestion, allowing the CAV to reroute proactively.
2. **Monitor:** It is the capacity to observe system performance and external conditions in real time to detect deviations from normal operations [Hollnagel 2015] through continuous health monitoring of sensors and actuators, and environmental sensing. For example, continuously checking the status of sensors. If a sensor is found to be malfunctioning, the system can switch to redundant sensors to maintain accurate environmental perception. Similarly, keeping track of external conditions such as road surfaces, weather, and traffic signals to adjust vehicle operations accordingly can increase the resiliency of the CAV system.
3. **Respond:** It is the ability to react effectively to regular and irregular changes or disruptions in the environment or system [Hollnagel 2015]. and implementing autonomous decision-making algorithms to adjust driving behaviour in response to sudden obstacles or system errors. An example is executing an emergency stop when a pedestrian is detected unexpectedly [Chen et al. 2015a].
4. **Learn:** It is the capacity to learn from past experiences to improve future performance [Hollnagel 2015] by analysing data from incidents to refine algorithms and system responses. For example, improving object recognition algorithms after a near-miss incident [Bojarski et al. 2016].

Table 14.3 summarizes the RAG components and their applications in CAVs.

TABLE 14.3
RAG Components and Applications in CAVs

RAG Component	Definition	Application in CAVs
Anticipate	Foresee potential events and prepare accordingly	Predictive analytics for traffic and hazards
Monitor	Observe system performance and external conditions in real time	Health monitoring of components and environmental sensing Autonomous decision-making and failover mechanisms
Respond	React effectively to changes or disruptions	Data analysis and algorithm refinement
Learn	Learn from past experiences to improve future performance	Data analysis and algorithm refinement

14.3.3.3 Case Study: Enhancing Resilience Against Cyber Attacks

CAVs rely heavily on interconnected systems and networks, making them susceptible to cyber attacks that can compromise safety, privacy, and functionality. By applying the RAG framework, we can enhance the resilience of CAVs against such threats through the following stages:

- **Anticipate:** Anticipation involves proactively identifying potential cyber threats and vulnerabilities that could affect CAVs. Developing comprehensive threat models is essential for understanding how an attacker might exploit system weaknesses [Humayed et al. 2017]. These models consider various attack vectors, such as
  - **Vehicle-to-Vehicle (V2V) Communication Exploits:** Attackers may intercept or manipulate messages exchanged between vehicles to cause collisions or disrupt traffic flow.
  - **Sensor Spoofing:** False inputs to sensors like GPS, LiDAR, or cameras can mislead the vehicle's perception of its environment.
  - **Malware Infiltration:** Malicious software introduced through external media or over-the-air updates can compromise vehicle control systems.

  By anticipating these threats, designers can prioritize security measures and allocate resources effectively.
- **Monitor:** Monitoring entails continuously observing the system's performance and network activities to detect anomalies indicative of a cyber attack [Khan and Salah 2018]. Implementing intrusion detection systems (IDS) and intrusion prevention systems (IPS) enable real time analysis of data flows and system behaviours. Key monitoring strategies include
  - **Network Traffic Analysis:** Monitoring data packets for unusual patterns, such as unexpected communication with unknown entities or abnormal data volumes.

  - **System Integrity Checks:** Verifying the integrity of software and firmware through checksums and digital signatures to detect unauthorized modifications.
  - **Behavioural Anomaly Detection:** Utilizing machine learning algorithms to establish normal operation baselines and identify deviations.

  For example, an IDS may detect repeated failed access attempts to critical modules, triggering alerts for potential brute-force attacks.
- **Respond:** Effective response mechanisms are crucial for mitigating the impact of detected cyber threats. Upon identifying an attack, the system must isolate affected components and maintain essential functions to ensure safety [Petit and Shladover 2015]. Response strategies include:
  - **Module Isolation:** Segregating compromised subsystems to prevent the spread of malicious activity. For instance, disconnecting a hacked infotainment system from the vehicle's control network.
  - **Fail-Safe Operations:** Transitioning the vehicle to a safe state, such as reducing speed or pulling over, if critical systems are compromised.
  - **Alerting and Reporting:** Notifying the driver, nearby vehicles, and relevant authorities about the security breach.

  By responding promptly, the system minimizes potential damage and maintains a level of operational continuity.
- **Learn:** The learning phase focuses on analysing cyber incidents to enhance future resilience. Post-attack assessments provide valuable insights into vulnerabilities and the effectiveness of response actions. Learning activities involve
  - **Attack Pattern Analysis:** Studying the methods and tools used by attackers to refine threat models.
  - **Security Protocol Updates:** Implementing patches, updating encryption keys, and strengthening authentication mechanisms based on identified weaknesses.
  - **Knowledge Sharing:** Collaborating with industry partners and cybersecurity communities to share findings and develop best practices.

  Continuous learning ensures that the system evolves to counter emerging threats and adapts to the changing cyber landscape.

14.4 CONCLUSION

The advancement of CAVs promises to revolutionize smart transportation by enhancing safety, efficiency, and convenience. However, integrating CPSs into CAVs introduces new complexities and vulnerabilities that must be addressed to realize these benefits fully. This chapter has explored the critical importance of resilience in CPS design for CAVs, highlighting the need for systems that can anticipate, withstand, respond to, and recover from various disruptions, including cyber attacks and component failures. By prioritizing resilience, we can develop vehicles that not only operate efficiently under normal conditions but also maintain safety and functionality in the face of unexpected challenges. Future work should focus on interdisciplinary collaboration, integrating advanced technologies like artificial intelligence and machine learning to enhance predictive capabilities, and developing comprehensive regulatory frameworks to ensure the resilience and security of CAVs in increasingly complex environments. Embracing these strategies will pave the way for a safer, more reliable, and efficient transportation future.

BIBLIOGRAPHY

Karam M. Abughalieh and Shadi G. Alawneh. Predicting Pedestrian Intention to Cross the Road, IEEE Access, doi: 10.1109/ACCESS.2020.2987777

Theyazn H. H. Aldhyani and Hasan Alkahtani. Attacks to automatous vehicles: A deep learning algorithm for cybersecurity. *Sensors*, 22(1), 2022. ISSN 1424-8220. doi: 10.3390/s22010360. URL www.mdpi.com/1424-8220/22/1/360.

Mariusz Bojarski, Davide Del Testa, Daniel Dworakowski, Bernhard Firner, Beat Flepp, Prasoon Goyal, Lawrence D. Jackel, Mathew Monfort, Urs Muller, Jiakai Zhang, Xin Zhang, Jake Zhao, and Karol Zieba. End to end learning for self-driving cars. *arXiv preprint arXiv:1604.07316*, 2016. URL https://arxiv.org/abs/1604.07316.

Ming-Fang Chang, John Lambert, Patsorn Sangkloy, Jagjeet Singh, Slawomir Bak, Andrew Hartnett, De Wang, Peter Carr, Simon Lucey, Deva Ramanan, and James Hays. Argoverse: 3d tracking and forecasting with rich maps. In *Proceedings of the IEEE/CVF Conference on Computer Vision and Pattern Recognition (CVPR)*, June 2019.

Chenyi Chen, Ari Seff, Alain Kornhauser, and Jianxiong Xiao. Deep-driving: Learning affordance for direct perception in autonomous driving. In *Proceedings of the IEEE International Conference on Computer Vision*, pages 2722–2730, 2015a. doi: 10.1109/ICCV.2015.312. URL https://doi.org/10.1109/ICCV.2015.312.

Chenyi Chen, Ari Seff, Alain Kornhauser, and Jianxiong Xiao. Deepdriving: Learning affordance for direct perception in autonomous driving. In *2015 IEEE International Conference on Computer Vision (ICCV)*, pages 2722– 2730, 2015b. doi: 10.1109/ICCV.2015.312. URL https://doi.org/10.1109/ICCV.2015.312.

Alexey Dosovitskiy, German Ros, Felipe Codevilla, Antonio Lopez, and Vladlen Koltun. CARLA: An open urban driving simulator. In *Proceedings of the 1st Annual Conference on Robot Learning*, pages 1–16, 2017.

Paolo Falcone, Francesco Borrelli, Jorge Asgari, Huei Peng, and David Hrovat. Predictive active steering control for autonomous vehicle systems. *IEEE Transactions on Control Systems Technology*, 15(3):566–580, 2007. doi: 10.1109/TCST.2007.894653. URL https://doi.org/10.1109/TCST.2007.894653.

Andreas Geiger, Philip Lenz, and Raquel Urtasun. *Are we ready for autonomous driving? The kitti vision benchmark suite*. In *Conference on Computer Vision and Pattern Recognition (CVPR)*, 2012.

Antoine Grall, Christophe Dieulle, Christophe Bérenguer, and Michel Rousignol. Continuous-time predictive-maintenance scheduling for a deteriorating system. *IEEE Transactions on Reliability*, 51(2):141–150, 2002. doi: 10.1109/TR.2002.1011518. URL https://doi.org/10.1109/TR.2002.1011518.

Erik Hollnagel. *Prologue: The Scope of Resilience Engineering*. Ashgate Publishing, UK, 2011.

Erik Hollnagel. Rag—The resilience analysis grid. In Erik Hollnagel, editor, *Safety-II in Practice: Developing the Resilience Potentials*, pages 29–47. Routledge, Abingdon, UK, 2015. ISBN 9781317059790.

Abdulmalik Humayed, Jingqiang Lin, Fengjun Li, and Bo Luo. Cyber-physical systems security—A survey. *IEEE Internet of Things Journal*, 4(6):1802–1831, 2017. doi: 10.1109/JIOT.2017.2703172. URL https://doi.org/10.1109/JIOT.2017.2703172.

Christos Katrakazas, Mohammed Quddus, Wen-Hong Chen, and Lenia Deka. Real-time motion planning methods for autonomous on-road driving: State-of-the-art and future research directions. *Transportation Research Part C: Emerging Technologies*, 60:416–442, 2015. doi: 10.1016/j.trc.2015.09.011. URL https://doi.org/10.1016/j.trc.2015.09.011.

Mohammad A. Khan and K. Salah. Iot security: Review, blockchain solutions, and open challenges. *Future Generation Computer Systems,* 82:395–411, 2018. doi: 10.1016/j.future.2017.11.022. URL https://doi.org/10.1016/j.future.2017.11.022.

Will Maddern, Geoff Pascoe, Chris Linegar, and Paul Newman. 1 Year, 1000 km: The Oxford RobotCar Dataset. *The International Journal of Robotics Research (IJRR)*, 36(1):3–15, 2017. doi:10.1177/0278364916679498. URL http://dx.doi.org/10.1177/0278364916679498.

Charlie Miller. Lessons learned from hacking a car. *IEEE Design & Test*, 36(6):7–9, 2019. doi: 10.1109/MDAT.2018.2863106.

Jonathan Petit and Steven E. Shladover. Potential cyberattacks on automated vehicles. *IEEE Transactions on Intelligent Transportation Systems*, 16(2):546–556, 2015. doi: 10.1109/TITS.2014.2342271. URL https://doi.org/10.1109/TITS.2014.2342271

S. M. Popa and G. M. Manea. Using traffic self-similarity for network anomalies detection, *2015 20th International Conference on Control Systems and Computer Science*, Bucharest, Romania (2015) pp. 639–644, doi: 10.1109/CSCS.2015.89.

Dhiraj K. Pradhan. *Fault-Tolerant Computer System Design*. Prentice Hall, Upper Saddle River, NJ, 1996.

Joseph H. Saleh, Karen B. Marais, Evangelos Bakolas, and Ravindra V. Cowlagi. Highlights from the literature on accident causation and system safety: Review of major ideas, recent contributions, and challenges, Reliability Engineering and System Safety., *Elsevier*, vol. 95(11), pp. 1105–1116. DOI: 10.1016/j.ress.2010.07.004

Xiaoqiang Sun, F. Richard Yu, and Peng Zhang. A survey on cyber-security of connected and autonomous vehicles (CAVS). *IEEE Transactions on Intelligent Transportation Systems*, 23(7):6240–6259, 2022. doi: 10.1109/TITS.2021.3085297.

Anand Sundaram, Swathi Pavithran, and S. Selvakumar. An efficient lightweight mutual authentication and key exchange protocol for healthcare RFID applications. In *2013 International Conference on Advanced Computing and Communication Systems*, pages 1–6, 2013. doi: 10.1109/ICACCS.2013.6938702. URL https://doi.org/10.1109/ICACCS.2013.6938702.

Peng Wang, Chenglu Zhang, Minghui Guo, and Jianwei Li. A robust multi-sensor fusion approach for simultaneous localization and mapping in autonomous vehicles. *2015 IEEE Intelligent Vehicles Symposium (IV),* pages 1300–1305, 2015. doi: 10.1109/IVS.2015.7225864.

Jae Hyung Yoon, Md Tawhid Hossain, and Jong Min Kim. Reliable automotive system design using multi-core processors. *IEEE Transactions on Industrial Informatics*, 12(5):1865–1875, 2016. doi: 10.1109/TII.2016. 2583442.

15 Application of AI in Secured Supply Chain Management

15.1 INTRODUCTION

In addition to embedded systems like machinery, structures, and modes of transportation, cyber-physical systems (CPSs) also include internet services, logistical, coordination, and management procedures. While actuators are used to respond to organizational or production changes and connect with other components, CPSs use sensors to gather, handle, and analyse data.

By strengthening the ability to recognize and evaluate the underlying causes and consequences of supply chain problems and to come up with innovative and workable solutions that satisfy stakeholders, CPSs may assist you in enhancing your supply chain management (SCM) performance. Additionally, CPSs may be used to quickly and economically deploy and evaluate solutions, as well as to learn from results and feedback and make necessary improvements.

15.1.1 Application of Smart CPS in Effective Supply Chain Management

- **Increased Warehouse Efficiency:** Route optimization and effective manpower allocation are achieved via the application of machine learning (ML) techniques. The manufacturing line, point-of-sale, and marketing departments' demand signals may be analysed by artificial intelligence (AI)-enabled CPSs, allowing for cost-effective inventory management.
- **Enhanced Safety and Security:** AI algorithms may be used to monitor dangerous workplaces for safety regulations. By continuously tracking workers' movements on the assembly line and sounding a warning anytime it anticipates potentially dangerous situations, the wearable CPS with AI technology protects worker safety.
- **Operations Optimization:** To facilitate operations across the logistics network, CPSs are installed in logistics trucks. These algorithms are able to identify any real-world obstructions in the path and to bring the operations under control, it is possible to optimize the routes.

DOI: 10.1201/9781003489986-20

- **Decreased Mistakes and Effective Waste Management:** AI-enabled CPS may assist in lowering human error and, through the use of clever algorithms, identify faulty items in the product process. In addition to lowering waste, this can aid in effective waste management.
- **Logistics Automation:** AI-enabled CPS in autonomous cars contributes to the automation of logistics as a whole. It can streamline the supply chain's transportation procedures and delivery routes.
- **Warehouse Automation:** The COBOTS, or collaborative robots, assist with automated warehouse operations. It facilitates effective product sorting, packing, and transportation.
- **Automatic Quality Verification:** AI-enabled computer vision systems have the ability to automate quality inspections on the delivery line. Without the need for human involvement, the sophisticated algorithms in these CPSs will keep an eye on and examine the items from several angles to determine their quality.
- **Automated Inventory Management:** AI-powered automated bots assist in automating inventory tracking and real-time scanning. These automated systems will be in at order to accurately and promptly maintain inventory. These bots will be stationed at warehouses, production facilities, and delivery units.
- **Demand Forecasting Using Predictive Analytics:** AI-enabled CPSs can aid in demand forecasting. The managers may use this to efficiently prepare for purchase tactics, inventory levels, and suitable scheduling.

Since various issues may call for various strategies and resources, using CPS in SCM is a customized procedure. It has the following step–by-step procedure.

- Describe and comprehend the problem and its context using SWOT analysis, stakeholder maps, fishbone diagrams, and problem statements.
- Use TRIZ, SCAMPER, mind mapping, and brainstorming to produce ideas.
- Assess and improve your concepts using a feasibility study, a Pugh matrix, a criterion matrix, and prototyping.
- Lastly, draft an action plan that includes duties, responsibilities, resources, due dates, and performance metrics.

Additionally, before implementing the solution across the supply chain, it is essential to evaluate it on a limited scale via pilot testing. Finally, the plan–do–check–act (PDCA) cycle may be used to plan, carry out, evaluate, and act upon your findings and feedback, as well as to make necessary modifications.

There are several benefits of using CPSs in SCM, including higher productivity and efficiency, better quality and customer satisfaction, increased profitability and competitiveness, innovation and learning, and improved teamwork and communication. These advantages are attained via cutting down on waste, mistakes, and delays; satisfying or surpassing client demands and expectations; adding value and cutting expenses; and incorporating and energizing team members and stakeholders.

15.1.2 Cyber-Physical Supply Chain Management Systems (CPS^2)

Flavio et al. [1] have presented a framework for cyber-physical supply chain management systems (CPS^2) for forward-thinking product supply planning. Three tiers make up their suggested CPS^2: database and integration (local server), App and algorithms (cloud server), and front-end (local client).

A dynamic data representation engine, shared by all apps, is incorporated into the front-end development with the sole purpose of analysing and driving the supply chain planning simulation in real time.

To gather, process, and store this real-world data into the CPS^2 database management system, various data sources, including ERP, PLM, and CRM, can be connected starting from the bottom. Additionally, these data can be supplied via internet of things (IoT) connections, which serve as the actual basis for the development of the CPS and allow direct communication with the edge computing layer (machine data collection). All of the gathered data is archived in a Big Data Service (BDS) and made accessible to the entire organization.

As shown in Figure 15.1, the SAAS distribution model at the top layer normally faces security threats as the finance transactions made at this layer are vulnerable to attacks by hackers. This leads to the need for secured framework in addition to the stated CPS^2.

In recent years, the integration of blockchain technology into the finance domain has shown immense potential for revolutionizing the way sensitive information is handled, stored, and secured. This research will explore how to combine blockchain technology and ML to create a stronger security system for financial systems.

This chapter helps to look at existing security frameworks to find weaknesses and develop a new solution that is more secure and meets the specific needs of the finance industries. The goal is to create a strong security system that can protect sensitive information in these industries as they continue to change in the digital age. Fraudulent financial transactions cause a loss of large sums of money and therefore has a negative impact on the economy as a whole. To prevent this, we must employ the latest technologies that we have at our disposal.

15.2 ML-INTEGRATED BLOCKCHAIN FRAMEWORK FOR CPS SECURE SUPPLY CHAIN FINANCE

In blockchain technology, ensuring security and reliability is a constant challenge, especially with the ongoing threat of fraudulent activities. This study introduces a novel framework specifically crafted to detect fraudulent transactions within a private blockchain.

The framework's effectiveness is validated by rigorously testing it with real-time transaction data from the private blockchain platform Ganache. Using this live transaction data, we evaluate various gradient boosting models such as XGBoost, random forest, LightGBM etc. for their ability to accurately classify transactions as fraudulent or legitimate. Ganache provides valuable real-world data, offering insights into transaction behaviour within a controlled blockchain environment.

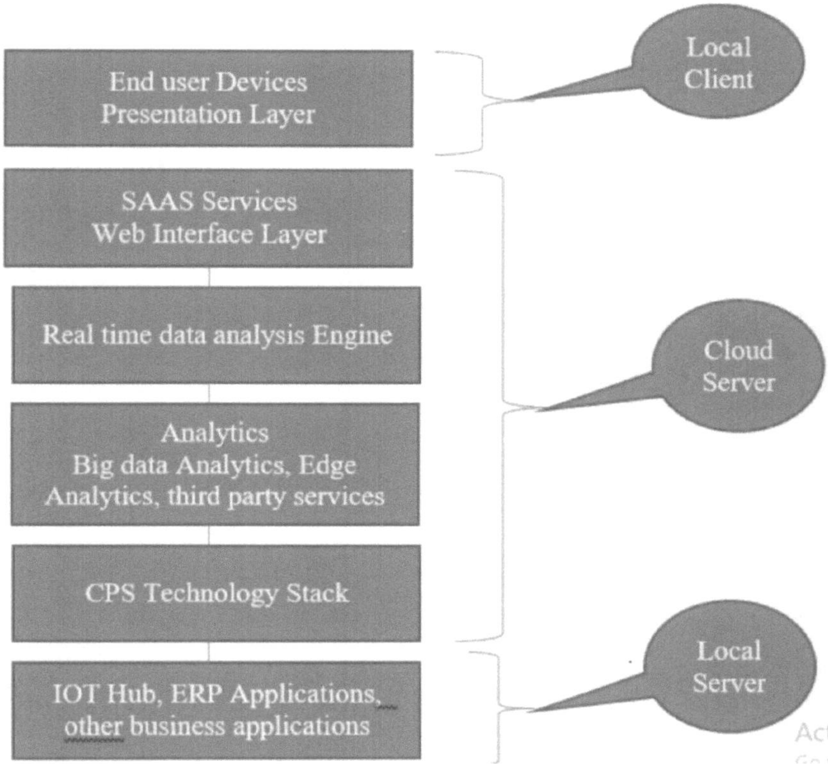

FIGURE 15.1 CPS^2 system architecture. (From Ref. [1].)

Note: On the left side the stacked collection of components from top to bottom as end-user devices, SAAS services, real-time data analysis engine, analytics components, CPS technology stack, and finally at the bottom the IoT hub, ERP architecture, and other business applications. On the right side, they are grouped and indicated with common names such as local client, cloud server, and local server.

Through careful experimentation and analysis, our goal is to determine the most efficient gradient boosting model for fraud classification, thereby enhancing security and trust in blockchain ecosystems.

The architecture of this proposed framework is presented in Figure 15.2.

The components of the framework, as shown in Figure 15.2, are explained further.

15.2.1 UI Module

Initially in the user interface module the transaction is registered with the help of specific account address and their private key to process the transaction.

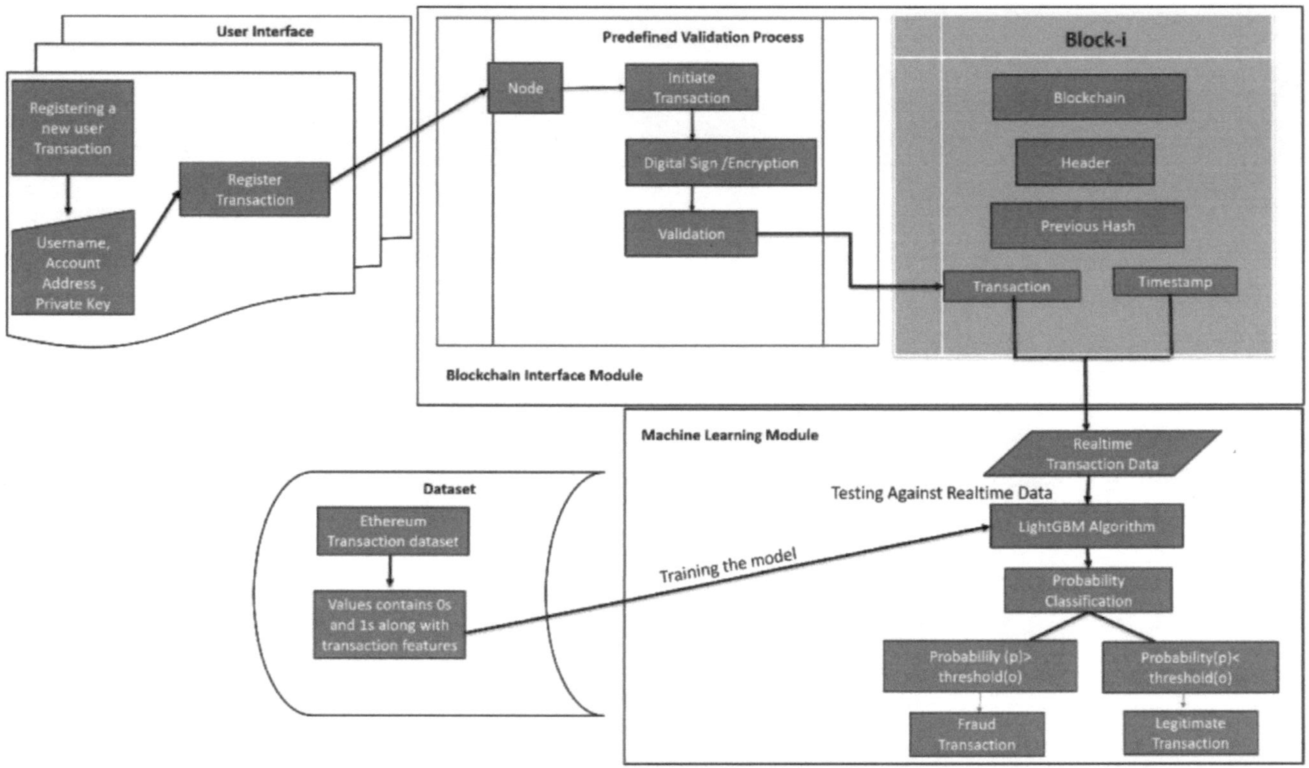

FIGURE 15.2 Framework for secured blockchain-based cyber-physical supply chain management system that has at the top-left User Interface component which sends the user-entered data to the Blockchain Interface module that is at the top right side and then at the below Machine Learning module which processes data from Dataset component at the left-bottom component.

15.2.2 Blockchain Interface Module

The blockchain interface module consists of a predefined validation process where the transaction is validated and verified by the miners, and the transaction is added to the blockchain. Once the block is created the block contains a transaction hash and timestamp. The transaction hash contains the details of the transaction that has been made. The transaction data is stored in a database for real-time use.

To interact with the blockchain module, a private blockchain server is used called Ganache server. The Ganache server is a private Ethereum blockchain used to perform transaction between accounts that are existing. This server is used to perform real-time transactions and get the transaction details for further fraud detection. The verification is done by a group rather than a single individual, and so the point of authority is decentralized.

The user's financial activity is stored in a blockchain using the hash method. A hash is a string that is produced involving the information as info. With a similar info, the result hash will constantly be something very similar. The main thing put away on the blockchain is the hash of the information. In contrast with the information, the hash is tiny, so the expense of an exchange is generally low.

15.2.3 Machine Learning Module

In the Machine Learning module the training and testing of the real-time data is performed for fraud classification. From the blockchain the transaction details are used to check for fraud. The ML model is trained with fraudulent and legitimate transactions and tested against the input transaction to detect fraud. Initially the training data is tested with various gradient boosting algorithms such as XGBoost, random forest classifier, decision tree classifier, and LightGBM classifier.

The aforementioned algorithms work based on probability classification where the probability is calculated and compared against the threshold probability. If Probability (p) > threshold (o), then it is detected as fraud; else if Probability (p) < threshold (o), then it is detected as legitimate.

A k-means algorithm is an unsupervised AI algorithm that separates a dataset into bunches called groups. The algorithm works iteratively and allots information focuses to one of the predefined number of classes (k), in view of the elements that are in the dataset. Information focuses are bunched in light of element likeness. Light gradient boosting, likewise called LightGBM, is a library of gradient boosting algorithms streamlined for information science issues. It leverages the boosting techniques. The k-means algorithm has the dual benefit of data clustering and outlier detection simultaneously. This means the correlation which would not be immediately obvious will be detected by the algorithm. Outliers will be easily detected which helps with quickly spotting a fraudulent transaction. When LightGBM encounters a missing value at a node, it tries both the left and right hand split and learns the way leading to higher loss for each node.

Overall, this combination proves to be efficient and quick, which is important for real-time execution. While endeavouring to carry out nonstop learning ML use cases, there might be an outrageous handling bottleneck on the grounds that the information is regularly put away in a standard disk- and value-based dataset (OLTP).

Performing examination on this information utilizing customary ML techniques regularly requires a concentrate, change, and burden (ETL) interaction to move the information into a logical dataset (OLAP), where the information can be dissected. Nonetheless, ETL is a clump cycle that might call for extensive investment to finish in light of the size of the information, and that implies the information in the OLAP framework is lifeless before the examination even happens.

15.3 EXPERIMENTATION AND RESULT ANALYSIS

15.3.1 Dataset Description

This is experimented on the existing dataset available on Kaggle (www.kaggle.com/datasets/gescobero/ethereum-fraud-dataset/data) that has benchmarks by several research works. Further this work has generated blockchain transactions using Ethereum framework and taken the real-time dataset for further validation. It is identified that the proposed algorithm (LightGBM) works well for both benchmark and real-time dataset (Table 15.1).

The average time between transactions appears infrequent, with an average of about 14 hours between sending transactions and 18 hours between receiving them. This could indicate low activity or a system designed for infrequent communication.

The data spans a long period of almost 49.6 days, suggesting a long-term observation. However, without additional context about the system, units used for time intervals, and the number of transactions, it's difficult to draw more specific conclusions. Table 15.2 consists of 10 records from Ethereum transaction with flag values 0 and 1 depicting the fraud and non-fraudulent transactions.

TABLE 15.1
Dataset Description

Attributes	Description
Address	The address of the Ethereum account
FLAG	Whether the transaction is fraud or not
Avg min between sent tnx	Average time between sent transactions for account in minutes
Avg min between received tnx	Average time between received transactions for account in minutes
Time Diff between first and last (mins)	Time difference between the first and last transaction
UniqueReceivedFrom_Addresses	Total unique addresses from which account received transactions
MinValueReceived	Minimum value in Ether ever received
MaxValueReceived	Maximum value in Ether ever received
MinValueSentToContract	Minimum value of Ether sent to a contract
MaxValueSentToContract	Maximum value of Ether sent to a contract
AvgValueSentToContract	Average value of Ether sent to contracts

TABLE 15.2
Benchmark Dataset Values

Index	Address	FLAG	Avg min between sent tnx	Avg min between received tnx	Time Diff between first and last (mins)
1	0x00009277775ac7d0d59e aad8fee3d10ac6c805e8	0	844.26	1093.71	704785.6
2	0x0002b44ddb1476db43c8 68bd494422ee4c136fed	0	12709.07	2958.44	1218217
3	0x0002bda54cb772d040f7 79e88eb453cac0daa244	0	246194.5	2434.02	516729.3
4	0x00038e6ba2fd5c09aedb9 6697c8d7b8fa6632e5e	0	10219.6	15785.09	397555.9
5	0x00062d1dd1afb6fb0254 0ddad9cdebfe568e0d89	0	36.61	10707.77	382472.4
6	0x000895ad78f4403ecd94 68900e68d6ee506136fd	0	9900.12	375.48	20926.68
7	0x000d63fc5df52b020437 4c2f5a3249779805d5d1	0	69.46	629.44	8660.35
8	0x000e001ab444fa8d6dc4a 402f8d7cfc88fe8c64d	0	1497.39	176.84	319828.1
9	0x0012cb699c836049a4bb eaac2d8c4d47c688e0e4	0	0	0	496.62
10	0x0012f247c9f980eea0a9a d06893bfd95c3145794	0	2570.59	3336.01	30572.7
……					
……					

TABLE 15.3
Experimentation Setup for Benchmark Data Analysis

S. no.	Hardware	Software	Test Environment
01	Intel(R) Core(TM) i5-8250U CPU @ 1.60GHz 1.80 GHz	Python, VSCode, Ganache(Private Blockchain)	VSCode, React, Google Collab

15.3.2 Experimentation Setup

The experimentation setup for benchmark data analysis and real-time data analysis are given Tables 15.3 and 15.4, respectively.

The heat map visualization of the features present in the training dataset is given in Figure 15.3, where each feature is listed against the other existing attributes. This heat map appears to visualize the interdependencies between features (variables) from the dataset using normalized mutual information (NMI).

TABLE 15.4
Experimentation Setup for Real-Time Data Analysis

Serial number	Hardware	Software	Test Environment
01	Intel(R) Core(TM) i5-8250U CPU @ 1.60GHz 1.80 GHz	Python, VSCode, Ganache(Private Blockchain)	VSCode, React, Google Collab

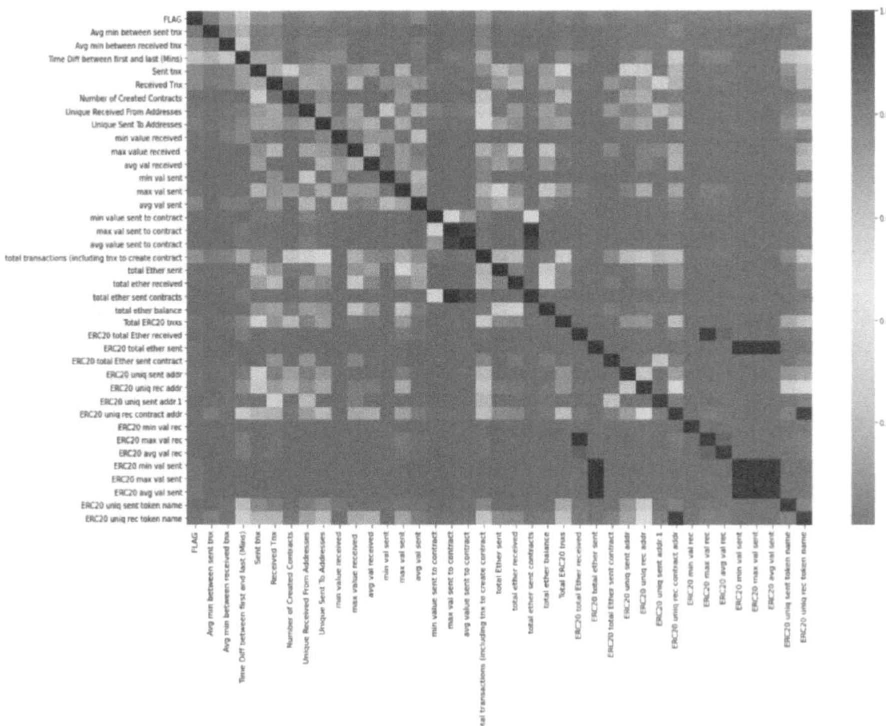

FIGURE 15.3 Heat map for the given dataset which has the features importance and the correlation between each feature with other feature and it shows colour to indicate highly important correlated features and the others are indicated in blue colour.

Looking at the heat map, we see several distinct squares along the diagonal where each feature has a high NMI with itself, as expected. Interestingly, there are also coloured blocks outside the diagonal, suggesting strong relationships between certain pairs of features. These potentially redundant features might be good candidates for removal during dimensionality reduction, a technique for simplifying data while preserving key information. It's worth noting that the heat map isn't symmetrical, meaning the NMI between feature A and B might not be the same as the NMI between B and A.

The training dataset is tested and evaluated with different ML models such as random forest, logistic regression, decision tree classifier, XGBoost classifier, and LightGBM classifier.

15.4 REAL-TIME IMPLEMENTATION OF PRIVATE BLOCKCHAIN

15.4.1 Real-Time Dataset Collection

A private blockchain server is established in Ganche server as shown in Figure 15.4. The ganache server consists of accounts that are predefined with several accounts with a balance of 100ETH in each account. To collect real-time Ethereum transaction data, the transactions are made within the existing accounts to test it against different models. The transaction details of the sample transactions that are made over the ganache server are collected and stored for evaluation.

Table 15.5 consists of 10 records from real-time Ethereum transaction that are made over ganache server consisting of account address, timestamp, and other attributes.

15.4.2 ML Model Implementation

The current research presents an advancement over existing studies by focusing on real-time data sourced from a private blockchain. The evaluation results of the best

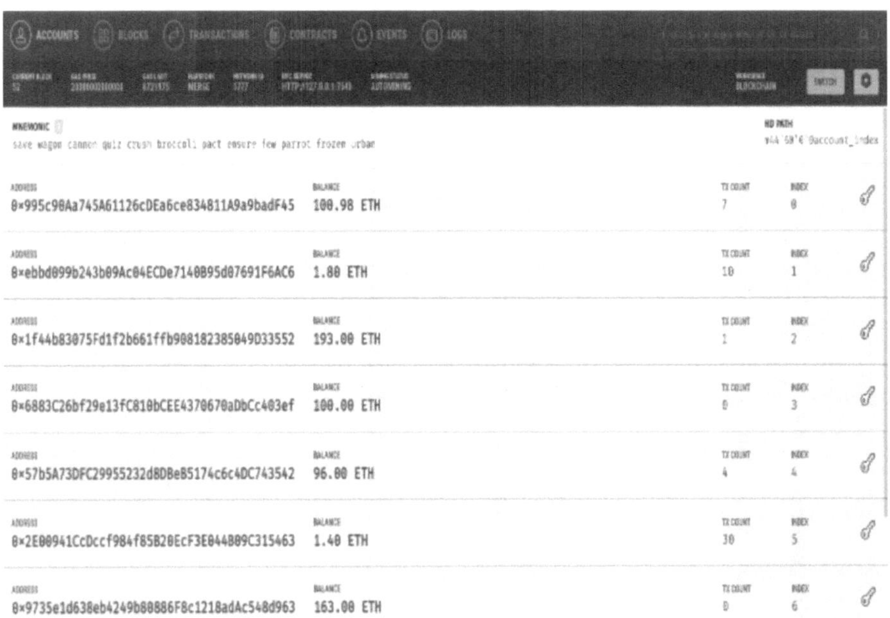

FIGURE 15.4 Private blockchain server (Ganache) based transactions generation which has the smart contract key information at the first column and then the Ethereum value etc.

TABLE 15.5
Real-Time Transaction Data

Ut id	Sender id	Send Address	Received Address	Transaction	Transaction Date
1	6	0xb2DFb8fE09d8fF761679e4A256B2ba83722B8b95	0x89b3E5C84cB94a481353E4B11648c0413f185695	154364B671AEA98FB51322A6DF8C4152DB0A3D1CF1D0C27A8D2F75F19F2DBA2A	13-02-2024 04:13
2	6	0xb2DFb8fE09d8fF761679e4A256B2ba83722B8b95	0x89b3E5C84cB94a481353E4B11648c0413f185695	E3B3C45B8F6D1E9CEF1DC1934B970AF1C2462FDE01B66BF7F8B0A42A0B9FD843	13-02-2024 04:13
3	7	0x35240934c130d40076E649F0fE3a28259BF29bff	0x89b3E5C84cB94a481353E4B11648c0413f185695	BB1FF0B51E447FC69EC7B37A66E95D180F75F8D12A1A24D5F22A34E90D628E04	13-02-2024 06:58
4	8	0x35240934c130d40076E649F0fE3a28259BF29bff	0x89b3E5C84cB94a481353E4B11648c0413f185695	2521AF1584768689931FF6E33399EF8CBCF4D6393C152A06358AE806448D47D0	13-02-2024 07:05
5	9	0x35240934c130d40076E649F0fE3a28259BF29bff	0x89b3E5C84cB94a481353E4B11648c0413f185695	4C5E53CA4562B0AFF93A679F5F1C7D66F8552DFB87B0107326748CA6F4EC81BB	13-02-2024 07:07
6	10	0x35240934c130d40076E649F0fE3a28259BF29bff	0x89b3E5C84cB94a481353E4B11648c0413f185695	76EEB378D631FCFB4399C3FBF2D864763ABCF4D37EC03D47822F1BC202E49CF9	13-02-2024 07:31
7	11	0x35240934c130d40076E649F0fE3a28259BF29bff	0x89b3E5C84cB94a481353E4B11648c0413f185695	F4A00AE8C9ED9A5B43E6B8F7EA636775DDCC4DD1B5D6C12BDE9F31D5B97C9416	13-02-2024 07:32
8	12	0x35240934c130d40076E649F0fE3a28259BF29bff	0x89b3E5C84cB94a481353E4B11648c0413f185695	471097D9A054B3F2B6B2BFC421577820736D6E53AA05661822CA398F042CB217	13-02-2024 07:40
9	13	0x35240934c130d40076E649F0fE3a28259BF29bff	0x89b3E5C84cB94a481353E4B11648c0413f185695	C313638D1744209C432ED043B418FC88C277230567645F9CF163EE7DF3F6874B	13-02-2024 07:48
10	14	0x35240934c130d40076E649F0fE3a28259BF29bff	0x89b3E5C84cB94a481353E4B11648c0413f185695	027E0AF2335BEB0FCDD733ED79B59CAB25DA59A978DE0D33FFD9728E4103D5FB	13-02-2024 07:52
		………			
		………			
		………			

TABLE 15.6
Evaluation Metrics of ML Models

Model	Accuracy	Precision	Recall	F1-Score
Random forest	98%	98%	92%	95%
Logistic regression	83%	87%	31%	46%
Decision tree	96%	92%	92%	92%
XGBoost	96%	97%	95%	96%
LightGBM (Best model)	98%	97%	95%	96%

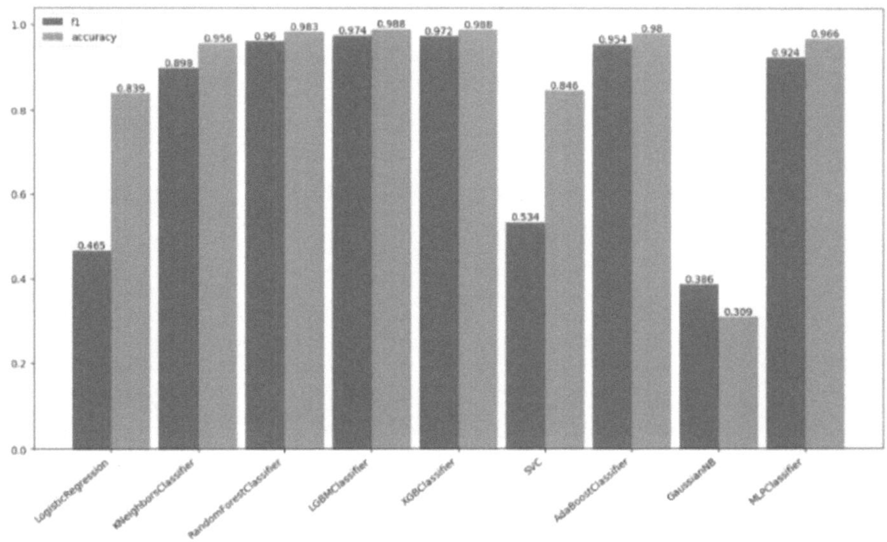

FIGURE 15.5 Comparison analysis of ML models in secured cyber-physical supply chain finance transactions.

model LightGBM for benchmark dataset with an accuracy of 98% is displayed in Table 15.6.

The bar graphs in Figure 15.5 shows the comparison of the different ML models. The graph suggests that LightGBM has the highest accuracy and F1 score of 0.98 and 0.97, respectively.

The evaluation results of the best model, LightGBM, for real-time data with an accuracy of 98% is displayed in Table 15.7.

The pair plot of the attributes present in the dataset is plotted in Figure 15.6. The variables appear to be continuous; they can take on any value within a certain range. There appears to be a positive correlation between the FLAG and Time Diff, as the values of the FLAG increases, the values of the Time Diff also tend to increase. There appears to be a negative correlation between the FLAG and Avg min between

TABLE 15.7
LightGBM Model Evaluation Metrics

	Precision	Recall	F1-Score	Support
0	0.98	1.00	0.99	170821
1	0.88	0.96	0.56	6341
Accuracy	–	–	0.98	177162
Macro average	0.93	0.71	0.78	177162
Weighted average	0.98	0.98	0.97	177162

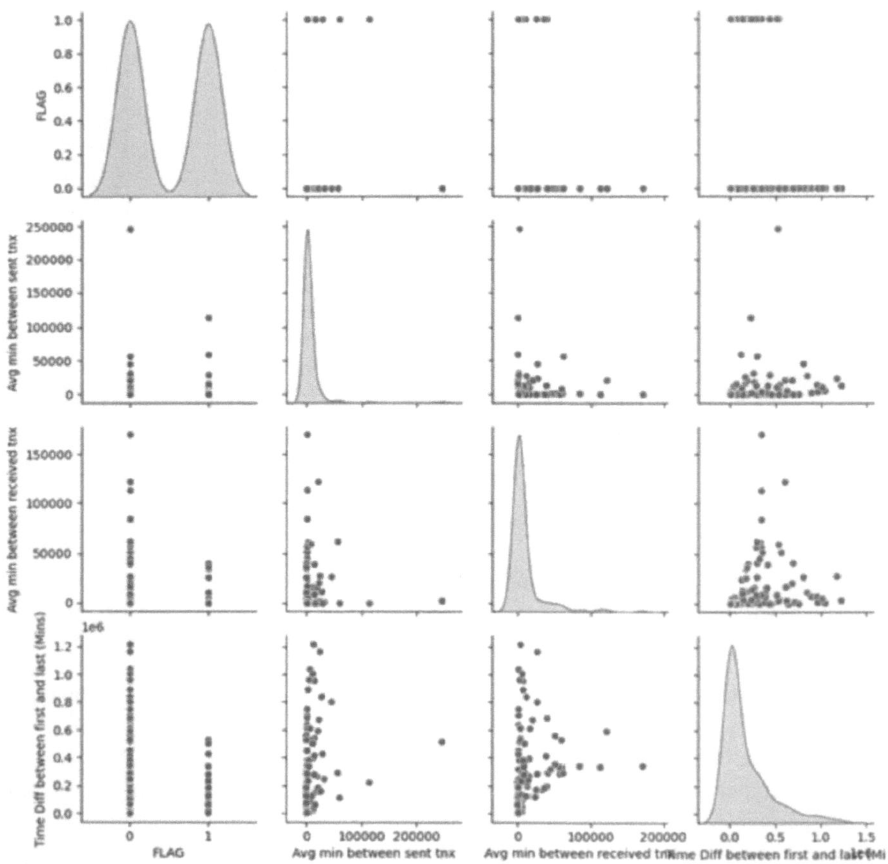

FIGURE 15.6 Pair plot of the correlation analysis as a 4×4 tabular format where each plot represents the correlation level of each feature with other feature in the dataset.

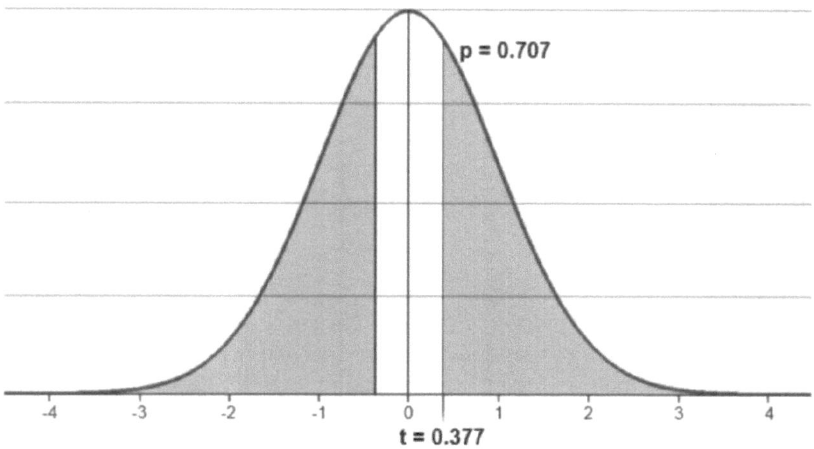

FIGURE 15.7 A bell curve that shows the *t*-test results of real-time data versus benchmark data is given to analyse the trustworthiness of the results.

received tnx, as the values of the FLAG increases, the values of the Avg min between received tnx tend to decrease.

15.5 STATISTICAL ANALYSIS

Statistical measures such as *t*-tests were performed which compared the prediction result of real-time data in each model against the proposed model and benchmark data. The results of *t*-test is displayed in Figure 15.7 and can help in interpreting how much each of the prediction result from standard ML classifiers vary when compared with the proposed model.

The line represents a normal distribution, which is commonly used in statistical tests to represent the sampling distribution of a statistic under the null hypothesis. In this case, the null hypothesis is likely that there is no difference between the predicted values and the test values.

The *p*-Value: The *p*-value is 0.707, which is a relatively high value. In most statistical tests, a *p*-value greater than 0.05 is considered to be evidence in favour of the null hypothesis. So, in this case, the *p*-value suggests that there is no statistically significant difference between the predicted values and the test values.

The *t*-Statistic: The *t*-statistic is 0.377. While the exact interpretation of the *t*-statistic depends on the specific test being used, it generally represents the standardized difference between the two groups being compared. In this case, the *t*-statistic is close to zero, which again suggests that there is no substantial difference between the predicted values and the test values. Along with it, ANOVA test is performed and the results of the test are displayed in Table 15.8.

TABLE 15.8
ANOVA Test Results

Source	DF	SS	MS	*F*-stat	*p*-Value
Between groups	1	0	0	0.02	0
Within groups	398	71.38	0.18		
Total	399	71.39			

The ANOVA test results presented in Table 15.8 suggest that there is no statistically significant difference between the real-time data and benchmark data.

The row indicates the source of the variation in the data. "Between groups" refers to the variability between the groups we are comparing, while "Within groups" refers to the variability within each group.

This represents the number of independent pieces of information used to estimate a parameter. In this case, there is 1 degree of freedom for "Between groups" (since we're likely comparing two groups) and 398 degrees of freedom for "Within groups" (likely the number of data points minus the number of groups).

SS (Sum of Squares): This measures the total variance in the data. "Between groups" SS reflects the variation due to the differences between the groups, while "Within groups" SS reflects the variation within each group due to random error or individual differences.

MS (Mean Squares): This is obtained by dividing the corresponding SS by its degrees of freedom (DF). It represents the average variance per source.

***F*-stat (*F*-statistic):** This is a test statistic used to compare the variances between groups and within groups. It is calculated by dividing the "mean squares between groups" by the "mean squares within groups". In this case, the F-statistic is 0.02.

***p*-Value:** This represents the probability of observing an F-statistic as extreme as the one calculated, assuming that there is no real difference between the groups (null hypothesis). A low p-value suggests that the observed difference is unlikely to be due to chance. In this case, the p-value is 0.

Since the F-statistic (0.02) is very low and the p-value is 0, we can reject the null hypothesis. However, this is because the F-statistic is so close to 0, indicating practically no difference between the variances of the groups.

15.6 OTHER RELATED RESEARCH WORKS

Ashfaq et al. [2] proposed a ML and blockchain-based extortion recognition component, exhibiting the coordination of cutting-edge innovations for productive misrepresentation location in monetary frameworks. Khan et al. [3] presented a superior financial application model using blockchain innovation, demonstrating the developing revenue in utilizing blockchain for upgrading financial tasks and security. Dashottar and Srivastava [4] examined a blockchain application-based way to deal with corporate financial gamble, displaying the capability of blockchain in tending to administrative and detailing necessities in the Indian financial area. Chakraborty et al. [5] introduced a blockchain-based credit examination structure pointed towards upgrading the productivity of monetary frameworks, featuring the job of blockchain in further developing credit evaluation processes. Garg et al. [6] directed a review estimating the apparent advantages of executing blockchain innovation in the financial area, giving experiences into the expected benefits of taking on blockchain arrangements in financial tasks. Dzhaparov [7] investigated the use of blockchain and computerized reasoning in bank risk, underscoring the cooperative energies between these advances for moderating dangers in financial conditions. Arjun and Suprabha [8] gave a scientometric view on the development and difficulties of blockchain in financial systems, offering an extensive examination of the exploration scene in this space. Sant and Tripathi [9] proposed profound blockchain-empowered security upgrades in exchange finance, featuring the meaning of blockchain in fortifying safety efforts in monetary exchanges. Islam et al. [10] tended to the identification of disseminated refusal of administration (DDoS) assaults in IoT-based financial checking frameworks utilizing ML models, tending to security worries in financial IoT applications. Chowdhury et al. [11] researched the use of blockchain in financial frameworks, adding to the comprehension of how blockchain innovation can be utilized to work on different parts of financial activities. Osterrieder et al. [12] examined progressed procedures for upgrading security in blockchain networks, zeroing in on identifying peculiarities and fakes to work on the trustworthiness of blockchain frameworks. Trivedi et al. [13] led a precise writing survey on the use of blockchain innovation in e-finance and monetary administrations, giving bits of knowledge into the present status of exploration in this field. Ajayi and Saadawi [14] concentrated on the discovery of insider assaults in blockchain networks, tending to security challenges related with insider dangers in blockchain conditions. Rathod and Motwani [15] talked about security dangers on blockchain and proposed countermeasures, adding to the comprehension of weaknesses and relief procedures in blockchain frameworks. Putz and Pernul [16] investigated procedures for distinguishing blockchain security dangers, zeroing in on proactive measures to recognize and address possible weaknesses in blockchain networks. Bhowmik et al. [17] directed a similar investigation of ML calculations for misrepresentation recognition in blockchain, assessing the exhibition of various calculations in recognizing false exercises. Taher et al. [18] proposed an outfit learning and reasonable man-made intelligence approach for cutting-edge extortion location in blockchain exchanges, meaning to upgrade the exactness and interpretability of misrepresentation discovery frameworks. Aziz et al. [19] presented a changed hereditary calculation with profound learning for recognizing extortion exchanges in Ethereum savvy contracts, exhibiting creative ways to deal with further developing security in blockchain applications. Atzei et al. [20] studied assaults on Ethereum shrewd agreements, giving bits of knowledge into normal weaknesses and assault vectors focusing on brilliant agreement stages. Samreen and Alalfi [21] led a study of safety weaknesses in Ethereum brilliant agreements, featuring the significance of tending to security worries in decentralized applications based on blockchain stages.

15.7 SUMMARY

The chapter presented a comprehensive exploration of various ML models, including random forest, decision tree, logistic regression, XGBoost, and LightGBM, applied to a specific dataset. LightGBM stands out as a preferred choice for fraud detection systems in financial systems based on its exceptional performance metrics derived from the evaluations performed.

With the highest accuracy score of around 97.7% and a robust AUC-ROC score of 0.93, LightGBM demonstrates superior discrimination ability, effectively distinguishing between legitimate and potentially fraudulent transactions. Its efficiency in handling imbalanced classes, reflected in improved precision, recall, and F1-score for the minority class, addresses a critical challenge in fraud detection.

Moreover, LightGBM's parallel processing capabilities contribute to its computational speed, ensuring swift analysis of transactions in real-time scenarios. The model's incorporation of regularization techniques and adept handling of missing values further enhances its reliability and generalization in the dynamic context of financial transactions, making it a compelling choice for accurate and timely fraud identification.

The analysis of the diverse range of existing works spanning and financial applications has provided valuable insights into the current state of research at the intersection of blockchain, ML, and security. This study suggests that LightGBM works well in detecting fraud transactions in both training and real-time dataset collected from a private blockchain. Along with it, the aforementioned tests conclude that the most important feature is "Time Diff between first and last (mins)", which refers to the difference in time between the first and last recorded during a transaction. This suggests that the model heavily relies on the time it takes to complete a transaction to classify it. Despite the advancements, there are identifiable literature gaps. While some chapters focus on specific applications or use cases, a holistic, unified approach is lacking. Additionally, the need for more empirical studies and real-world implementations is apparent to validate the effectiveness of proposed frameworks and models. The future work focuses on using Hyperledger as the private blockchain for integration and real-time use.

PRACTICE QUESTIONS

I. Short Answers – Remember/Understand
   1. Which are the most common areas of cyber-physical systems requirement?
   2. List the three tiers in cyber-physical supply chain systems.
   3. Brief the need for blockchain in supply chain.
   4. What do you mean by fraudulent banking transactions? How it affects the supply chain management?
   5. List the modules present in the blockchain-integrated CPS architecture.

6. For the blockchain-based CPS supply chain management, which ML classifiers are experimented here?

II. Descriptive Questions – Apply/Analyse
1. Analyse the CPS^2 framework and examine the issues related with the SAAS layer.
2. Apply the secured blockchain-based cyber-physical supply chain management system.
3. If the system to be developed don't have Ethereum framework, then examine the effects of it both positive and negative from the deployment perspective and suggest any other alternative to it.
4. The AUC shows 0.93 for both random forest and LightGBM models, which would be preferred for the CPS^2 system.
5. Analyse the application of any one of the ML algorithms in CPS-based supply chain management.

III. Higher Order Thinking Questions – Analyse/Create/Evaluate
1. Explore the application of ML in securing the CPS^2.
2. Showcase the application of AI in CPS^2 route optimization.
3. Examine the evaluation metrics applied to select the best ML models for the CPS^2.
4. Indicate the OLAP and ETL operations to be performed in SAAS layer of the CPS-based supply chain management system.
5. Critically analyse the application of AI models in CPSs development for agricultural supply chain.

REFERENCES

1. Flavio Tonelli, Melissa Demartini, Massimo Pacella, Roberta Lala, Cyber-physical systems (CPS) in supply chain management: From foundations to practical implementation, *Procedia CIRP*, vol. 99, pp. 598–603, 2021. https://doi.org/10.1016/j.procir.2021.03.080.
2. T. Ashfaq et al., A machine learning and blockchain based efficient fraud detection mechanism, *Sensors*, vol. 22, no. 19, p. 7162, 2022.
3. M. K. Khan et al., *An Improved Finanacial Application Model Using Blockchain,* 2021.
4. S. Dashottar and V. Srivastava, Corporate financial—Risk management, regulatory and reporting framework in India: A blockchain application-based approach, *Journal of Finanacial Regulation*, vol. 22, pp. 39–51, 2021.
5. S. Chakraborty, S. Aich, S. J. Seong, and H. C. Kim, A blockchain based credit analysis framework for efficient financial systems, in *2019 21st International Conference on Advanced Communication Technology (ICACT)*, February 2019, pp. 56–60.
6. P. Garg et al., Measuring the perceived benefits of implementing blockchain technology in the finanacial sector, *Technological Forecasting and Social Change*, vol. 163, p. 120407, 2021.
7. P. Dzhaparov, Application of blockchain and artificial intelligence in bank risk management, *Икономика и управление*, vol. 17, no. 1, pp. 43–57, 2020.
8. R. Arjun and K. R. Suprabha, Innovation and challenges of blockchain in finanacial: a scientometric view, *IJIMAI*, vol. 6, no. 3, pp. 7–14, 2020.

9. T. G. Sant and V. Tripathi, Deep blockchain-enabled security enhancement in trade finance, *International Journal of Systematic Innovation*, vol. 7, no. 8, pp. 48–64, 2023.
10. U. Islam et al., Detection of distributed denial of service (DDoS) attacks in IoT-based monitoring system of finanacial sector using machine learning models, *Sustainability*, vol. 14, no. 14, p. 8374, 2022.
11. M. U. Chowdhury et al., Blockchain application in finanacial system, *Journal of Software Engineering and Applications*, vol. 14, no. 7, pp. 298–311, 2021.
12. J. Osterrieder, S. Chan, J. Chu, Y. Zhang, B. H. Misheva, and C. Mare, Enhancing security in blockchain networks: Anomalies, frauds, and advanced detection techniques, 2024. *arXiv preprint arXiv:2402.11231*.
13. S. Trivedi, K. Mehta, and R. Sharma, R. Systematic literature review on application of blockchain technology in E-finance and financial services. *Journal of Technology Management & Innovation,* vol. 16, no. 3, pp. 89–102, 2021.
14. O. Ajayi and T. Saadawi, Detecting insider attacks in blockchain networks, in *2021 International Symposium on Networks, Computers and Communications (ISNCC)* (pp. 1–7), 2021. IEEE.
15. N. Rathod and D. Motwani, Security threats on blockchain and its countermeasures. *International Research Journal of Engineering and Technology*, vol. 5, no. 11, pp. 1636–1642, 2018.
16. B. Putz and G. Pernul, Detecting blockchain security threats, in *2020 IEEE International Conference on Blockchain (Blockchain)* (pp. 313–320), November 2020. IEEE.
17. M. Bhowmik, T. S. S. Chandana, and B. Rudra, Comparative study of machine learning algorithms for fraud detection in blockchain, in *2021 5th international conference on computing methodologies and communication (ICCMC)* (pp. 539–541), April 2021. IEEE.
18. S. S. Taher, S. Y. Ameen, and J. A. Ahmed, Advanced fraud detection in blockchain transactions: An ensemble learning and explainable AI approach. *Engineering, Technology & Applied Science Research,* vol. 14, no. 1, pp. 12822–12830, 2024.
19. R. M. Aziz, R. Mahto, K. Goel, A. Das, P. Kumar, and A. Saxena, Modified genetic algorithm with deep learning for fraud transactions of ethereum smart contract, *Applied Sciences*, vol. 13, no. 2, p. 697, 2023.
20. N. Atzei, M. Bartoletti, and T. Cimoli, A survey of attacks on ethereum smart contracts (sok), *in Principles of Security and Trust: 6th International Conference, POST 2017, Held as Part of the European Joint Conferences on Theory and Practice of Software*, ETAPS 2017, Uppsala, Sweden, April 22–29, 2017, Proceedings 6 (pp. 164–186). Springer: Berlin Heidelberg.
21. N. F. Samreen and M. H. Alalfi, A survey of security vulnerabilities in ethereum smart contracts, 2021. *arXiv preprint arXiv:2105.06974*.

Index

Note: Page numbers in *italic* indicate figures and in **bold** indicate tables.

C

D

T

U

V